RISK MANAGEMENT:
VALUE AT RISK AND BEYOND

The Isaac Newton Institute of Mathematical Sciences of the University of Cambridge exists to stimulate research in all branches of the mathematical sciences, including pure mathematics, statistics, applied mathematics, theoretical physics, theoretical computer science, mathematical biology and economics. The research programmes it runs each year bring together leading mathematical scientists from all over the world to exchange ideas through seminars, teaching and informal interaction.

RISK MANAGEMENT: VALUE AT RISK AND BEYOND

edited by

M.A.H. Dempster

University of Cambridge

CAMBRIDGE UNIVERSITY PRESS
Cambridge, New York, Melbourne, Madrid, Cape Town, Singapore,
São Paulo, Delhi, Dubai, Tokyo, Mexico City

Cambridge University Press
The Edinburgh Building, Cambridge CB2 8RU, UK

Published in the United States of America by Cambridge University Press, New York

www.cambridge.org
Information on this title: www.cambridge.org/9780521781800

First published 2002

A catalogue record for this publication is available from the British Library

ISBN 978-0-521-78180-0 Hardback

CONTENTS

Contributors

Philippe Artzner, Department of Mathematics, Université Louis Pasteur, 7 rue Rene Descartes, F 67084 Strasbourg Cedex, France

Freddy Delbaen, Department of Mathematics, ETH-Zentrum, Raemistrasse 101, CH-8092 Zürich Switzerland.

M.A.H. Dempster, Centre for Financial Research, Judge Institute of Management, Trumpington Street, Cambridge CB2 1AG, UK.

Jean-Marc Eber, LexiFi Technologies, 17, Square Edouard VII, F-75009 Paris, France

Paul Embrechts, Department of Mathematics, ETH-Zentrum, Raemistrasse 101, CH-8092 Zürich, Switzerland.

David Heath, Department of Mathematical Sciences, Carnegie Mellon University, Pittsburgh, PA 15213, USA.

R. Kiesel, Department of Statistics, London School of Economics, Houghton Street, London WC2A 2AE, UK.

Paul H. Kupiec, International Monetary Fund, 700 19th Street, NW, Washington DC 20431, USA.

M.N. Kyriacou, Group Operational Risk Management, BNP Paribas, 10 Harewood Avenue, London, NW1 6AA, UK.

Alexander J. McNeil, Department of Mathematics, ETH-Zentrum, Raemistrasse 101, CH-8092 Zürich, Switzerland.

E.A. Medova, Centre for Financial Research, Judge Institute of Management, Trumpington Street, Cambridge CB2 1AG, UK.

W. Perraudin, School of Economics, Mathematics & Statistics, Birkbeck College, 7–15 Gresse Street, London W1P 2LL, UK.

Evan Picoult, Managing Director, Head of Risk Methodologies and Analytics, Risk Architecture, Citigroup, 399 Park Avenue, 11th Floor/Zone 1, New York, NY 10043, USA.

Richard L. Smith, Department of Statistics, University of North Carolina, Chapel Hill, NC 27599-3260, USA.

Sanjay Srivastava, Graduate School of Industrial Administration, Carnegie Mellon University, Pittsburgh, PA 15213, USA.

Daniel Straumann, Department of Mathematics, ETH-Zentrum, Raemistrasse 101, CH-8092 Zürich, Switzerland.

A.P. Taylor, Centre for Financial Research, Judge Institute of Management, Trumpington Street, Cambridge CB2 1AG, UK.

G.W.P. Thompson, Centre for Financial Research, Judge Institute of Management, Trumpington Street, Cambridge CB2 1AG, UK.

Introduction

The modern world of global finance had its antecedents in two significant events which occurred approximately thirty years ago: the breakdown of the post-war Bretton Woods system of fixed exchange rates between national currencies and the (re-) introduction of option trading in major financial markets emanating from the creation of the Chicago Board of Trade Options Exchange.

The latter coincided with the Nobel Prize-winning work of Black, Scholes and Merton who produced both a formula for the 'fair' valuation of stock options and an idealised prescription for the option seller to maintain a self-financing hedge against losing the premium charged – the famous delta hedge – which involved trading in the underlying stock only. The essence of their argument involved the concept of perfectly replicating the uncertain cash flows of European options. This argument, which required a continually rebalanced portfolio consisting only of the underlying stock and cash, applied more generally to other financial derivatives products whose introduction followed rapidly and at a rate which is still accelerating today. The new concepts were soon applied to futures and forwards and to the burgeoning market in foreign exchange in terms of derivatives written on currency rates, as FX market makers and participants attempted respectively to profit from, and to employ the hedging capabilities of, the new contracts in order to protect cross border cash flows in domestic terms in a world of uncertain exchange rates.

The market for derivative products in the fixed income sphere of bills, notes and bonds – although the basic theoretical foundations were established early on by Vasicek – has been much slower to develop, not least because fixed income instruments, even those issued by major sovereigns such as the US, Japan or the UK, are subject to multiple risk factors associated with their different multiyear tenors so that they are considerably more complex to value and hedge. Nevertheless, in less than twenty years the global market for swaps – in which two cash flows are exchanged for a specified period between counterparties – has grown from a single deal between IBM and the World Bank to over a trillion US dollar market accounting for about 40% of the global value of the derivatives markets. When the credit risk involved in similar instruments issued by less creditworthy sovereigns or public corporations must be factored in, derivative product valuation and hedging becomes even more complicated. Only recently a rough consensus on at least the alternative approaches to credit migration and default risk valuation has begun to emerge. Further, the derivatives markets are currently attempting to meet head on the risk inherent in all banking intermediation by using the new derivative tools and techniques both to securitize all types of risky cash flows such as mortgages, credit card payments and retail and commercial loan repayments and to create a global market in credit derivatives.

In the meantime, the use of derivative products in risk management is also spreading to such virtual commodities as energy, weather and telecommunications bandwidth. While futures contracts have been in use for agricultural commodities for over two centuries and for oil products and minerals for more than a hundred years, the markets for forward, futures and option contracts written on kilowatt hours of electricity, heating or cooling degree days and gigabits of fibre optic transport, like their traditional commodity predecessors, introduce a spatial location element that adds to valuation complexity. Moreover, the nature of the asset price processes underlying these new areas often results in a very poor fit to the classical diffusion processes used to model the equity, FX and major sovereign treasury worlds. Arising originally from the impacts of credit events on fixed income asset valuation, research continues unabated into valuation models and hedging schemes involving jumping diffusions, extreme value processes and the unpriced uncertainties of so-called incomplete markets.

Although often denied, it was a maxim of nineteenth century commodity and futures markets that speculative trading led to excessive price fluctuations – today termed volatility. A new development is that investment banks currently operating in the major financial markets have switched from being comfortable fee earners for assisting the equity and bond flotations of major corporations, together with giving them advice on mergers and acquisitions, to deriving a considerable portion of their profits from derivative product sales and trading on own account. Like the development of modern derivatives trading, the subsequent introduction of formal risk management techniques to cope with the effects of increased volatility in financial markets can be traced to two relatively recent events.

The first of these was the 1988 recommendation of the Bank of International Settlements in Basle of a flat 8% capital charge meant to be appropriate to all financial institutions to cover all types of risks - market (due to price changes), credit (due to counterparty defaults), liquidity (due to market imbalance), etc. This Capital Adequacy Accord was a more or less direct reaction to credit problems following the equity market crash of October 1987 and was subsequently refined in an attempt to cover off-balance-sheet derivatives and enacted into law in many of the world's economies with varying lags. In the absence of a global financial regulator this so-called 'soft law' has been remarkably effective in the leading economies. Indeed, the current BIS proposals to revise the Accord and to explicitly cover the risks inherent in banking operations is enjoying heated debate largely in recognition of the fact that the lags in national enforcement are likely to be much shorter this time around.

The second, more technical, event occurred on Wall Street about seven years ago at J.P. Morgan in response to an earlier demand by the Chairman for a 4:15 report each day on the potential trading earnings at risk overnight due to global market price movements. The result was the concept of *Value at Risk* (VaR) which figures in the title of this volume, together with a formal model for the evaluation of the such market risks for portfolios and trading desks over short periods of several trading days. This concept has been taken up by financial regulators in the 1996 Basle Accord supplement and has subsequently been extended – more controversially – to measuring credit risks over much longer horizons. Moreover, it has led to the Risk Metrics spin-off which markets data and software systems based upon its previously published approaches and has become a major player in the rapidly growing market for so-called enterprise-wide risk management solutions appropriate to the world's financial institutions at all levels. This market trend will no doubt continue under the pressure of the new BIS Capital Adequacy Accord and it is hoped that the present book can play some small role in helping to clarify the complex issues revolving around the future stability of the global financial system.

We now turn to a brief description of the contributions to this volume which are based to a greater or lesser degree on a very successful Workshop on Risk Management held at the Isaac Newton Institute for Mathematical Sciences on 2–3 October 1998, organized by its Director, Professor H.K. Moffat FRS, and attended by both practitioners and academics. The contents of the volume reflect the mix of theory and practice which is required for survival in today's capital markets.

The opening chapter by Picoult, the senior risk analyst at Citicorp, the world's largest and arguably most global bank, sets the practical context for the rest of the book. In a clear and parsimonious style the author discusses in some detail techniques for three of the four most important risks of trading: valuation risk, market risk and counterparty credit risk. (The fourth, operational risk, will be discussed in the last chapter of this volume, where the impact of the Russian Crisis of late summer and early autumn 1998 upon trading profits of an anonymous European bank will be analysed.) Chapter 1 begins by describing the important features of (expected) discounted cash flow models used for the valuation of financial instruments and portfolios. The author points out that valuation error can stem not only from the model error beloved of quantitative analysts, but also from erroneous or misused data and human misunderstanding, and he goes on to clarify the factors required to establish market value. The next two sections of the chapter discuss in detail the methods used to 'measure, monitor and limit' market and counterparty credit risk respectively. The principal approaches to statistical analysis of market risk – *parametric* (Gaussian or mean-variance), *historical* (empirical) and full *Monte Carlo* VaR analysis and *stress testing* – are described precisely. Analysis of credit risk is as indicated above usually more complex, and techniques for the measurement of both pre-settlement and settlement risks are set out next. Finally the main attributes of market and credit risk are compared and contrasted.

In Chapter 2, Srivastava uses parametric VaR analysis based on a binomial tree implementation of the popular Heath–Jarrow–Morton model for forward interest rates to provide a succinct dissection of one of a string of celebrated derivative fiascos of the early 1990s – the fixed-floating five year semi-annual swap between Bankers Trust and Proctor and Gamble (P&G) initiated in November 1993. The author's step-by-step exposition demonstrates that had P&G carried out such a straightforward analysis using modern risk management tools, they would have seen that the VaR of the contract was about seven times its value. In the event this so-called *unexpected loss* amount – $100M – was actually lost. Using the expected excess loss over the VaR limit – a *coherent* risk measure as introduced in Chapter 6 and applied in subsequent chapters – a factor of about ten times the market value of the contract would have been found.

Kupiec proposes in Chapter 3 a methodology to parametrize extreme or *stress test scenarios*, as used by many banks to evaluate possible market value changes in a large portfolio *in addition* to VaR analysis, in a context which is completely consistent with VaR. The author shows how assuming multivariate normal return distributions for all risk factors leads to automatic consideration of value changes due to the non-stressed factors which are commonly ignored in stress testing. He demonstrates on data for the period of the 1997 Asian crisis that his conditional *Gaussian Stress VaR* (95%) approach to stress testing leads to historically accurate estimated value changes for a global portfolio with instruments in the US, European and Asian time zones. The chapter concludes with a detailed discussion of the practical problems involved in stressing the correlations and volatilities needed in any Gaussian analysis.

In the last chapter to deal primarily with market risk, Chapter 4, Dempster and Thompson return to the fundamental Black–Scholes concept of accurate trading strategy replication of risk characteristics in the context of dynamic portfolio replication of a large target portfolio by a smaller self-financing replicating portfolio of tradable instruments. Two applications are identified: *portfolio compression* for fast portfolio VaR calculation and *dynamic replication* for hedging by shorting the replicating portfolio or for actual target portfolio simplification. The first (virtual) application involves no transaction costs and is shown to be a promising alternative to other portfolio compression techniques such as multinomial factor approximations to a full daily portfolio revaluation using Monte Carlo simulation. With or without the use of variance reduction techniques such as low-discrepancy sequences, using full Monte Carlo simulation to value large portfolios for VaR analysis is for many institutions barely possible overnight. The authors demonstrate that the use of stochastic programming models and standard solution techniques for portfolio compression can produce an expected average absolute tracking error of the easily evaluated replicating portfolio which (at about 5% of the initial target portfolio value) is superior to both more static replicating strategies and target portfolio delta hedging and within acceptable limits for fast VaR calculations.

Chapter 5, by Kiesel, Perraudin and Taylor, turns to an integrated consideration of market and credit risks for VaR calculations. Reporting on part of a larger comparative study of credit risk models for US corporate bonds supported by the Bank of England, the authors emphasize the very different horizons needed for market and credit risk VaR calculations – respectively several days and one or more years over which the time value of money clearly cannot be ignored. They note that interest rate risk should always be included in long horizon credit VaR calculations if interest rates and credit spreads are less than perfectly correlated and they set out to study this correlation and its analogue for ratings transition risks. They find – somewhat counter intuitively but in agreement with some previous studies – that interest rate changes and both credit spreads and ratings transitions are negatively correlated even over one year horizons. Recently it has been suggested that such effects may be explained by the empirical fact that expected default rates – and *a fortiori* possible credit transitions – account for a surprisingly small proportion of so-called credit spreads, the bulk of which may be due to state tax effects and premia for nondiversifiable systemic risk in the bond markets analogous to equity premia.

The remaining four chapters of this volume take the reader well beyond the concepts of VaR analysis. The first, Chapter 6 by Artzner, Delbaen, Eber and Heath, is a classic. The authors *axiomatize* the concept of financial risk measurement in terms of the risk or economic capital required to neutralize potential losses from the current position and relate such *coherent* risk measures to existing VaR and stress testing techniques. They show by example that VaR is *not* a coherent risk measure in that it fails to possess the subadditivity – i.e. portfolio diversification – property. This property assures that the risk capital required to cover two risky positions is never more than the sum of those required to cover each individually. It has the important demonstrated consequence that individual coherent risk measures for classes of risk factors – for example relevant to market and credit risk individually – can be combined into an overall conservative coherent risk measure based on all risk factors present. The abstract approach to risk measurement is applied in the chapter both to improve the stress testing schemes for margining proposed by the Chicago Mercantile Exchange and the US Securities and Exchange Commission and to demonstrate that the *expected excess* over a VaR level added to the VaR yields a coherent measure – an idea with its roots in nearly 150 years of actuarial practice.

Embrechts, McNeil and Straumann provide in Chapter 7 a thorough *primer* on the measurement of static statistical dependencies from both the actuarial and financial risk management viewpoints. They demonstrate, both by theory and illuminating example, that the concept of linear correlation is essentially valid only for the multivariate Gaussian and other closely related spherical distributions. Correlation analysis is based on second moments, breaks down for fat-tailed and highly stressed distributions and is not defined for many *extreme value* distributions. From the risk management perspective, these facts constitute a different criticism of VaR analysis to that studied in the previous chapter: namely correlation matrices calculated from data non-spherically distributed but used in practice for parametric Gaussian VaR calculations can lead to highly misleading underestimates of risk. As well as classical rank correlation and concordance analysis, the use of the *copula* function, appropriate to the study of dependencies amongst the coordinates of *any* multivariate distribution, is proposed and its basic properties set out. Much work remains to be done in this area – particularly with respect to practical computational multivariate techniques – but this chapter provides among many other things a basic grounding in the copula concept.

Following its extensive use by insurance actuaries, possible uses of *extreme value theory* (EVT) in risk management are discussed by Smith in Chapter 8. After a brief exposition of EVT and maximum likelihood estimation of extreme value parameters, these concepts are illustrated on both fire insurance claims and S&P500 equity index data. Next the author introduces the Bayesian approach to the predictive EVT distributions with unknown parameters which are needed for risk management in the presence of extreme loss events. He goes on to describe the limited progress to date in handling multivariate extreme value distributions and then to propose a dynamic changepoint model to attack the volatility clustering of the S&P500 index data. The latter allows the extreme value parameters to change at a fixed number of timepoints, which number is estimated from the data along with the other parameters using hierarchical Bayesian methods. The posterior distributions of all parameters are simultaneously estimated using reversible jump *Markov chain Monte Carlo* (MCMC) sampling. The suggested conclusion of this analysis is that the NYSE enjoys periods of (short-tailed) normal returns but exhibits extreme value behaviour in periods of high volatility.

The themes of Chapter 8 are continued in the final chapter by Medova and Kyriacou in the context of extreme operational risks in financial institutions. The authors set the scene by describing current definitions of operational risk and the recently proposed Basle Accord revisions to cover it and then argue that all risks must be considered in an integrated framework in which extreme operational risks correspond to events in the unexpected loss tail of the appropriate integrated profit and loss (P&L) distribution *whatever* their underlying source. Next they provide an overview of the relationship between the classical limit theory for stable distributions and extreme value theory and then set up a time homogeneous version of the Bayesian hierarchical model of the previous chapter, again estimated by MCMC sampling, in the context of extreme operational risk measurement and capital allocation in terms of the coherent expected excess loss measure. Bayesian techniques are appropriate to the measurement of operational risk in financial institutions in that, although such data is scarce, using median posterior distribution extreme value parameter estimates – i.e. absolute value loss functions – and the *peaks over threshold* (POT) model, stable accurate estimates are produced for very small sample sizes (10–30). The methods are illustrated using five quarters of proprietary data on the daily P&L of four trading desks of a European investment bank through the Russian crisis of 1998. The authors demonstrate that for this data their techniques could have been used to get relatively accurate estimates of the risk capital required to cover actual losses throughout the period and that this capital provision enjoyed a portfolio diversification effect across trading desks in spite of the presence of extreme events.

MAHD Cambridge

Quantifying the Risks of Trading

Evan Picoult

Abstract

This article defines and describes methods for measuring three of the prominent risks of trading: valuation risk, market risk and counterparty credit risk. A fourth risk, operational risk, will not be discussed. The first section of the article describes the essential components of discounted cash flow models used for valuation, identifies the sources of valuation error and classifies the types of market factors needed to measure market value. The second section of the article describes the nature of and the methods that can be used to measure, monitor and limit market risk. A similar analysis of counterparty credit exposure and counterparty credit risk follows. Finally, the nature of and methods for measuring market risk and counterparty credit exposure will be compared and contrasted.

1 Introduction

The term 'risk' is used in finance in two different but related ways: as the magnitude of (a) the potential loss or (b) the standard deviation of the potential revenue (or income) of a trading or investment portfolio over some period of time.[1]

Our discussion and analysis of market risk and counterparty credit risk will almost exclusively focus on risk as potential loss. That is, we will describe methods for measuring, in a specified context, the potential loss of economic value of a portfolio of financial contracts. The context that needs to be specified includes the time frame over which the losses might occur (e.g. a day, a year), the confidence level at which the potential loss will be measured (e.g. 95%, 99%) and the types of loss that would be attributed to the risk being measured (e.g. losses due to changes in market rates vs. losses due to

[1] The quantitative relationship between risk as potential loss and risk as uncertainty in future revenue is a function of the estimated probability distribution of future revenue. For example, if the estimated probability distribution of potential revenue is normally distributed around an expected value of zero then the potential loss at some confidence level can simply be expressed in terms of the standard deviation of potential revenue. In many cases the expected total revenue from a trading business is not zero (else the firm would not be in the business) and the probability distribution of future revenue may not be symmetric about its expected value.

the default of a counterparty). Part of the context for measuring the potential loss, whether due to market or credit risk, is the distinction between an economic perspective and an accounting perspective.

The distinction arises for market risk because the income from financial contracts may be accounted for in one of two ways: by accrual accounting (e.g. as is typically done for a portfolio of deposits and loans) or by mark to market accounting (e.g. as is typically done for a trading portfolio). The primary difference between the two approaches is in the timing of their recognition of financial gains or losses.[2] Only the mark to market approach is equivalent to the continual measurement of economic value and change in economic value. The relative merits and demerits of measuring the income and risk of a particular business on an accrual basis will not be evaluated here This article is focused on the risks of trading and will analyze and describe market risk from an economic perspective.

A similar issue arises for credit risk. One example of this is the potential difference between the loss in the market (economic) value of a loan caused by the default of the borrower and the timing of the recognition of the loss in the income statement.

A more important example of this issue for credit risk is the treatment of the deterioration of a borrower's credit worthiness. Consider as an example a corporate loan. Assume that in the period after the loan was made the only relevant factor to undergo a material change was a deterioration of the credit worthiness of the borrower. In more detail, assume that one year ago a bank made a three year loan to a corporation for which the corporate borrower was required to make periodic interest payments and to pay back the principal and the remaining interest payment on the maturity date of the loan. Assume further that both the public credit rating and the bank's internal credit rating of the borrower has deteriorated since the loan was initially made. Finally, assume that general market rates have remained unchanged since the loan was made and that the borrower has made all interest payments on time.

[2] As a simple example of the difference consider two portfolios. Portfolio A is a standard deposit and loan portfolio. It consists of a ten year \$100 million loan to firm X at a fixed semi-annual rate of 10.00% and a one year \$100 million deposit from firm Y, at a semi-annual rate of 9.50% interest. Portfolio B is a trading portfolio. It consists of a long position in a ten-year debt security issued by firm X at a fixed semi-annual rate of 10.00% and a short position in a one-year debt security issued by firm Y at a fixed semi-annual rate of 9.50%. If both portfolios were viewed from a marked to market perspective, they would have identical market risks. However, under standard accounting practices, the effect of a change in market rates on the reported revenues of the portfolios will differ. Assume the only change in market rates is a 1% parallel increase in the risk free rate at all tenors. On a marked to market basis the net value of the securities in Portfolio B would fall in value. In contrast, the accrued interest earned by Portfolio A is locked in for the year and is independent of the level of interest rates. If interest rates should continue at their higher level the accrued interest earned by Portfolio A will only be affected after its one year deposit matures and has to be replaced with a deposit at a higher interest rate.

Under standard accrual accounting the loan would be recorded on the balance sheet of the bank at its par value. The bank would only record a loss if the borrower defaulted on a payment. The bank would not in general recognize any loss due to the deterioration of the credit quality of the borrower. At most the bank could establish a general loan loss reserve for the expected credit loss of the portfolio.

In contrast, the market value of the loan would fall if the borrower's credit worthiness deteriorated. To appreciate the reality of this loss, assume the bank took action to actively managing its credit risk to this corporate borrower, after the borrower's credit risk had deteriorated. For example, if the bank were to sell the loan in the secondary loan market, then *pari passu*, it would suffer an economic loss – i.e. the market value of the loan would be less than its par amount because of the increased credit riskiness of the borrower. Or, if the bank tried to hedge its credit exposure to the borrower by buying a credit derivative on its underlying loan it might have to pay an annual fee for that credit insurance that was higher than the net interest income it was earning on the loan. Both of these examples of active portfolio management illustrate that a deterioration in the credit quality of the borrower, all other things held constant, will cause the market value of the loan to decrease, even if the borrower had not defaulted.

From the accrual accounting perspective no credit loss would occur without a default by the borrower. From the economic perspective, the increased riskiness of the borrower would cause the economic value of the loan to decrease.

This article will not focus on loan portfolio credit risk. It will however analyze another form of credit risk, the risk that the counterparty to a forward or derivative trade could default prior to the final settlement of the cash flows of the transaction. This form of credit risk is called *counterparty pre-settlement credit risk* and will be analyzed in detail below.

We will describe methods for measuring four aspects of the risks of trading:

- Methods for measuring and controlling *valuation uncertainty* and valuation error.
- Methods for measuring *market risk*. These methods measure the potential decrease in the economic value of contracts caused by potential future changes in market rates.
- Methods for measuring a *counterparty's pre-settlement credit exposure*. These methods measure the potential future replacement cost of the forward and derivative contracts transacted with a counterparty, should the counterparty default at some time in the future before all the contracts mature. The potential credit exposure will depend on the potential future market value of the contracts transacted with the counterparty, on any risk mitigating agreements (such as netting) that have

been contracted with the counterparty and on the legal enforceability of such agreements.

- Methods for measuring *counterparty credit risk.* These methods measure the probability distribution of loss due to counterparty default and rest, in part, on measurements of the potential future credit exposure to the counterparty, the future default probability of the counterparty and the potential loss in the event of counterparty default.

Our measurements of market risk, counterparty credit exposure and counterparty credit risk all rest on our ability to measure the current and the potential future economic value of financial contracts. At the end of the article we will summarize and contrast each type of risk measurement. Because of the crucial connection between methods for valuing contracts and methods of risk measurement, we shall begin our discussion with a review of revaluation models, valuation errors and market factors.

2 Market valuation and valuation uncertainty

2.1 Discounted cash flow formula

Marking to market is the activity of ascertaining the market value of each financial instrument in a trading portfolio. Market value is ascertained in one of two ways: directly, by observing the market price of identical (or nearly identical) instruments or indirectly, by using a discounted cash flow revaluation model. When a discounted cash flow model is used, it is necessary to periodically calibrate the model against the market to ensure that the model's valuation corresponds to the market's.

Very liquid, cash-like financial instruments such as spot FX, equities and simple debt securities are marked to market by discovering the prices or rates at which identical (or nearly identical) instruments are traded in the market. For example, the market value of a portfolio of US Treasury securities of different maturities and coupon rates would be calculated simply by discovering the unit market price of each security in the portfolio and by multiplying the unit price by the number of units owned (positive for long, negative for short). Nothing beyond simple arithmetic would be needed to calculate the mark to market value of the portfolio.

In contrast, forward and derivative contracts are revalued in terms of discounted cash flow models (reval models). In essence, a *reval model* calculates the net present value of the expected future cash flows of the contract. It does this by representing the economic value of a contract as a function of its terms and conditions, basic market rates, and the current date:

$$PV(t)_k = f(T\&C_k, \{X_j(t)\}, t), \tag{2.1}$$

where:

$PV(t)_k$ is the present value of contract k, at time t;

$T\&C_k$ are the terms and conditions of contract k;

$X_j(t)$ is the value of market factor j, at time t;

$\{X_j(t)\}$ are the values of the complete set of market factors j, at time t, needed to value contract k;

t is today's date.

As an example, the terms and conditions of a simple interest rate swap would include: (a) a description of which party is paying the fixed cash flow and which is paying the floating; (b) the specification of the dates (and time of day) at which each floating rate will be set and the reference rate (e.g. three month US$ LIBOR) for setting the floating rate; (c) the specification of the dates at which fixed and floating cash flows will settle; and so forth.

Market factors $\{X_j(t)\}$ are the fundamental market prices and rates in terms of which contracts are valued. Examples include the term structure of interest rates, spot FX rates, spot equity and equity index prices, spot and forward commodity prices and the term structures of implied volatilities of the preceding market factors.

For the simple interest rate swap, described above, the relevant market factors needed to value of the contract are the term structure of LIBOR interest rates. Given the terms and conditions of the interest rate swap, the current term structure of LIBOR interest rates and today's date, the simple swap revaluation formula estimates the forward interest rate as of each floating rate reset date and then calculates the discounted present value of all the fixed and expected floating cash flows from their settlement date to today.

More generally, given the terms and conditions of a financial contract, the current values of the relevant market factors and today's date, reval models estimate all the floating and contingent cash flows; discount all fixed, floating and contingent cash flows to present value; and convert cash flows denominated in different currencies into some base currency.

2.2 Types of revaluation models

There are two basic type of revaluation models: *arbitrage-free, risk-neutral* models and *behavioral* models. The most common type of financial model described in textbooks are the former. The essential feature of such models is their derivation by means of an arbitrage-free analysis. That is, an analysis which demonstrates that if the value of the contract were different from that given by the model it would be possible to risklessly make a return higher than the risk-free rate by constructing a portfolio consisting of the contract and some other financial contracts with suitable weights.

Examples of risk-neutral, arbitrage-free models are the standard formula for valuing a forward FX contract in a liquid market or the Black–Scholes valuation formula for a European equity option.

The second basic class of models are behavioral models, such as those for mortgage-backed securities. Instruments that require behavioral models have some option-like feature (e.g. a prepayment option) whose exercise is only partly a function of the level of market rates (e.g. some people with mortgages may sell their home, and therefore prepay their mortgage, for reasons other than the level of interest rates). The likelihood that the pre-payment option will be exercised in the future can be approximately modeled as a function of the future level of interest rates on the basis of the historical patterns of the pre-payment behavior of borrowers. Another feature of behavioral models is that given the uncertainty in modeling the potential future cash flows as a function of interest rates, the expected value of the future cash flows are not discounted to present value at a risk-free rate (e.g. LIBOR). Instead, they are discounted to present value at some spread ('option adjusted spread') over LIBOR to take into account both the uncertainty in modeling the prepayment option and the uncertain liquidity of the market.

Examples of financial assets and liabilities that require behavioral models to value them include mortgages (and mortgage backed securities), credit card balances and demand deposit account balances.

When contracts are marked to market with a model, whether it is of risk-neutral arbitrage-free or behavioral type, the model ultimately function as complex interpolation tool.[3]

2.3 Revaluation systems and valuation error

Revaluation models function within revaluation systems. To understand potential errors in valuation and to understand the steps of calculating market risk, we need to describe the essential components of a revaluation system.

[3]As an example of how a model functions as an interpolation tool, consider a portfolio of European FX options (i.e. simple puts and calls) on the US Dollar/Japanese Yen. The spot exchange rate and the yield curves of both currencies are easily observable. We can observe the prices of a set of options in the market and from these observed prices, using the standard FX option formula, we can infer the corresponding implied volatilities. For a sufficiently large set of observed option prices (i.e. for a sufficient range of strike prices and option tenors) we can construct an implied vol surface from which any particular FX option could be valued. In effect, for each currency pair (e.g. US$/JPY), the model allows us to transform the observed prices of options that have standard strike prices and tenors into an implied volatility surface. The model then enables us to value an FX option on that currency pair, with any particular tenor and strike price (within a range), simply by interpolating in the implied vol space and calculating the corresponding value of the FX option.

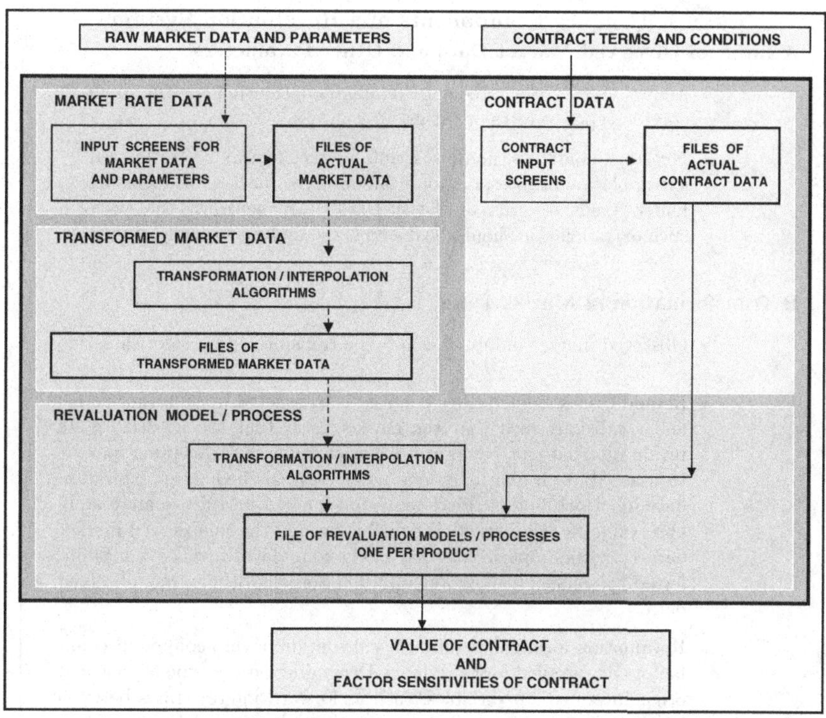

Figure 1: Revaluation system: components and flow of information

Table 1 and Figure 1 depict the essential components and flow of information in a revaluation system.

2.4 Sources of valuation differences and valuation error

A given contract might be valued differently by two different valuation systems. The difference in valuation may be large (as a percentage of total valuation) or small. There are several potential causes of valuation difference between systems: (a) the set of observed market factors chosen as input; (b) the set of transformation algorithms applied by each system to the observed market data; and (c) the revaluation algorithm as such.

For a given contract (or class of contracts) a valuation system might calculate a value that differs from the market's valuation. This can arise for all the reasons that would cause two systems to calculate different valuations, as discussed in the previous paragraph. In addition, the market's and the system's valuation may differ because there may be causes or factors affecting the price in the market place that are not captured or incorporated in the revaluation model. That is why it is important to periodically calibrate the system's valuation to the market's.

Table 1. Primary Components of a Revaluation System

A Input of Observed Market Data and Other Parameters

- Market data are entered into templates formatted for the input of specific types of rates or prices.
- For some contracts, non-observable market factors are estimated by extrapolation or interpolation from observed market data (see below, under 'Transformation of Market Data'). All algorithms that perform such extrapolation should be documented and periodically evaluated.

B Transformation of Market Data

- Observed market data will usually be transformed by algorithms:
- Illiquid Prices. Sometimes a contract is transacted with unusual terms and conditions such that the market rates that are needed for its revaluation can not be observed. For example, an FX option may be transacted at an unusually long tenor, beyond the longest expiration date for which option prices or option implied volatilities are readily observable in the market. When this occurs, the unobserved market data (e.g. the implied volatility of the long dated option) will be inferred by some extrapolation/interpolation algorithms from observed data.
- Revaluation formulae require the calculation of zero coupon discount factors at specified forward dates. Depending on the type of contract, other intermediate variables, such as forward interest rates between specified pairs of forward dates, may also be needed for the valuation. These intermediate variables are derived from the observed and inferred market factors by transformation algorithms.

C Input of Contract Terms and Conditions

- The terms and conditions of each contract, for each form of contract, are entered via product specific templates and are stored in a suitably formatted database.

D Valuation Formula/Process

- Each contract is revalued according to the revaluation formula (or process) specific to the particular form of the contract, its detailed terms and conditions and the transformed market data.
- Factor sensitivities (which are critically important for measuring market risk, see below) are usually calculated as part of the revaluation process by revaluing each contract many times, under different scenarios in which some market rates are changed from its current level. For example, during the end-of-day batch process, each contract might be revalued $N+1$ times. One revaluation will be at the current set of market factors. The other N additional revaluations will be done, one for each of the N factor sensitivity scenarios. These N additional revaluations enable us to calculate N factor sensitivities.

As an example of how valuation differences can occur even for a relatively simply product in a liquid market, consider a simple fixed/floating US$ LIBOR interest rate swap. The value of a simple, 'plain vanilla' US$ LIBOR swap of specified terms and conditions is determined once the appropriate set of LIBOR zero coupon discount factors (zcdfs) are specified.[4] What observable market data are available to define a continuous set of US$ LIBOR zcdfs?

Information about US$ LIBOR interest rates appears in the market in several different forms: There are: (a) *LIBOR money market rates* from one to twelve months (e.g. the London Interbank Offer Rate at which banks offer to lend to each other); (b) market prices for about *forty Eurodollar future contracts*, one for each quarter for approximately the next ten years (a LIBOR future contract has a final value which is set on some specified future date and is determined by the then current three month LIBOR interest rate); and (c) *LIBOR interest rate swap spreads* over a range of tenors (e.g. the difference between the fixed rate of a US$ LIBOR swap and the US Treasury yield-to-maturity of the same tenor) and the *US Treasury yield curve* for a range of tenors (note that the current US Treasury yield, of a given tenor, plus the current swap spread, for that tenor, equals the current fixed rate of the swap, for that tenor). The different forms of market data overlap. Consequently there are choices in selecting the observed market data that will be used to derive the LIBOR zero coupon discount factors needed to value a general US$ LIBOR swap. For example, a particular valuation system might have as input: (a) US$ LIBOR money market rates for some specific set of tenors (e.g. overnight, 1 week, 1 month, 2 months, 3 months); (b) a set of three-month Eurodollar future contracts (e.g. the first sixteen contracts); (c) a set of US Treasury yields at specified maturities (e.g. two years, three years, five years to thirty years) and a set of corresponding US$ LIBOR swap spreads to Treasuries.

After the observed market data are selected, algorithms must be employed to transform them into a consistent set of zero coupon discount factors (zcdfs), one for each future date needed for valuation. There are several methods for making the transformation (e.g. the transformation could assume continuously compounded zero coupon rates or could assume constant forward rates; the transformation could make one of several different assumptions about the relationship between the price of a US$ LIBOR Eurodollar future and the corresponding US$ LIBOR forward rate, etc.).

Consequently, the set of zcdfs used for valuation are not uniquely specified by the market because there are alternative sets of observed market data to choose from and there are alternative methods for transforming a given set of observed data into a set of zcdfs.

[4] I am assuming that the counterparty's credit risk can be ignored. In principle the yield curve used to discount expected future cash flows should take into account the relative credit risk of the counterparty to the trade.

Simple interest rate derivatives (e.g. interest rate swaps, simple interest rate caps or floors) have standard revaluation formula. Consequently, almost all of the differences in valuation of simple interest rate derivatives between different systems will be caused by differences in the choices of observed market data and differences in the transformation and interpolation algorithms used on the observed market data. The ultimate test of the validity of the entire valuation system is a comparison of the value obtained from the valuation system to the value observed in the market place. For the liquid US$ swap market the differences in valuation caused by different choices of input and different transformation algorithms, as a percentage of notional principal, should be relatively very small. In less liquid markets the difference in valuation may be quite large and material.

In summary, the market value calculated by a valuation system depends on: (a) the observed market factors that are selected; (b) the algorithms used to transform and interpolate the observed data into inferred market data and intermediate market rates (e.g. zcdfs); and (c) the revaluation model or process itself. To the extent there is a reasonable range of choice in any of the components of valuation, the valuation system could generate a range of valuations. The magnitude of valuation uncertainty is called *valuation risk*. It is a misnomer to call this 'model' risk because doing so attributes the entire uncertainty of the valuation to the revaluation model and misdirects attention from the other components of the valuation process.

A valuation error occurs when there is a difference between the market's valuation of a contract and the value assigned by a valuation system. In a liquid market, with small bid/offer spreads, the range of potential valuation is constrained (usually tightly constrained) by observable rates in the market.

To prevent or minimize valuation losses, a firm should take two actions:

- It should set aside a reserve for valuation uncertainty, whenever there is material uncertainty in the value assigned by the valuation system, given the choices or uncertainty in each step of the valuation process. Setting aside such a reserve requires careful scrutiny of accounting and regulatory standards.
- It should periodically calibrate the valuation system by comparing the system's valuations to the market's, to the degree that is possible, to ensure they are essentially the same.

My experience is that most of the large valuation losses that have been reported by firms in recent years can be attributed to either (a) errors in the values of the market factors used for valuation or (b) errors in the algorithms used to extrapolate from observed market rates to inferred, unobserved rates. Valuation losses caused by a problem with the revaluation formula/process as such are rarer. That is why it is more precise to speak of a 'valuation system error' rather than 'model error'.

3 Market errors

3.1 General and specific market risk factors

A very large commercial or investment bank might require more than twenty thousand market factors (including specific equity and debt security prices) to revalue all the contracts in its portfolio.

Market factors are the basic market prices and rates in terms of which contracts are valued. Market factors can be classified as general or specific. Examples of *general market factors* are:

- The term structure of Treasury or LIBOR interest rates
- Spot FX rates
- Spot equity indices
- Spot and forward commodity prices
- The term structure of the implied volatilities of options on each of the above.

Examples of *specific market factors* are:

- Spot price of an equity. The specific risk component is more precisely defined as the component of the change in the spot price of the equity that is idiosyncratic (or specific) to the firm that issued the equity and that can not be explained by changes in more general prices. As a simple example, the component of the change in the price of an equity that has zero correlation with the change in the primary equity index of that country.
- Spot price (or, equivalently, the yield) of a corporate debt security. The specific risk component is more precisely defined as the component of the yield (and the change in the yield) that is idiosyncratic to the issuer and that can't be accounted for by more general yields. As an example, the total yield of a fixed income corporate debt security of a specific tenor could be decomposed into a general base rate (e.g. five year LIBOR rate), a general spread to the base rate by risk rating and tenor (e.g. the average spread for corporate debt securities with a risk rating of 'A' and tenor of five years) and an idiosyncratic component, specific to the issuer of the debt security. In this example only changes in the latter (idiosyncratic or issuer specific) spread would be labeled as specific risk.

- Spot price (or equivalently the yield) of a sovereign debt security issued in another country's currency. More precisely, the specific risk is the component of the yield of the security that is idiosyncratic to the issuer and that can't be accounted for by more general yield components (e.g. for a South American government bond denominated in US$, the base yield would be the US Treasury yield while the spread to Treasury for that issuer and tenor could be considered to be idiosyncratic and hence specific). Alternatively, the total yield of the bond might be decomposed into a US Treasury yield for that tenor, a general spread for that tenor for South American government debt issued in US$ and finally an idiosyncratic spread for that tenor for that particular issuer).
- The term structure of the implied volatilities of options on each of the above.

General market factors reflect the current and the market's expectations and uncertainties about future macro-economic conditions (e.g. level of economic growth, inflation, etc.). Specific market factors reflect the current and the market's expectations about both general macro-economic conditions and the financial conditions and risks of specific issuers.

4 Introduction to market risk measurement

Valuation depends on the market data that is input, the algorithms used to transform the input data and the valuation formula or process as such. Uncertainties in any of these components will give rise to uncertainties in valuation. In the remainder of this section, unless otherwise explicitly stated, I will put aside the issue of valuation uncertainty and assume that there is sufficient information to unambiguously value contracts. I will focus on the risk of a change in market value due to potential changes in observed market rates.

The value of market factors tomorrow, or at any other future date, may change from their current values – e.g. yield curves, spot exchange rates, implied volatilities may change. Consequently the future marked to market value of contracts may differ from their current value. The potential future loss that can occur because of a decrease in market value is called *market risk.*

Although the market value of a debt security should be assessed directly from its market price, its market risk is usually analyzed by means of a discounted cash flow model. This enables one to ascertain the sensitivity of the market value of the debt security to changes in the term structure of interest rates (and interest rate spreads). The benefit of this analysis is that it facilitates the integration of the interest rate risk of debt securities with the interest rate risk of a wide range of other types of contracts.

4.1 Types of market risk

Two broad categories of market risk are directional risk and relative value risk. A trader intentionally takes *directional risk* when he expects that the value of a particular market factor (e.g. the price of an equity, the US$ value of Japanese Yen currency) or a class of market factors (e.g. the US Treasury yield curve, the level of implied volatilities of Pound Sterling LIBOR caps) will change in a particular direction in the near future. If the market factor(s) changes in the direction the trader expects, his position will gain in value, if not, it will lose value.[5] A trader intentionally takes *relative value risk* when he expects the relative value of two market factors to change in a particular direction (i.e. the relative difference in value will get either smaller or larger). An example of relative value risk would be buying equity A and selling equity B under the expectation that A will appreciate in value more than B or, equivalently, that B will depreciate in value more than A.

Market risk is sometimes differentiated into two related risks: price risk and liquidity risk. *Price risk* is the risk of a decrease in the market value of a portfolio of contracts. *Liquidity risk* refers to the risk of being unable to transact a desired volume of contracts at the current market price – i.e. without materially affecting the market price. For example, assume I own a thousand shares of equity of company X and assume that on average one million shares of company X's equity trades per day. In that context I should be able to sell all my shares at the current market price without materially affecting the market price. In contrast, assume I own one million shares of equity in company Y and assume that on average one hundred thousand shares of company Y's equity trades per day. In this latter case I will not be able to sell all of my shares quickly, or even in one day, without depressing the price per share. This can be described as ordinary liquidity risk – the risk of having a position in an asset (or type of financial contract) that is materially large with respect to the average daily trading volume of that asset (or type of financial contract). Another type of liquidity trading risk is a form of event risk and can be called *liquidity event risk* – the sudden and unexpected decrease in demand for an asset (or type of financial contract) relative to supply in a market.

Liquidity and price risk are obviously intertwined. The total price risk of a position will depend on the potential change in market rates and prices over the time needed to liquidate (or fully hedge) the position. During a liquidity

[5]More precisely, a trader takes positive directional risk if he expects a market factor to appreciate greater than its forward value, suitably calculated. For example, assume an equity pays no dividend, assume taxes can be ignored and assume it costs the trader 5% per annum to fund the purchase of shares of equity. The forward price of the equity for settlement in one year would be 5% higher than the spot price. If the trader held the equity for one year and it's value only appreciated by 5% then, given his funding cost, his return on the position would be zero.

credit event, the (often) temporary decrease in demand relative to supply will cause the relevant prices to decrease significantly.

4.2 Types of market risk measurement

There are two broad methods for measuring market risk: *scenario analysis* and *statistical analysis*. These two methods of risk measurement are complementary. Neither form of measurement alone is sufficient to monitor and limit market risk. One reason is that a statistical measure of potential loss always rests on some assumptions about the future behavior of markets (e.g. if markets behave in the near future like they have in the past then there is only an $X\%$ probability that the loss in this portfolio over the next day will exceed $\$Y$). Markets however sometimes exhibit 'regime shifts' or 'event risks', which are sudden discontinuous changes in volatilities, correlations or liquidity that often are difficult to model. Scenario analysis is a tool to estimate the potential loss if the market were to jump into such a state.

4.3 Scenario measurements of market risk

A *scenario analysis* measures the change in market value that would result if market factors were changed from their current levels, in a particular, specified way. The scenario analysis might be applied to a single contract, to a portfolio of contracts or to an entire trading business. The specified scenario may entail a change in only a single market factor (all others held constant) or may entail a concurrent change in many market factors. A scenario analysis calculates the change in market value for a specified scenario, without necessarily specifying the probability that the particular scenario would occur. The market factors that are changed may be observed factors (e.g. money market rates, Eurodollar future prices) or derived, intermediate factors (e.g. forward rates or zero coupon rates).

There are several important sub-types of scenario analysis:

- A *Factor Sensitivity* is the measurement of the change in the value of a contract (or a portfolio) for a specified change in a particular market factor, all other market factors held constant. A factor sensitivity requires the specification of the *particular market factor* that is changed (e.g. the three month forward US\$ LIBOR interest rate, starting twelve months from now), the *type of change* (e.g. a relative change, such as a +1% increase in an equity price, or an absolute change, such as a +6 basis point change in the five year Treasury yield), the *magnitude and direction* of the change (e.g. +1 basis point, −10 basis points, +1 standard deviation etc.) and the specific portfolio of contracts subject to the market factor change. A factor sensitivity is not identical to a first derivative and is a broader concept than the standard 'Greeks' (e.g. delta, gamma) described in textbooks.

- A *one-dimensional grid* (or *ladder* or *table*) *of Factor Sensitivities* is a way of representing the functional relation between the *change in value* of a contract, or portfolio, and the change in level of a *particular market factor*. A one-dimensional grid of factor sensitivities is the set of changes in the market value of a contract, or portfolio corresponding to a specified set of instantaneous changes in the level of particular market factor. For example, the set of changes in the market value of a portfolio of FX option contracts corresponding to a specific set of instantaneous changes (e.g. −10%, −5%, −1%, 0%, +1%, +5%, +10%) in the spot US$/Yen exchange rate from its current level. The set of sensitivities to an array of changes in one market factor, with all other market factors held constant, is called a *one-dimensional grid of sensitivities.* As I will explain later, a one-dimensional grid is a better way of representing the nonlinearity of a portfolio than using the first few terms of a Taylor series expansion (e.g. delta, gamma, etc.). In the limit of a very large set of small incremental changes in the value of the market factor, the one-dimensional grid of Factor Sensitivities becomes a continuous function describing the change in portfolio value as a function of changes in the market factor. An example of a one-dimensional grid for a linear and a nonlinear portfolio are illustrated in Figure 2.

- A *two-dimensional grid of Factor Sensitivities* is a way of representing the functional relation between a change in the value of a contract, or portfolio, and the concurrent change in the level of two market factors, e.g. the changes in the value of the FX options portfolio which would result from a specified set of instantaneous, concurrent changes in the spot US$/Yen exchange rate and the US$/Yen implied volatility. In the limit of small incremental changes, the two-dimensional grid of factor sensitivities becomes a factor sensitivity surface.

A *Stress Test* is a measurement of the change in the market value of a portfolio that would occur for a specified unusually large change in a set of market factors. The specified changes may correspond to an historical stressful event (e.g. the change in market rates that actually occurred during the 1987 stock market crash) or may be a set of hypothetical changes corresponding to some possible future stressful market event. A large financial firm may have sensitivities to over 20,000 market factors. Consequently it would be impossible to define every potential combination of stressful market changes. Judgement must be used in selecting stress tests both with regard to the plausibility of stress event occurring and the materiality of its consequence.

The second major form of measurement of market risk is a *statistical measurement* of potential loss. In essence the potential loss at some confidence

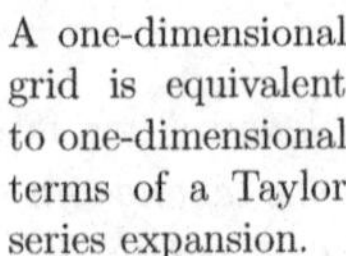
A one-dimensional grid is equivalent to one-dimensional terms of a Taylor series expansion.

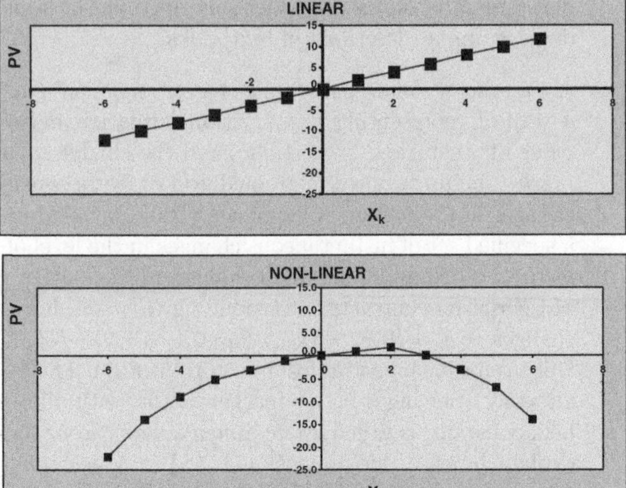

Figure 2: Grids of factor insensitivities. One-dimensional: changes in portfolio value for specific changes in only one market factor, all others held constant.

level is derived by first simulating many scenarios of changes in the underlying market factors, then deriving the corresponding probability distribution of changes in portfolio value and finally calculating a potential loss at some confidence level. There are many (usually an unlimited number of) potential scenarios that correspond to a loss at a particular confidence level.

In summary, a scenario measurement of market risk is a calculation of the potential loss for a given scenario of changes in market factors, without necessarily specifying the probability of that scenario occurring. A statistical measurement of market risk is an estimate of potential loss at some confidence level, corresponding to a countless number of potential scenarios. These are complementary ways of measuring market risk.

4.4 Statistical measurements of market risk

Value at Risk (VAR) is an example of a statistical measurement of price risk. It is an order statistic in which the potential loss of a portfolio is represented by a single number, its value at some confidence level.

Most measurements of VAR assume a *static portfolio.* That is, they measure the potential loss in the value of the portfolio at some confidence level, given a set of potential changes in market rates, calculated under the condition that the composition of the portfolio remains unchanged. This should be contrasted with the assumptions and processes that underlie the accurate calculation of Pre-Settlement Counterparty Credit Exposure (discussed below).

4.5 Holding period and assumption of static portfolio

VAR is usually measured as the potential loss in value of a static portfolio due to changes in market rates. For many, if not most portfolios of traded products, an accurate or realistic measurement of market risk over a time period longer than twenty four hours would require more complex modeling than is typically done for a VAR calculation. To calculate the potential loss in portfolio value over a period longer than one day, it would be necessary to model:

- The structural changes in the portfolio that could contractually occur during the period. For example, to compute VAR over a two week period, it would be necessary to simulate, for each day of the period: (a) the potential daily changes in all market factors; (b) the concurrent setting each day of floating rates for some of the forwards, swaps and options, as required by the contract's terms and conditions; and (c) the concurrent daily settlement of fixed, floating or contingent cash flows, as contractually specified. As floating rates are set and cash flows settle, the portfolio's remaining unrealized value and the sensitivity of its value to changes in market rates will change.

- Depending on the purpose of calculating VAR, it might be necessary to also simulate over some period of time the dynamic adjustment to the portfolio's hedges that would correspond to the simulated changes in market rates (and the resultant simulated changes in the portfolio's factor sensitivities). Given some rules for hedging, given the simulated changes in market rates and given assumptions about or models of market liquidity, the dynamic hedging of the portfolio could in principle be simulated. A prudent VAR calculation would assume that no adjustment to hedges would occur (or that adjustment to hedges could not occur because of market illiquidity).

Even the Bank of International Settlement Internal Models Approach effectively assumes a static portfolio, although it nominally refers to a 'ten day holding period'. In actuality, that approach requires a firm to calculate the potential loss in portfolio value at the 99% confidence level, assuming a static portfolio and the simulation of market rate scenarios corresponding to a ten-day change in market rates.[6]

[6] If one assumes that market rates are log-normally distributed, that daily changes in market rates are independent of one another and finally that the portfolio is linear then the ten-day 99% shock in market rates is equivalent to a 7.36 standard deviation measure of VAR.

4.6 Limitations on VAR

VAR measures the potential loss in the market value of a portfolio of contracts at some confidence level, based on an N-day shock in market rates. The standard calculation of market risk VAR assumes a static portfolio and an instantaneous shock in market rates. The actual change in the profit or loss of a trading portfolio over one day will be affected by several activities not captured in the standard VAR calculation:

- *Intra-day adjustments to hedges.* Traders may adjust their hedges more frequently than once a day.
- *Intra-day trading.* In principle, traders could go home at the end of the day with no directional or relative value positions (and consequently have an end-of-day VAR of zero) while taking large positions in the market intra-day. In practice, the relative amount of intra-day risk taking will vary by desk and will depend on many considerations (e.g. the objectives of the trading desk, the liquidity of the market, etc.).
- *Customer flow.* A large commercial bank may have a large corporate customer base. The spreads earned on customer business will normally be included in the daily profit and loss of the trading desk and may be difficult to cleanly separate from the profit and loss attributable to daily position taking.
- *Net interest revenues or expenses.* This includes the net cost-of-carry of assets (i.e. funding costs minus interest earned), the cost of funding unrealized gains (and the corresponding interest income from unrealized losses) and the drip from any deferred income.
- *Adjustments for expected cost of counterparty credit risk, liquidity risk and valuation risk.* Some financial firms may adjust the mark to market value of positions, that are initially calculated at a bid-offer midpoint, to reflect the expected costs of the risks of the position.

The consequence of the above is that depending on the relative importance of these various factors (e.g. customer flow vs. proprietary position taking, intra-day risk taking vs. longer term position taking, etc.) the hypothetical probability distribution of revenue that underlies the calculation of VAR will more or less correspond to the actual probability distribution of the revenue of the trading business.

Even if we only consider a static portfolio with no intra-day hedging or trading and even if we ignore the revenue impact of customer flow and net interest revenue, the calculation of VAR will likely have certain limitations. One of the most common limitations of VAR models is their poor ability to correctly model event risk, i.e. their inability to incorporate into the VAR

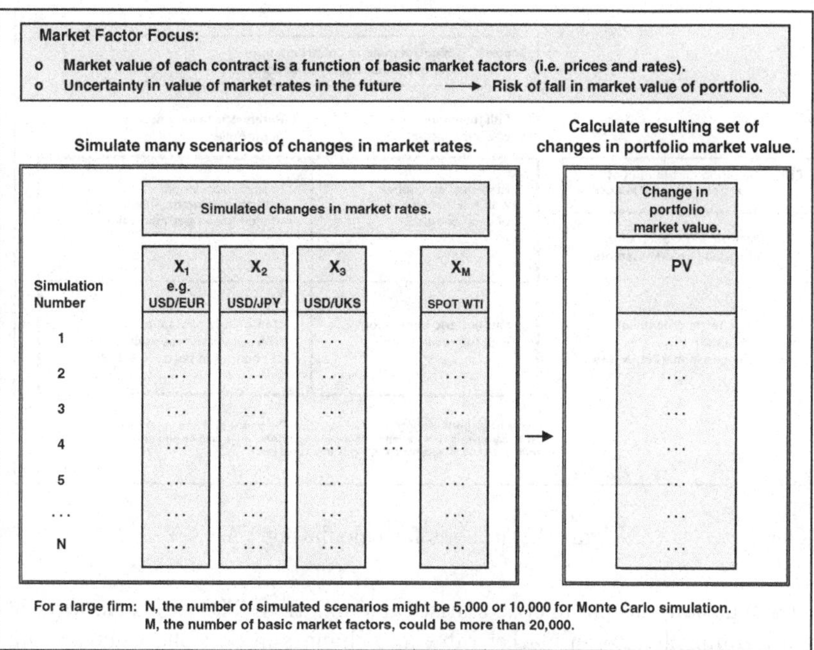

Figure 3: Steps in calculating VAR

calculation the magnitude and the probability of very large changes in market factors and the correlations that correspond to these very large change in market factors.

4.7 The two basic components of VAR calculation

The two basic steps in the calculation of VAR are the simulation of changes in market rates and the calculation of the resultant change in portfolio value. In Figure 3 the simulation of market factors is represented by a two-dimensional array of numbers. Each row of numbers corresponds to a particular scenario of simulated changes in market factors. Each column of numbers corresponds to the simulated changes in a particular market factor. There should be a column for each market factor on which the value of the contracts in the portfolio depends. A very large commercial bank might need to simulate changes in over 20,000 market factors.

VAR methodologies differ both with respect to how market rates are simulated and with respect to how changes in market rates are transformed into changes in portfolio value. The basic alternatives are described in Figure 4. The two basic methods for simulating changes in market factors are histor-

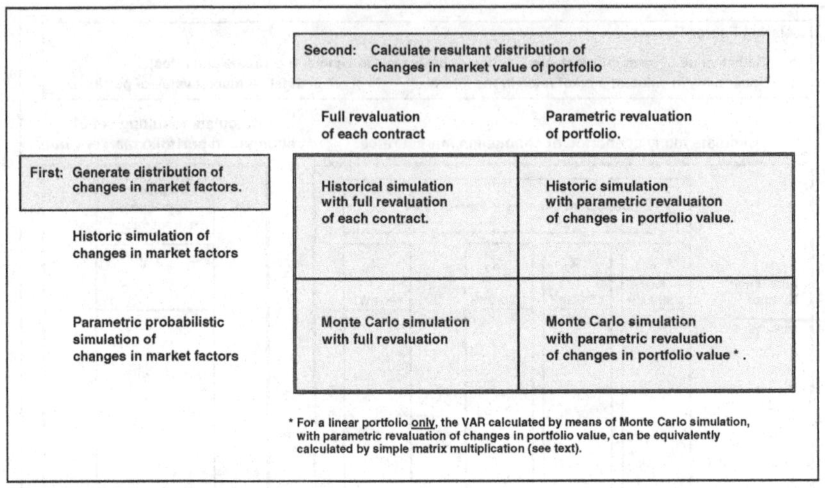

Figure 4: Methods for calculating VAR

ical simulation and parametric statistical simulation. The two basic ways of transforming changes in market rates into changes in portfolio value are full contract revaluation and parametric portfolio revaluation. As we shall see, matrix multiplication is a special, limited case of combining statistical parametric simulation with parametric portfolio revaluation.

We shall first describe methods for simulating changes in market rates and then describe methods for transforming these simulated changes into changes of portfolio value.

5 VAR: Simulating changes in market factors

The different forms of simulation can be most easily understood as different methods for populating the rows of Figure 3 – i.e. different methods of generating a scenario of simulated changes in market rates.

5.1 Historical simulation

In *historical simulation* the rows in Figure 3 are populated by the actual changes in market factors that occurred between specified pairs of dates in the past. As an example of this method, one scenario for the potential change in each market factor between today and tomorrow would be the actual changes in those market factors that occurred between a consecutive pair of dates in the past. By this method a thousand simulations of the potential changes in market factors between today and tomorrow could be obtained from a thousand pairs of past dates. Historical simulation has some strengths and some weaknesses.

Its strengths are:

- It does not make any explicit assumptions about the shapes of the probability distribution of changes in market rates.
- It does not assume that correlations are stable and in fact makes no explicit assumptions about correlations (although it implicitly assumes that past changes in market rates, whatever they may have been, are useful for estimating the potential loss from future changes in market rates).
- Therefore, historical simulation can incorporate fat tails, skewness and dynamic correlations (correlations that are a function of the magnitude of the change in market rates) if the historical sample period had such features.

It weaknesses are:

- A problem with obtaining a sufficient number of historical simulations – e.g. to generate a thousand independent simulations of a one day change in rates, it would be necessary to obtain a time series of four years (roughly two hundred fifty business days per year times four). To generate 1,000 independent historical simulations of an N-day change in rates, it would be necessary to obtain a time series of N thousand days – e.g. 1,000 independent simulations of a ten-day changes in rates (as required by BIS) would require a time series of forty years! This would be impossible to obtain for many market factors and would be meaningless for all market factors, even if the data could be obtained, because of the structural changes in markets that would have occurred over that time period.
- A problem with trends for short sample periods. For example, assume that historical simulation is done on the basis of the last 100 business days. Assume that on most days during that period the Japanese Yen fell in value against the US$. Consequently, a large spot FX position that was short Yen against the US$ would gain in value for most of the historical simulations based on the last 100 days. Under this condition the VAR might be small. This might significantly understate the actual risk – the potential loss that would occur should the trend in the market (a falling Yen) reversed itself.

5.2 Parametric statistical simulation

The essence of *parametric statistical simulation* is the use of statistical parameters (e.g. standard deviation, correlation, etc.) derived from time series of past changes in market rates to simulate future changes in market rates. In parametric statistical simulation the rows of Figure 3 are populated by generating random changes in market factors based on historically derived statistical parameters.

The simplest form of parametric simulation (implicitly) assumes that near term changes in market factors are normally (or lognormally) distributed and that correlations of changes in market rates are stable. On these assumptions future changes in market rates are simulated in terms of the standard deviations and correlations of past changes in market rates. The statistical parameters are calculated for some historical sampling period. Even in its simplest form, parametric statistical simulation entails many choices, such as: (a) the length of the historical sample period (e.g. three months, three years); and (b) whether the volatilities and correlations are derived by giving the terms of the historical time series equal or unequal weights (e.g. exponential weights or GARCH analysis).

In a more complex form, statistical simulation would depend on more complex assumptions than normality and stable correlations. It would incorporate kurtosis and skewness in the probability distributions of changes in market rates and dynamic correlations of changes in market rates (e.g. the correlations could depend on the magnitude of changes of market rates). Incorporating fat tails, skewness and dynamic correlations is particularly important if the VAR measurement is to accurately represent the potential loss that could occur at a very high confidence level. It is non-trivial to incorporate such features into the statistical simulation of a large portfolio.

Monte Carlo simulation is a general method of modeling stochastic processes (i.e. processes involving human choice or processes for which we have incomplete information). It simulates such a process by means of random numbers drawn from probability distributions which are assumed to accurately describe the uncertain components of the process being modeled. Monte Carlo simulation is extensively used in physics and engineering as well as in finance.

5.3 Technical issues in simulating changes in market factors

The calculation of VAR at a very large global financial institution will require the simulation of thousands of general and specific market factors. There are many mathematical and practical issues associated with the simulation of such a large number of factors, and a thorough discussion of all the technical issues involved in such a simulation would make this article far too lengthy. Consequently I have only highlighted some of the issues below. A more detailed technical discussion of some of the general issues of simulating market factors can be found in another paper by the author.[7]

5.4 Covariance matrix

The calculation of the VAR for a general portfolio of transactions (i.e. a portfolio which includes options) whose market value depends on N market factors, requires the simulation of changes to each of these N factors. Under the simplest form of parametric statistical simulation, this simulation would be based on N volatilities and $N \times (N-1)/2$ correlations. These statistical parameters can be combined in an $N \times N$ covariance matrix V, for which $V_{j,k} = \sigma_j \times \sigma_k \times \rho_{j,k}$, where σ_k is the standard deviation of the daily change in market factor k (i.e. historical daily volatility of X_k) and $\rho_{j,k}$ is the correlation of the daily changes in market factors j and k.

A full parametric simulation of the N market factors would start with the $N \times N$ covariance matrix, but in practice the number of statistical elements required for simulation can be reduced by various simplifications and approximations. For example, market factors can be grouped into classes of similar factors (e.g. all spot exchange rates, all term structures of interest rates, all spot and forward commodity prices, etc.). Under some conditions the correlations between market factors in different classes are weak and to a good approximation can be set to zero. Setting such cross factor correlations to zero results in a block diagonalized covariance matrix.

5.5 Positive definiteness

The standard techniques for simulating correlated changes in N market factors are *Cholesky decomposition* and *Principal Component Analysis*.[8] Both start with the $N \times N$ covariance matrix V. Each of these standard techniques requires that V be positive definite. A matrix is *positive definite* if all of its eigenvalues are greater than zero. In other words, an $N \times N$ covariance matrix V can be used directly for statistical simulation only if an $N \times N$ matrix C can be found for which CVC^{-1} is a diagonal matrix L with positive diagonal elements, i.e. $L_{k,k} = (\lambda_k)^2 > 0$, $L_{j,m} = 0$ for $j \neq m$. The calculation of the eigenvalues of V can be done with *Singular Value Decomposition*.[9] The covariance matrix V may not be positive definite for several reasons.

[7]See Evan Picoult, 'Calculating Value At Risk with Monte Carlo Simulation', in *Risk Management for Financial Institutions*, Risk Publications, 1997. The same essay with a minor change regarding BIS rules appears in two other books by Risk Publications: *Monte Carlo*, 1998 and *Internal Modeling and CAD II*, 1999.

[8]For a description of using Cholesky decomposition or Principal Component Analysis to simulate correlated changes in market factors, see Picoult, 1997 *op. cit.*

[9]For a description of Singular Value Decomposition and Cholesky decomposition, see, for example, W. Press, S. Teukolsky, W. Vetterling and B. Flannery, 1992, *Numerical Recipes in C, The Art of Scientific Computing*, Second Edition, Cambridge University Press.

- *Eigenvalues of Zero* The covariance matrix V may be degenerate and have zero eigenvalues. One cause of zero eigenvalues could be that the elements of the historical time series of one market factor were a linear combination of the elements of the time series of other market factors, e.g. if the time series for the 7-year Treasury yield was inferred at each date as a weighted average of the 5- and 10-year Treasury yields on that date. Zero eigenvalues can also occur if the number of sample dates in the time series is less then the number of market factors.
- *Negative Eigenvalues* Negative eigenvalues represent a material problem in that they correspond to market factors whose standard deviation is an imaginary number! A negative eigenvalue whose magnitude is extremely small relative to the largest positive eigenvalue is likely due to a rounding error. Techniques can be used to effectively treat a very small negative eigenvalue as equivalent to a zero eigenvalue. Large negative eigenvalues on the other hand indicate a set of time series with a material lack of internal coherence and consistency.

Techniques can be followed to perform a simulation using covariance matrix V if its only problem is zero eigenvalues or extremely small negative eigenvalues.

5.6 Completeness of market factors

The number of market factors needed for simulation depends on the types of contracts that might be traded and the types of risks that might be taken by the trading desk. Consider, as an example, the US Treasury yield curve.

One issue in simulating changes in the US Treasury yield curve concerns the *representation of the curve.* The term structure of interest rates can be represented by a set of yields-to-maturity, by tenor; or by a set of forward rates, by forward tenor; or by a set of zero coupon discount rates, by tenor. As each representation can be transformed into the other, it is optional which representation is used for the very short term simulations entailed in calculating VAR. Similar issues occur in the simulation of implied volatility curves or surfaces.

A second issue concerns the *number and type* of discrete elements to simulate. For example, assume a trading desk only trades US Treasury securities. Assume we choose to represent the term structure of interest rates by means of thirteen Treasury yields-to-maturity, comprising a set of tenors from 1 month through thirty years. Once we have selected the form of representation (e.g. yields-to-maturity) and the number, M, of discrete elements (e.g. 13 points on the yield curve from 1 month to 30 years) that are needed for simulation, we can construct an $M \times M$ covariance matrix. In our example the $M \times M$ matrix will be created by the historical volatilities of 13 points on the yield curve and the correlations of changes in each of these points. For a large trading business we would need a much larger matrix than the 13×13 matrix just described. To make the following points however, I will limit my discussion to only the US Treasury curve, as if the trading business had no sensitivity to any other yield curve.

If it is positive definite, the $M \times M$ covariance matrix can be transformed into 13 principal components (or eigenvectors). The eigenvalues of these components can be represented as a spectrum from most material to least material. For the US Treasury curve, most of the daily changes in the yield curve can be explained by just the first three principal components (i.e. a change in the 'level' of the yield curve, a change in the 'tilt' of the yield curve and a change in the 'bending' of the yield curve).

Some people have mistakenly argued that since the first three principal component are sufficient to account for most of the daily change in the US Treasury yield curve, the calculation of VAR for a US Treasury desk should also only require the simulation of the first three components. The error in this view is that it confuses two issues: (a) the number of factors needed to account for most of the daily changes in the US Treasury yield curve with (b) the number of factors needed to explain the risk of the position.

To see the difference in these issues, consider the following example. Assume a trading desk had a very large relative value position: a large 'butterfly' position in which they were long twenty year Treasury bonds, short 25 year bonds and long 30 year bonds. A trading desk would take such a position if it thought the longer tenor portion of the yield curve would become more negatively convex – i.e. if the desk expected the 25 year Treasury yield to increase relative to the 20 and 30 year yields. This position would lose money if the longer tenor portion of the yield curve became less negatively or more positively convex – i.e. if the yield at 25 years decreased relative to the yields at 20 and 30 years. The position might have been structured so that it had virtually no sensitivity or very little sensitivity to changes in any of the first three principal components – i.e. virtually no sensitivity to a change in the level or 'tilt' or 'bend' of the yield curve (where the 'bend' would have an inflection point under 15 years). Consequently, if the VAR for this position was calculated by simulating changes in only the first three principal components of the treasury yield curve, no material risk would be measured.

That is why the number of market factors needed to fully model the risk of the portfolio needs to take into account the kind of contracts that might be traded and the types of positions the trading desks might hold.

5.7 Simulation of market risk with specific risk

Specific risk refers to the issuer risk of equity and most debt securities (the exception being debt securities issued by a sovereign in its own currency). In terms of simulation, specific risk refers to the fact that (a) the change in the yield of a corporate debt security, issued in some currency, is not perfectly correlated to the change in the Treasury yield of that currency and (b) that the change in price of an equity, issued in some currency, is not perfectly correlated to the change in the value of an equity index within the same country. The need to model specific risk is a form of the larger issue of the completeness of the market factors used in simulation. This is because not all the risk of the portfolio would be captured if the simulation were based on the crude assumptions that all yield curves within a country were perfectly correlated and all equity prices within a country were perfectly correlated.

One way of modeling specific risk is to treat each corporate yield curve as a separate market factor that has some correlation with each point on the Treasury yield curve and each point on every other corporation's yield curve. A more parsimonious method would model each corporate yield in terms of factors. For example a three factor approach could model a fixed rate corporate yield of a given remaining tenor issued by a firm with a particular risk rating as the sum of three components (a) the yield to maturity, at the given tenor, of a base yield curve selected for that country (e.g. the LIBOR curve or the Treasury curve), (b) a general (average) market spread above the base yield for all corporate debt issued in that country by firms with the particular risk rating and the given, remaining tenor and (c) an idiosyncratic spread (which could be positive or negative) for that particular corporate issuer. Changes in the fixed rate corporate yield would then be modeled as the sum of changes in each of the three components.[10]

6 VAR: simulating changes in portfolio value

The second stage in calculating VAR is to translate the simulated change in market factors into the corresponding simulated change in the market value of the portfolio. There are two basic ways of doing this transformation: (a) *Full Valuation* or (b) *Parametric Portfolio Valuation.*

6.1 Full valuation

In a full valuation method each contract in the portfolio is revalued for each simulated change in market factors (i.e. for each row of Figure 3). This would be a costly process in several ways. For example, a costly infrastructure would have to be built if the full valuation were to be done centrally. A central process would require (a) a database with detailed information on every transaction and (b) a central collection of every revaluation algorithm for every product. In addition it would be costly in time in that each transaction would have to be revalued thousands of times.

[10]For more details see Picoult, 1997, *op. cit.*

6.2 Parametric portfolio revaluation

A more efficient process than full valuation would be the calculation of the Grids of Factor sensitivities described above and illustrated in Figure 2. For each of the one-dimensional grid of sensitivities that is shown in Figure 2, each contract in the portfolio would be revalued thirteen times – once at the current level of market rates and twelve additional times, under the condition that the single market factor was varied over a range above and below its current market level.

For a linear portfolio the result would of course be a grid of factor sensitivities that formed a straight line, as illustrated. Consequently, for portfolios that only have linear instruments, it is sufficient and most efficient to simply calculate a single factor sensitivity per market factor. The information in a multiple point factor sensitivity grid is redundant.

For a nonlinear portfolio (i.e. one with various put and call options bought and sold, at different strike prices) the grid of factor sensitivities will form some curve that may have a complex shape – i.e. that may not be described simply by means of only the first few terms of a Taylor series expansion.

The grids of factor sensitivities that will be generated, for each market factor that a portfolio has sensitivity to, function as look-up tables by which one can calculate the change in the value of a portfolio for a given scenario of simulated changes in market factors. Grids of factor sensitivities eliminate the need to perform a full revaluation of every transaction in the portfolio for each scenario of simulated changes in market factors.

Under this method, for each portfolio of transaction on each revaluation system, it would be necessary to calculate a grid of factor sensitivities for each and every market factor to which the transactions were sensitive. Depending on context one might calculate one-dimensional grids of factor sensitivities (in which only one market factor was varied from its end of day value) or two-dimensional grids (in which a pair of market factors were concurrently varied from their end of day value).

6.3 Grids of factor sensitivities and the terms of a Taylor series expansion

The total market value of a portfolio $PV_{\text{portfolio}}$ is the sum of the values of each of its component contracts. Given the revaluation formula PV_m of each contract m, we have:

$$PV_{\text{portfolio}} = \sum PV_m(T\&C_m, X_j(t), t)$$

where $T\&C_m$ are the terms and conditions of each contract, m and $X_j(t)$ are the values of each of the market factors, j, on which the contract's value depends. For the Taylor series expansion, the $X_j(t)$ are the independent variables. Consequently, we have:

$$\begin{aligned}\Delta PV_{\text{portfolio}} &= \sum_j \left(\frac{\delta PV}{\delta X_j}\right)\Delta X_j + \frac{1}{2}\sum_j \left(\frac{\delta^2 PV}{\delta X_j^2}\right)\Delta X_j^2 \\ &\quad + \frac{1}{2}\sum_{j\neq k}\left(\frac{\delta^2 PV}{\delta X_j \delta X_k}\right)\Delta X_j\,\Delta X_k + \cdots \\ &\quad + \frac{1}{6}\sum_j \left(\frac{\delta^3 PV}{\delta X_j^3}\right)\Delta X_j^3 + \frac{1}{6}\sum_{j\neq k}\left(\frac{\delta^3 PV}{\delta X_j^2 \delta X_k}\right)\Delta X_j^2\,\Delta X_k + \cdots \\ &\quad + \frac{1}{6}\sum_{j\neq k\neq l}\left(\frac{\delta^3 PV}{\delta X_j \delta X_k \delta X_l}\right)\Delta X_j\,\Delta X_k\,\Delta X_l + \cdots,\end{aligned}$$

where j, k, l, etc. are summed over all market factors. The portfolio's partial derivatives are summed over each contract's partial derivatives.

This polynomial is one form of parametric portfolio revaluation. The parameters in the polynomial are the instantaneous partial derivatives of the change of the portfolio's value with respect to changes in its market factors. These parameters are derived for each contact from the contract's exact revaluation formula and then summed across all contracts in the portfolio. The first set of parameters are the instantaneous deltas. The second set are the instantaneous gammas. The third set are the instantaneous cross gammas. And so on. An *unlimited set of parameters*, representing higher order derivatives, could potentially be included.

The minimum number of terms of a Taylor series expansion that are needed to achieve a specified degree of accuracy in the VAR calculation depends on the composition of the portfolio, the confidence level at which VAR is to be measured and the relationship between the set of market factors used for simulation and the market factors on which the contract's value depends. This latter point is not obvious and will be explained below.

For reasons that will be clearer below, we can reorganize the terms of the Taylor series expansion into the following subset of terms:

$$\begin{aligned}\Delta PV_{\text{portfolio}} &= \text{One-Dimensional Terms} + \text{Two-Dimensional Terms} \\ &\quad + \text{Three-Dimensional Terms} + \cdots.\end{aligned}$$

Let us now define each of these terms and explain what we mean by dimensionality. We will begin with the 'one-dimensional' terms:

$$\begin{aligned}\text{1D Terms} &= \sum_j \left(\frac{\delta PV}{\delta X_j}\right)\Delta X_j + \frac{1}{2}\sum_j \left(\frac{\delta^2 PV}{\delta X_j^2}\right)\Delta X_j^2 \\ &\quad + \frac{1}{6}\sum_j \left(\frac{\delta^3 PV}{\delta X_j^3}\right)\Delta X_j^3 + \cdots \\ &= \sum_j \text{polynomial function of } (\Delta X_j).\end{aligned}$$

The one-dimensional terms are the sum of polynomial functions of the change of each single market factor. The Taylor series expansion will consist of only one-dimensional terms if the revaluation formula of each contract can be expressed as a separable function of each market factor, that is, if the sensitivity of the value of a contract to changes in each market factor is independent of the magnitude of the change of any other market factor.

$$\begin{aligned}\text{2D Terms} &= \frac{1}{2}\sum_{j\neq k}\left(\frac{\delta^2 PV}{\delta X_j\,\delta X_k}\right)\Delta X_j\,\Delta X_k + \frac{1}{6}\sum_{j\neq k}\left(\frac{\delta^3 PV}{\delta X_j^2\,\delta X_k}\right)\Delta X_j^2\,\Delta X_k\cdots\\ &= \sum_{j\neq k}\text{polynomial function of }(\Delta X_j, \Delta X_k).\end{aligned}$$

The two-dimensional terms are the sum of polynomials of pairs of market factors. The Taylor series expansion will depend on two-dimensional terms if the sensitivity of the market value of a contract to changes in a market factor is conditional on the magnitude of the change in one other market factor. An example would be a spot exchange rate and the corresponding implied volatility of an FX option. Under certain conditions the change in the value of an option corresponding to a change in the spot exchange rate (particularly for a large change) will depend on the magnitude of the change in the implied volatility.

$$\begin{aligned}\text{3D Terms} &= \frac{1}{6}\sum_{j\neq k\neq l}\left(\frac{\delta^3 PV}{\delta X_j\,\delta X_k\,\delta X_l}\right)\Delta X_j\,\Delta X_k\,\Delta X_l + \cdots,\\ &= \sum_{j}\text{polynomial function of }(\Delta X_j, \Delta X_k, \Delta X_l).\end{aligned}$$

The three-dimensional terms will be important if the sensitivity of a contract to a change in a market factor is conditional on the magnitude of changes in two other market factors.[11]

The one-dimensional grid of factor sensitivities to changes in market factor X_j corresponds to the one-dimensional terms of the Taylor series expansion – i.e. a one-dimensional grid of factor sensitivities to changes in market factor X_j corresponds to a polynomial function of ΔX_j . A one-dimensional grid of factor sensitivities is sufficient if the cross-derivative terms of the Taylor series expansion are relatively small.

The two-dimensional grid of factor sensitivities to market factors X_j and X_k corresponds to a combination of the one-dimensional and two-dimensional terms of the Taylor Series expansion terms, i.e. corresponds to a polynomial function of ΔX_j, a polynomial function of ΔX_K and a polynomial function of $(\Delta X_j, \Delta X_K)$.

[11]For more details see Picoult, 1997, *op. cit.*

6.4 Parametric portfolio revaluation and the representation of yield curve

In discussing simulations of changes in interest rates, I pointed out that there were choices in how the yield curve could be represented (e.g. in terms of yields-to-maturities, forward interest rates, etc.) and asserted that for purposes of short term simulation it was optional which representation was used.

When calculating grids of factor sensitivities for linear instruments (e.g. a simple debt security, an interest rate swap) it is also optional which representation is used. However, for interest rate options (e.g. caps, swaptions) the choice of representation will effect the materiality of the cross-derivatives.

For example, consider a simple cap. It can be viewed as a set of options on forward interest rates (i.e. a set of 'caplets'). Consider a caplet on the six month forward interest rate between six and twelve months forward. Assume that interest rate factor sensitivities are calculated in terms of yields-to-maturity. The caplet under discussion would have sensitivity to changes in the six month interest rate and the twelve month interest rate. However, it would also have a large sensitivity to the cross-derivative formed by concurrent change of both the six-month and the twelve-month yield. This is because the forward interest rate between six and twelve months depends on changes in both the six-month and twelve-month rate. Consequently if factor sensitivities were measured in terms of perturbing points on the yield curve, an accurate calculation of VAR would require that two-dimensional grids be calculated for each pair of points on the yield curve.[12] If only one-dimensional grids were used, the error in VAR could be quite large when VAR was calculated for a large shock in market factors.[13]

However, if the grids of interest rate factor sensitivity for caps were calculated in terms of forward interest rates there would not be as much need for the inclusion of the two-dimensional grids of sensitivities because the cross-terms for different forward rates would not be material. For each type of interest rate product there will be some optimal way of collecting grids of factor sensitivities. Consequently, for a large trading business with many types of interest rate options, grids of factor sensitivities might be calculated in several different representations (e.g. some grids will be in terms of points on a yield curve, some grids will be in terms of forward interest rates, etc.). A mechanism will be needed for integrating the grids of interest rate factor sensitivities calculated in different representations with the particular representation chosen for simulating changes in market factors. That is, the simulated states of the market will have been done for some representation of the yield curve (i.e. yields to maturity) while some of the grids of interest rate factor sensitivities might be in other representations (e.g. sensitivities to changes in forward rates).

[12] For other examples, see Picoult, 1997, *op. cit.*

[13] E. Epperlein, internal Citigroup calculation, 1999.

The mechanism for integration is straight forward. The first step is to transform each simulated change in state of the yield curve in the base representation (e.g. in change in the yield-to-maturity at various tenors) into the corresponding change in another representation (e.g. the change in each forward rate). The change in each forward rate can then be mapped into the corresponding change of portfolio value by using the grids of factor sensitivities to changes in that particular forward rates. Because of the nonlinearity of an option portfolio one *can not transform* grids of factor sensitivities to changes in forward rates into grids of factor sensitivities to changes in yields-to-maturity. One can only transform the simulated change in market factors and then look up the corresponding change in portfolio value.

Now that we have explored some of the issues involved in calculating VAR we will turn out attention to another form of trading risk, counterparty credit risk.

7 Pre-settlement counterparty credit exposure

7.1 Forms of credit risk

Credit risk is the risk that the obligor to a financial contract will be unable or unwilling to perform. An obligor to a firm is another firm, a government or an individual who is contractually obligated to pay the firm some net value. Credit risk takes several forms:

- *Lending Risk* Lending risk is the risk that a borrower could be unable or unwilling to pay interest and/or principal when he is contractually obligated to do so. This is the standard accounting view. From a more general economic perspective, lending risk includes that risk that the market value of a loan could decrease because the market's evaluation of the credit quality of the borrower has deteriorated. (See above for a comparison of the accounting and economic perspectives.)

- *Issuer Credit Risk* Issuer credit risk is the risk that the issuer of a debt (or equity) security could default or become insolvent. More generally, it is the risk that the market value of the security could decrease because the market's evaluation of the credit quality of the issuer has deteriorated. Issuer credit risk is one component of the specific market risk of a security, which was discussed above. The specific market risk of a security includes both issuer credit risk and issuer liquidity risk.

- *Counterparty Risk* Counterparty risk is the risk that the counterparty to a trade could default on his obligations. Counterparty risk occurs in two forms:

- *Settlement Risk* Settlement risk is the risk of a 'one-sided trade' – that is, it is the risk that the counterparty to a trade could fail to perform on its contractual obligations at settlement while one's firm performs on its obligations. For example, assume my firm had entered into a forward currency trade to buy (i.e. to receive) 10 million Pound Sterling in exchange for selling (i.e. paying) 15 million US dollars. My firm would have a credit loss due to settlement risk if we transmitted to the counterparty the amount of US Dollars we had agreed to sell, but did not receive from the counterparty the amount of Pounds Sterling we had agreed to buy.
- *Pre-Settlement Risk* Pre-Settlement Risk is the risk that a counterparty to a trade will default[14] *before* the final settlement of the transaction's cash flows. My firm would experience an economic loss if the counterparty defaulted and if, at the time of default, the contract (or portfolio of contracts) had a positive economic value to my firm.

7.2 Defining pre-settlement exposure

As it may not be obvious how an economic loss could occur prior to the final settlement of cash flows, consider a simple example involving two contracts, each with only one exchange of assets. Assume my firm has a trading desk that is a market maker in gold, buying and selling gold from customers for settlement at spot and forward dates. Assume my firm has entered into two transactions: (a) we agreed to *buy* 1,000 ounces of gold from counterparty X, for settlement in one year, at a forward price of \$349/ounce; (b) we agreed to *sell* 1,000 ounces of gold to counterparty Y, for settlement in one year, at a forward price of \$351/ounce. Assume the settlement dates of the two contracts are identical (to make this more concrete, assume we entered into the both contracts on September 1, 2000 and that both settle on September 1, 2001). Assume further that the two transactions were done at the prevailing forward market price, within a small bid/offer spread – i.e. assume the bid/offer midpoint of the forward market price was \$350/ounce.

The consequence of these two trades is that my firm will have locked in a forward bid/offer spread revenue of \$2,000 and will have no incremental market risk – i.e. the net market value of the two trades will have zero sensitivity to changes in the forward price of gold and *de minimus* sensitivity to changes in the yield curve.

The dominant market factor effecting the value of each contract is the forward price of gold, as of the settlement date of the two contracts. A secondary market factor effecting the market value of each contract is the yield curve

[14] The precise definition of the what constitutes default will depend on the specifics of the financial agreement entered into and the applicable legal code.

– more specifically the interest rate needed to discount the expected future cash flows to present value. To simplify the analysis of the potential change in the value of each contract, let us focus only on changes in the forward price of gold while keeping in the back of our mind that a full analysis of market value would also take into account changes in the yield curve.

If the forward price of gold increases, the market value of the forward buy contract with counterparty X will increase while the market value of the forward sell contract with counterparty Y will decrease and *vice versa* for a decrease in the forward price of gold. Although my firm will have no net market risk for the two trades, each trade will create pre-settlement credit risk.

Consider the consequences to my firm if counterparty X were to default before the final settlement of our forward trade. If counterparty X defaulted at some future date and, if at the time of default, the market value of our forward contract were *zero*, my firm would be able to replace it with another contract transacted with another counterparty, at a cost of zero (or, at most, a very small processing cost) – i.e. my firm should be able to find another firm from whom we could buy 1,000 ounces of gold for settlement on September 1, 2001, at a forward price of $350/ounce (ignoring the bid/offer spread).

Alternatively, if counterparty X defaulted at some future date and, if at the time of default, the forward buy contract had a *positive marked to market* value to my firm, we would incur an *economic loss* in replacing the contract. The magnitude of the loss would equal the mark to market value of the defaulted contract. The loss can be viewed from several equivalent economic perspectives:

- Because the contract had a positive market value at the time of the counterparty default, its terms and conditions were 'off market' – i.e. we had contracted to buy gold at a forward price that was now below the market's forward price. To replace the defaulted contract with an identical contract (with the same terms and conditions) but transacted with some other counterparty, my firm would have to pay the new counterparty the current market value of the contract. That replacement cost would be a loss to my firm.

- If we simply replaced the contract with another contract transacted at current market rates, we would have lost the positive unrealized value of the defaulted contract.

- If the defaulted contract were hedging the market risk of some other contract (or portfolio of contracts), which is the most likely case, the positive market value of the defaulted contract would have been offsetting the negative market value of the contracts it was hedging. If my firm replaced the defaulted contract with a new contract transacted at current market rates we would be locking in a net loss on the trading portfolio that would be equal to the forgone positive market value of the defaulted contract.

All three examples illustrate that *the immediate economic cost of replacing the defaulted contract is its market value*, calculated as if there were no default. This assumes, of course, that the size of the transaction with counterparty X is not large with respect to the liquidity in the market and that the transaction could therefore be replaced by another transaction with another counterparty at prevailing market rates.

Finally consider a scenario in which the contract with counterparty X has a negative mark to market value to my firm at the time counterparty X defaults. Naively, one might think my firm would benefit from this situation because we would be able to replace the contract we had transacted with X with a new contract, with another counterparty, under terms more favorable to us. However, in the scenario just painted, the contract has a negative value to us and a positive market value (i.e. it is an asset) to counterparty X. If counterparty X were in bankruptcy, the bankruptcy court would want all contracts that were assets to X to perform in order to improve X's ability to pay off his liabilities. Consequently, if the contract had a negative mark to market value to my firm, we could not replace it and would not benefit if the counterparty were to default.

The analysis of the credit exposure when there are multiple contracts with a counterparty and the function of legal netting agreements will be given below.

In summary, my firm's credit exposure to counterparty X for a single contract is asymmetric. If X defaults we will have a loss when the contract has a positive value but will have no gain if it has a negative value. Let us define the *immediate credit exposure* of a contract as the cost of replacing the contract at its current market value in the event of an immediate counterparty default:

$$\text{Immediate Credit Exposure}(t) = \text{Max}[PV(t), 0],$$

where $PV(t)$ is the current market value of the contract. A similar analysis can be done for the transaction with counterparty Y, the firm to which my firm is selling gold forward. My firm would experience a credit loss if firm Y defaulted when the forward sell contract had a positive mark to market value to my firm (which would occur if the forward price of gold for settlement on September 1, 2001 decreased).

The forward gold transactions with counterparty X and counterparty Y have equal and opposite sensitivities to a change in the forward price of gold. As a consequence, if we ignore the initial small bid/offer spread, it is clear that my firm could not experience a credit loss on both contracts if both counterparties defaulted on the same date and if my firm replaced each contract at the same time. However, my firm could experience a credit loss on each contract if X and Y were to default in the future on different dates and if each contract had a positive market value at the time of the default. For example, consider the scenario in which the forward price of gold appreciated, counterparty X defaulted and my firm replaced the contract with X at an economic loss. As part of the same scenario, assume that months later the forward price of gold dramatically fell and counterparty Y defaulted, causing me to replace the contract with counterparty Y at an economic loss. In this scenario I would have an economic loss on both contracts.

Consequently my firm will have potential credit exposure to both counterparties X and Y because we potentially could have a loss on each contract. The relative likelihood of my firm experiencing an economic loss on both the forward buy from X and the forward sale to Y would be taken into account in the proper assessment of the credit risk of the transactions, as I will explain in the next section where I differentiate credit exposure from credit risk.

A fuller discussion of pre-settlement credit risk needs to address several topics:

1. The distinction between credit exposure and credit risk.
2. The need to measure credit exposure in terms of both the current market value and the potential future market value of a transaction.
3. A description of methods for calculating pre-settlement exposure when there are multiple contracts with the counterparty, some which may have positive and some negative market value.
4. The effects on the measurement of exposure of risk mitigants, such as a netting agreement or a margin agreement.
5. A description of methods for measuring pre-settlement risk.

We will now discuss each of these topics briefly.

7.3 Credit exposure and credit risk

To manage and limit credit risk a firm should measure, monitor and limit two related but different quantities: credit exposure and credit risk.

Credit exposure is the *potential loss due to the default of the obligor, ignoring the probability of the default occurring and assuming no recovery value.* Consider the definition of credit exposure for each of the major types of credit risk:

- *Credit Exposure of Lending* From a standard accrual accounting perspective, the credit exposure of a loan is the outstanding amount of the loan. From an economic perspective the credit exposure of a loan is its market value.

- *Credit Exposure of Issuer Risk* The credit exposure of a security is the market value of the security.
- *Pre-Settlement Credit Exposure* A firm's immediate pre-settlement exposure to a counterparty for a single transaction is the larger of the current market value of the transaction or zero. 'Immediate' exposure refers to the economic loss if the contract had to be immediately replaced at its current market value. A fuller measure of pre-settlement exposure must take into account not only the current market value of a contract but also its potential future value, for reasons explained below. More generally, the immediate and potential pre-settlement exposure for multiple contracts with a counterparty depends not only on the current and potential future market value of the contracts but on the legal enforceability of any risk mitigating agreements that had been entered into, such as netting agreements, margin agreements or option for early termination agreements, as will be discussed below.

Credit Risk is a statistical measure of risk. It is similar to the VAR measurement of market risk. It is derived from the probability distribution of economic loss due to credit events, measured over some time horizon, for some large set of obligors. As illustrated in Figure 5, two properties of the probability distribution of economic loss are particularly important: (a) *The Expected Credit Loss*, over some time frame; and (b) *The Unexpected Credit Loss.* The latter is the difference between the potential loss at some very high confidence level (e.g. 99.0%, 99.9% or 99.97%) and the Expected Loss. These statistical measures have several uses. For example, a market maker with a large portfolio of transactions exposed to credit risk should earn at least enough income from customer spreads to cover the *cost of credit.* The cost of credit can be defined as the sum of the Expected Loss plus the *cost of Economic Capital* for credit risk, where Economic Capital for credit risk is equal to the Unexpected Loss. When defined in this way, economic capital for credit risk functions as a cushion to absorb unexpected credit losses, at a high confidence level, to avoid insolvency.

The probability distribution of loss will be based on several factors in addition to the current and potential future credit exposure. The loss distribution critically depends on the definition of credit loss, the time frame over which the potential loss is measured and the relative diversification of credit risk within the portfolio. As an example, consider a portfolio of loans to many borrowers. Assume that loss is defined only with respect to a borrower defaulting sometime over the life of a loan. The calculation of the probability distribution of credit loss for this portfolio would be derived from the credit exposure to each borrower at each forward period of time, the probability of a borrower defaulting within each forward period of time, the correlation between borrower defaults and the probability distribution of *Loss In the Event of Default* for each type of borrower. Loss In the Event of Default (LIED –

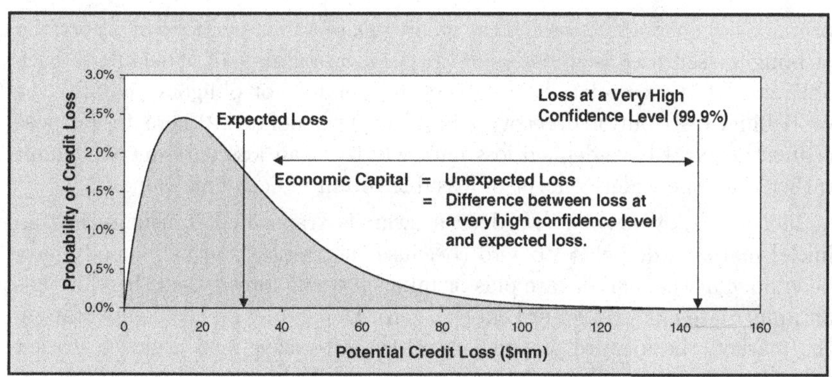

Figure 5: Probability distribution of potential credit loss for a set of obligors.

also known as Loss Given Default) is the percent of the exposure that is actually lost when the borrower defaults, after taking into account the recovery value and the costs of recovering that value. The LIED will be affected by several factors including the seniority of the credit claim on the obligor and on whether any of the loans were collateralized.[15]

If we had defined the loss differently (e.g. as the potential decrease in the economic value of the loan portfolio over a one year horizon instead of the loss due to default over the life of the portfolio) the probability distribution of potential loss would have differed.

The relationship between Credit Exposure and Credit Risk is similar to that between a Scenario Analysis of Market Risk and a calculation of VAR. Credit Exposure is a form of scenario analysis in that it answers a what-if question: what is the potential loss if an obligor were to default, given the assumption of no recovery. The calculation of the probability distribution of potential credit loss is similar to the calculation of the probability distribution of potential market loss in that it rests on information about both potential exposure and the probability of different risky scenarios occurring.

7.4 Contrasting the exposure of Lending Risk, Issuer Risk and Pre-Settlement Risk

A loan or a bought security is always an asset and always has credit exposure. The magnitude of the credit exposure is straightforward to calculate. The

[15]Note the different functions of margin for counterparty risk and collateral for lending risk. In the appropriate legal context, assets posted as margin may reduce the credit exposure whereas, under the appropriate legal context, assets posted as collateral for loans typically do not reduce the measured exposure. Instead, the assets posted as loan collateral function by *increasing* the expected recovery value of the loan (and thereby reducing the Loss in the Event of Default).

primary difficulties in assessing the credit risk of a loan portfolio or a portfolio of bought securities is in assessing: (a) the probability of a default of each obligor; (b) the correlations between the defaults of obligors; and (c) the probability distribution of recovery in the event of default. Where, for purpose of illustration, I have defined loss narrowly to mean loss only due to default rather than the wider concept of loss due to fall in economic value.

In contrast, when a forward or a swap is transacted it usually has an initial market value close to zero (because the transaction will usually have been done at the market rate plus or minus a small spread) and therefore has an immediate credit exposure close to zero. Depending on the future state of the market, the forward or swap may have a positive or a negative market value (i.e. it potentially could be an asset or a liability). Consequently it is not certain if the forward or swap will have any immediate credit exposure in the future, nor is it certain, should it have such exposure, what the magnitude of the exposure would be. It will depend on the future state of the market. The stochastic nature of pre-settlement credit exposure is a salient feature of this form of credit risk.

If the credit risk of the issuer of a security deteriorates, the owner of the security can sell it at the market price, assuming of course that the size of his position is small relative to the liquidity of the market. In contrast, after a firm enters into a forward or derivative transaction with a counterparty, it may have no means of subsequently terminating the transaction or hedging the counterparty credit exposure. It may have credit exposure to the counterparty until the final cash flow settles. Even if the derivative transaction could be assigned to a third party, the selling price (in a rational market) would take into account the potential exposure of the transaction over its remaining life. *Consequently, a prudent measurement of pre-settlement exposure should look at both the current market value of the contract (its immediate replacement cost) and its potential future value over the lifetime of the contract, taking into account any legally enforceable risk mitigants.*

7.5 Pre-settlement exposure to options

A bought option will have a market value to the buyer that is either zero or a positive amount. A sold option will have a market value to the seller that is either negative or zero. Consequently the buyer of the option has credit risk that the seller won't perform. The seller has an obligation to perform and has no credit risk to the option buyer, once the buyer pays the option premium.

Options that settle in terms of an underlying derivative are more complex. Consider an option to enter into a simple fixed/floating interest rate swap (a swaption). The buyer of the option has credit exposure on the option and, if the option is exercised, on the underlying swap that is created. The seller of the option has no credit risk as such on the option. Consider the credit exposure to the seller if the buyer exercises the option to create a swap. Initially the swap will have a positive market value to the option buyer (which is why the option was exercised) and a negative market value to the option seller. Longer term, however, interest rates may change and the swap might have positive value to the option seller. Consequently a sold option on a swap can create credit exposure for the option seller once the swap is created.

7.6 Pre-settlement exposure of multiple contracts and netting

The pre-settlement exposure for multiple contracts with a counterparty critically depends on whether any legally enforceable netting agreements have been entered into with the counterparty. If there is no enforceable *netting agreement*, a bankruptcy court may view each contract separately and 'cherry pick' in the event the counterparty defaults. That is, the court may require the non-defaulting firm to perform on all its contracts that have a positive mark to market value (i.e. are assets) to the defaulting firm (i.e. that have negative value to the non-defaulting firm). For all contracts that have a negative mark to market value (i.e. are liabilities) to the defaulting firm (i.e. that have a positive value to the non-defaulting firm) the non-defaulting firm may have a credit loss.

In contrast, under a legally enforceable netting agreement, the immediate exposure, in the event of counterparty default, is the *net market value* of all the transactions within the netting agreement. *All* transactions includes those with positive as well as those with negative market values. All transactions also includes any options *sold* to the counterparty that are covered by the netting agreement. The negative market value of the sold options could offset the potential positive market value of other transactions covered by the netting agreement.

Let us again define the immediate pre-settlement exposure as the cost of replacing contracts transacted with counterparty in default, at current market rates.

No Netting Agreement: Immediate Exposure$(t)_{M,\text{non-netted}} = \sum \text{Max}[PV(t)_k, 0]$, for all non-netted contracts k counterparty M. That is, there is immediate credit exposure only for contracts that have a positive market value.

A Netting Agreement: Immediate Exposure$(t)_{M,p}$ = Max$[\sum PV(t)_j, 0]$, for all contracts j, under a particular netting agreement p, with counterparty M. That is, there is immediate credit exposure only if the net market market value of all the contracts under the netting agreement is positive. As defined above, $PV(t)_k$ is the market value of contract k at time t.

A large, multinational financial institution might trade forwards and derivatives with another large multinational financial institution at many places around the world. The transactions may have been done between multiple pairs of each firm's booking centers around the world (e.g. firm A, New York with firm B, New York; firm A, New York with firm B, London; firm A, Hong Kong with firm B, Tokyo; etc.). As a consequence, some of the transactions between the two firms might be covered by one netting agreement, a second set of transactions might be covered by a second netting agreement, a third set of contracts by a third netting agreement, and so forth. In addition, some of transactions between the two firms might be covered by no netting agreement.

The legal enforceability of a netting agreement will depend on the home country of each of the parties to the trade and the country in which each has booked the transaction. For example, if the Singapore branch of a US bank entered into a set of forward currency exchanges with the Bangkok branch of a Japanese bank, the enforceability of the netting agreement could, on the face of it, depend on American, Singaporean, Japanese and/or Thai law.

The full measure of portfolio credit exposure requires that we take into account both the current and the potential future market value of contracts. A method for measuring the potential pre-settlement exposure for a portfolio of transactions under a netting agreement is described below.

7.7 Pre-settlement exposure and potential future replacement cost

As we have explained, the immediate credit replacement cost of a transaction is the larger of its current marked to market value or zero. The potential market value of the transaction at a future date may be significantly different from its current value. Two factors can cause the market value of a transaction to change over time: (a) for contracts with multiple cash flows, the contractual fixing of floating rates, expiration of options and settlement of cash flows over time will change the set of remaining unrealized cash flows and therefore the market value of those cash flows; and (b) changes in market rates will change the market value of the transaction's remaining unrealized cash flows.

Measurements of market risk normally assume a static portfolio of contracts and measure the effects on market value of potential changes in market rates. In contrast, the measurement of the potential future pre-settlement exposure of a transaction, or a portfolio of transactions with a counterparty, requires that we simulate potential changes in market rates over time and also simulate the contractual setting of floating rates, the expiration of options and the settlement of cash flows over time.

If the pre-settlement credit exposure on a forward or derivative were only defined in terms of its current market value, a firm would have no effective way to limit its potential future pre-settlement exposure. This is because the initial market value of a forward or swap will be very close to zero and the incremental change in the immediate credit exposure to the counterparty from each new forward or swap would be close to zero. Consequently, a limit on current exposure would not limit the volume of swaps and forwards, transacted with a single counterparty, that could be added to the portfolio. If market rates materially changed at a later date the potential credit exposure to the counterparty could be enormous. To avoid the surprise of a potentially large exposure to a counterparty at a future date, a firm should define pre-settlement exposure in terms of both the current replacement cost of a contract (or portfolio) and its potential future replacement cost.

The two most common ways of measuring pre-settlement exposure are a simple *transaction methodology* and a more precise and sophisticated *portfolio methodology.*

7.8 Simple transaction exposure method of measuring pre-settlement exposure

The simple transaction exposure method defines the *Pre-Settlement Exposure* (PSE) of each transaction as a *single number*, a prudent measure of its potential future market value, defined at some high confidence level. In this method, the PSE of a transaction is the sum of two terms, the transaction's current market value and its potential increase in value, measured at a very high confidence level:

$$PSE(t)_k = \text{Max}[PV(t)_k + PIV(t)_k, 0],$$

where: $PSE(t)_k$ is the pre-settlement exposure of transaction k, at time t; and $PV(t)_k$ is the current mark to market value of transaction k, at time t (this could be positive, zero or negative, depending on context – i.e. depending on market rates and the transaction's terms and conditions); $PIV(t)_k$ is the potential Increase in the Value of transaction k, at time t, estimated at a high confidence level over the remaining lifetime of the transaction – this is always a positive number.

In the transaction exposure method, the formula for the PSE of a transaction has some similarity in form to the value of a bought option. The similarity arises because the magnitude of the pre-settlement exposure can be thought of approximately in terms of the cost of buying a American Call option on the underlying transaction – i.e. the cost of buying an option, capable of being exercised at any time, to create the underlying transaction should the counterparty default. In contrast to the value of an option, which is measured in terms of the expected present value of its future cash flows, PSE is measured at a very high confidence level. It is a 'worst case' potential replacement cost rather than the expected potential replacement cost.

In a simple transaction exposure method, the total pre-settlement exposure of a portfolio of many transactions is calculated as simply the sum of each transaction's pre-settlement exposure.

To facilitate the implementation of a simple transaction method, the *Potential Increase in Value* is defined in terms of a *Credit Exposure Factor* (CEF):

$$PIV(t)_k = NP_k \times CEF(t,T)_k,$$

where NP_k is the notional principal of transaction k, and $CEF(t,T)_k$ is the credit exposure factor for transaction k, which matures at time T, as of time t.

The CEF of a transaction is its potential increase in value, over its remaining lifetime, per unit of notional principal. The potential increase in value is defined at some high confidence level, as explained below. Over the life of the transaction its CEF will tend to decrease. For a contract with multiple cash flows (such as an interest rate swap) one reason the CEF will tend to decrease over time is that unrealized cash flows periodically settle (which reduces the contract's remaining unrealized value) and floating rates get set (which reduces the factor sensitivity of the contract to changes in that market rate). In addition, for any transaction, as time goes by the potential change in market rates over the remaining life of the transaction decreases, assuming constant volatility of market factors. Corresponding to this reduction in the potential change in market rates is a reduction in the transaction's potential change in value.

Tables of CEFs can be defined for all standard forms of forward and derivative contracts in terms of: (a) the *form of the contract* (i.e. forward, European option, swap, etc.); (b) the primary type of *underlying market factor* (e.g. a currency exchange rate, an interest rate, etc.) that determines the market value of the transaction; (c) the *volatilities and correlations of the underlying market factors* of the transaction; and (d) the *remaining tenor* of the transaction. For non-standard transactions, deal specific CEFs can be calculated.

Tables of CEFs may change over time because the best estimate of the future volatility of market factors may change over time.

The best way of calculating the CEF of a transaction is by means of Monte Carlo simulation. For purposes of illustration it is easier to describe a CEF by a scenario simulation.

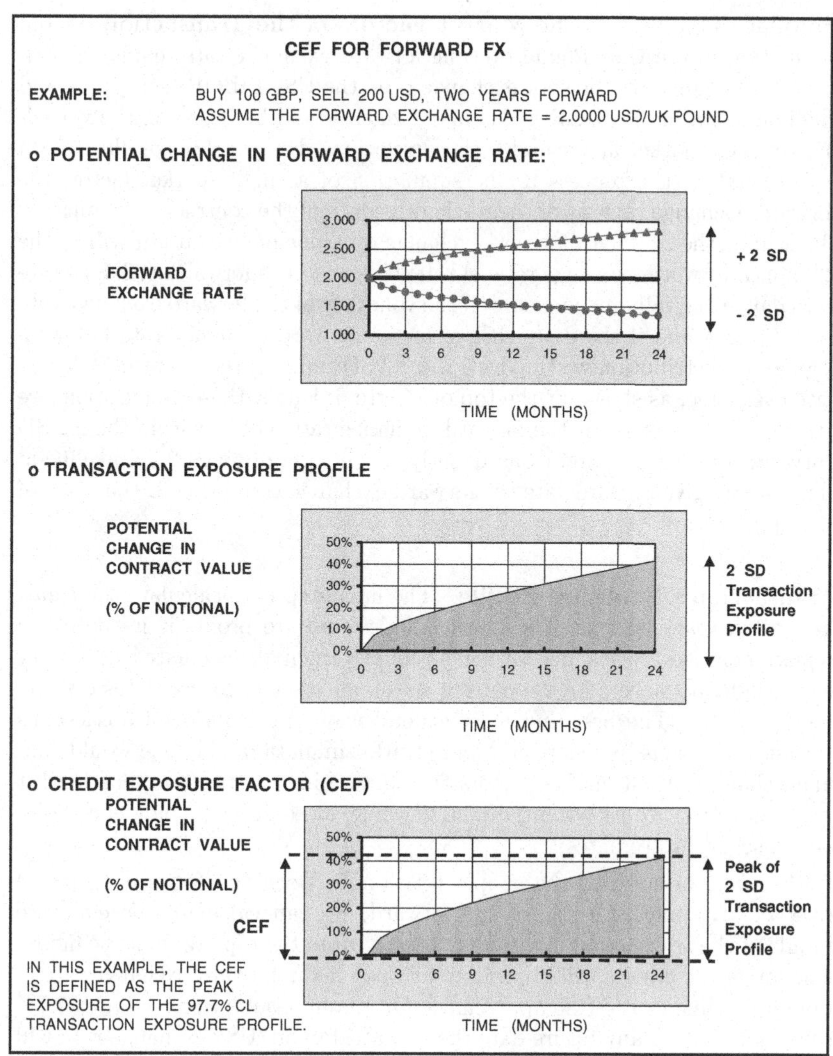

Figure 6: PSE for a Single Contract.

We illustrate the calculation of a CEF for a forward FX contract in Figure 6. The contract is a forward purchase of 100 Pounds Sterling for 200 US$, for settlement in two years. Assume the initial forward exchange rate is 2.00 US$ per Pound Sterling. There are three steps in calculating the CEF:

Simulate Changes in the Market Factors of the Transaction A full simulation in the US$/Pound Sterling forward exchange rate requires a simulation of changes in the spot exchange rate, the US$ LIBOR yield curve and the Pound Sterling LIBOR yield curve, which would be based on the volatilities of each market factor and their correlations. For purposes of illustration we can reduce the process to the simulation of a single market factor: the forward exchange rate, as of the settlement date of the contract – i.e. initially this will be the 24 month forward exchange rate, one month later it will be the 23 month forward exchange rate, etc. If changes in the forward exchange rate each day are serially independent, if the volatility of the forward exchange rate is stationary and if the daily change in the forward exchange rate follows a lognormal distribution we can draw the ± 2 SD range of the forward exchange rate over time, as shown at the top of Figure 6. Under these assumptions we are 95.4% confident that the forward exchange rate will lie within the ± 2 SD curves at any future date. Consequently, there will only be a 2.3% likelihood that on any given future date the forward exchange could exceed the value of the upper curve.

Transaction's Exposure Profile The next step is to calculate the *transaction's exposure profile.* The transaction's exposure profile is its potential replacement cost, as a percent of notional principal, calculated at a very high confidence level, for the current date and a set of future dates over its remaining life. The best and most general way to calculate a transaction's exposure profile is by means of Monte Carlo simulation – i.e. one would simulate changes in all market factors that determine the transaction's market value and measure its potential market value, at some high confidence level, over a set of future dates.

However, because of the simple relation between the forward exchange rate and the market value of the forward, we can calculate the exposure profile of the transaction in Figure 6 directly from the top graph of the figure. The forward contract will appreciate in value if the forward exchange rate of Pound Sterling to the US$ appreciates. Given our assumptions, we are 97.7% confident that on any future date the forward Pound/US$ exchange rate will not exceed the upper graph. Consequently, we can easily transform the set of forward exchange rates at future dates, at the 97.7% confidence level, into the 97.7% confidence level potential value of the forward contract over the same set of future dates.

The transactions profile of the forward FX contract is shown in the second graph of Figure 6. It has a shape that is typical of single cash flow contracts (such as forwards or options that cash settle).

Credit Exposure Factor (CEF) The *Credit Exposure Factor* is defined by reducing the transaction's exposure profile to a single number. This re-

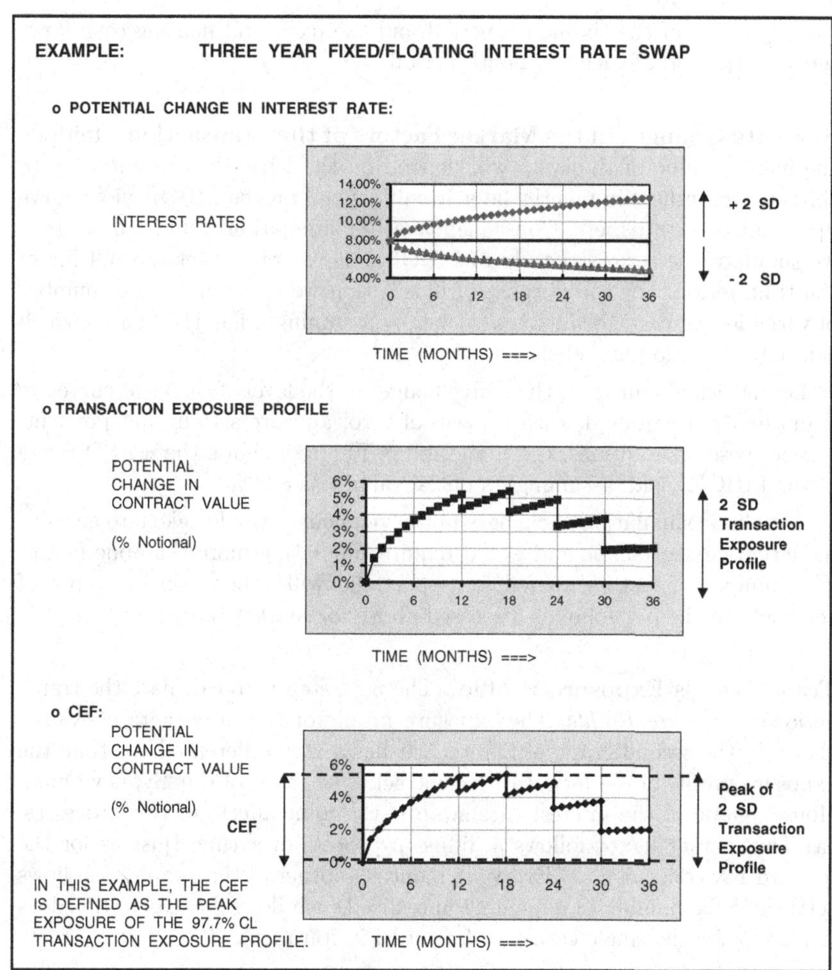

Figure 7: Single Contract.

duction can be made by selecting some aspect of the transaction's exposure profile (e.g. its peak, its average) and using that single aspect to represent the transaction's potential increase in value. If the CEF is defined with respect to the peak of the exposure profile it will equal the difference between the peak of the profile and its initial value – the latter is simply the transaction's current market value, per dollar of notional principal.

A second example of a CEF is shown in Figure 7 for a three-year fixed/floating interest rate swap for which the firm is paying the fixed and receiving the six-month LIBOR floating rate. Assume the floating six-month rate is set at the beginning of the six-month period and the fixed and floating cash flows settle at the end of each six-month period.

Simulate Changes in the Market Factors of the Transaction Initially the market value of the swap will depend on the LIBOR yield curve up to thirty-six months. Six months later it will depend on the LIBOR yield curve up to thirty months, etc. Consequently a full simulation would require that we simulate the term structure of LIBOR interest rates over the full life of the transaction. For the purpose of illustration we can simplify the number of variables we need to simulate to one, by assuming a flat yield curve which can only undergo parallel shifts.

Let us also assume that the daily changes in the level of the yield curve are lognormally distributed, with a mean of zero, and are serially independent. Under these assumptions, the top graph of Figure 7 shows the ±2 SD range of the LIBOR yield, assuming its initial value was 8.00%.

A realistic simulation of changes in the yield curve would take into account its actual current shape and would require simulating more than one factor. The number of factors needed for a portfolio will depend on the types of contracts in the portfolio, as discussed above for market risk simulation.

Transaction's Exposure Profile The next step is to calculate the transaction's *Exposure Profile.* The exposure profile of the interest rate swap is shown in the second graph of Figure 7. It has a very different shape than the exposure profile of the forward FX contract. The cause of the 'Sydney Opera House' shape of the interest rate swap is the joint affects of two processes: (a) The market factor follows a diffusion process over time (just as for the forward FX contract); (b) Every six months another of the swap's cash flows settle and the number of remaining unrealized cash flows decreases (this does not occur for the single cash flow forward FX contract).

Credit Exposure Factor (CEF) The *Credit Exposure Factor* of the three-year interest rate swap is shown in the bottom graph of Figure 7.

7.9 BIS requirements for pre-settlement credit risk

The 1988 Basle Accord on Bank Supervision (including more recent amendments) requires that regulated banks hold capital for pre-settlement credit risk. As of the writing of this paper, the amount of regulatory capital needed for pre-settlement risk is calculated by a procedure that is similar to the simple transaction methodology described above. The first step of the procedure requires that a bank calculate the loan equivalent exposure of each counterparty. According to the BIS algorithm, the *gross exposure* (i.e. ignoring the effects of netting)[16] per counterparty k is equal to sum of the positive market value of contracts with that counterparty (i.e. the 'gross' market value) plus an 'add-on' for every contract, independent of its current value. The 'add-on' is intended to represent the potential increase in the contract's value and is calculated by multiplying the contract's notional principal by a BIS exposure factor:

$$\text{BISExposure}_k = \sum \text{Max}[PV(t)_j, 0] + \sum (P_j \times \text{BISFactor}_j),$$

where: the sum is over all transactions j of counterparty k; $PV(t)_j$ is the current market value of contract j; P_j is the notional principal of contract j; BISFactor_j is the BIS equivalent of a CEF for contract j.

The BIS factors are similar to CEFs in intended function, i.e. as an approximate estimate of a contract's potential increase in value per unit of notional principal. BIS factors are defined for five very broad product types (e.g. interest rate contracts, forward currency exchange and forward gold contracts, etc.) and three very broad tenor buckets (i.e. under one year, one to five years, over five years). The very broad BIS product categories make no differentiation with respect to a contract's form (e.g. forward, swap, option, structured derivative, etc.) nor do they differentiate with respect to the volatility of the underlying market factor. For example, the same BIS factor is used for a forward exchange of two currencies whose exchange rate has a very low historical volatility and an exchange of two currencies with a very high historical volatility.

In contrast, when the CEF method was first implemented at Citibank in 1990, they were specified as a function of the *primary type* of underlying market factor (e.g. interest rate, FX, equity, etc.); the volatility of the underlying market factors; the *form of contract* (e.g. forward, swap, cross currency swap, etc.); and the *remaining tenor* of the transaction (i.e. usually the CEF was expressed as a continuous function of the time remaining until contract maturity). Consequently thousands of standard CEFs were specified at Citi in contrast to the fifteen BIS factors. However, in spite of this finer specificity, the simple transaction method had several shortcomings, some serious.

[16]The BIS method allows netting to be taken into account in a very approximate way. For contracts under an enforceable netting agreement, the first term in the formula for calculating the BIS exposure is the net, rather than the gross, current market value. However the second term (the sum of the 'add-ons') is only partially and approximately reduced by having a netting agreement. To fully measure the benefits of netting on both current and potential exposure it is necessary to implement a portfolio simulation method such as described in this text.

7.10 Limitations of simple transaction method

There are two classes of shortcomings inherent in the simple transaction method.[17]

One difficulty concerns the trade off between ease of implementation and precision. As described above, any implementation of the simple transaction method will rest on the calculation and firm wide dissemination of tables of credit exposure factors. The credit exposure factors will be used by transactors to calculate the potential exposure of a single transaction and by risk systems to calculate the total exposure to a counterparty. At issue is the question of how many factors need to be specified to achieve a desired amount of precision in the calculation of exposure. At Citi in 1990 we initially implemented thousands of factors, specified by the four categories described above. We knew at the time that this method had approximations:

- The probability distribution of the potential values of a market factor at a future date is not symmetric around the expected value for that date. This naturally follows if the distribution is lognormal. It also can occur under other assumptions about the future distribution of market factors. As a consequence, particularly for transactions with longer tenors, it is necessary to calculate separate CEFs for forward buys versus forward sells, for a swap in which the firm receives the floating rate versus one for which the firm pays the floating rate; etc.

- For transactions with multiple cash flows (e.g. an interest rate swap, a cap, a cross currency swap) the actual height of the transaction's exposure profile will depend on the shape of the yield curve (or, for a commodity swap, on the shape of the term structure of forward prices) and the terms and conditions of the contract.

- For transactions with multiple cash flows, an *in-the-money* contract will tend to have a smaller potential increase in value than an *at-the-money* contract of the same form, underlying market factor and tenor. This occurs because the multiple cash flow contract will experience 'time decay' as its unrealized positive cash flows are realized over time. Similarly, a transaction with multiple cash flows that is *out-of-the-money* has a higher potential increase in value than an *at-the-money* transaction.

[17]The portfolio method of calculating pre-settlement credit exposure and the representation of exposure as a portfolio exposure profile were developed in 1991 by David Lawrence and me with critical technical assistance from Byron Nicas. See D. Lawrence and E. Picoult, 1991, 'A New Method for Calculating Pre-Settlement Risk', *The Tactician* **4**, (3) (internal Citibank publication); D. Lawrence, 1995, in *Derivative Credit Risk: Advances in Measurement and Management*, Risk Publications; E. Picoult, 1996, 'Measuring Pre-Settlement Credit Risk on a Portfolio Basis', in *Risk Measurement and Systemic Risk, Proceedings of a Joint Central Bank Research Conference, November 1995*, Board of Governors of the Federal Reserve System.

- The potential change in value of an option depends on the degree to which it is out of the money, at the money or in the money – i.e. it is a function of the contract's current value and its time to expiration. For example, a deeply out of the money option will have a much smaller potential increase in value than a deeply in the money option.

- Many transactions are not 'plain vanilla'. They have extra terms and conditions (e.g. amortization of notional principal, structured derivatives) which materially affect their potential increase in value.

Consequently, a precise calculation of the potential increase in the value of a contract requires more specificity than simply categorizing the transaction in terms of its primary market factor, volatilities of underlying market factors, form of contract, and remaining tenor. These deficiencies in principle could be rectified by defining tens of thousand (or hundreds of thousands) additional CEFs, with much finer specificity. However, a second, more fundamental problem with the simple transaction method argues against the expenditure of the effort that would require.

A more fundamental limitation of the simple transaction exposure method is its inability to accurately *calculate* and *represent* the potential pre-settlement exposure of a portfolio of multiple contracts with a single counterparty. *Under the simple transaction exposure method the total exposure of a portfolio of many contracts with a counterparty is simply the sum of each contract's potential exposure.* There are several flaws in such a calculation of portfolio exposure:

- The transactions with the counterparty may have different tenors. The peak exposure of each transaction will consequently occur at different times. The sum of peak transaction exposures overestimates the peak exposure of the portfolio.

- The market factors underlying the value of all the transactions with the counterparty are unlikely to be perfectly correlated. Adding up the potential exposure of each transaction, calculated in isolation, ignores the diversification of having transactions with sensitivity to different market factors.

- The counterparty may have done transactions in offsetting directions, all which can't increase in value at the same time.

- The simple portfolio method can not properly calculate the effect of netting.

In summary, the actual counterparty portfolio exposure will tend to be less, and potentially dramatically less, then the sum of the potential exposure of each transaction. Additionally, the shape of the exposure profile to the counterparty will tend to be very different than the arithmetic sum of the shapes of the potential exposure to each transaction.

7.11 Counterparty portfolio simulation method and counterparty exposure profile

Citibank developed a portfolio simulation method for calculating a counterparty's pre-settlement exposure in 1991. The essence of the simulation method is described below:

- *Simulate thousands of paths of changes in market factors over time.*

  For each simulated path, begin with the current level of market factors. It is necessary to simulate as many factors as are required to value the contracts in the portfolio. For a large financial firm this may require simulating thousands of market factors

  For each simulated path, simulate market factors at a set of future dates. For example: start with today's market rates and simulate market factors at each day over the next week, each week over the next month and each month over many years (depending on the types of contracts traded this might be ten or more years).

  The simulation of long term changes in market factors can be done with varying levels of sophistication and subtlety. At a minimum it should take into account the long term volatilities and correlations of all simulated market factors and should make some assumption about how spot rates drift towards their expected forward value.

- *Calculate the potential market value of each transaction at each future date of each simulated path.*

  For each simulated path: calculate the simulated market value of each contract at the current date and at each future date for which market factors are simulated. The simulated market value of each contract at each future date will depend on the revaluation algorithm appropriate for the contract, the contract's terms and conditions, the number of remaining unrealized cash flows for the contract and the particular simulated path of market rates that was generated (which will effect how floating rates were set, etc.).

- *Calculate the potential exposure of each counterparty at each future date of each simulated path.*

  For each simulated path, at each simulated future date, employ the appropriate aggregation rules to transform the simulated market value of each contract into the simulated exposure of the portfolio of transactions with the counterparty. The aggregation rules need to take the legal context into account including the effect on exposure of any enforceable

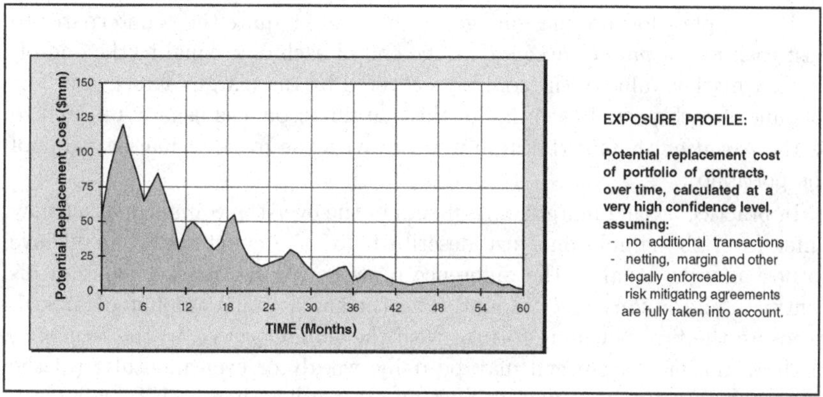

Figure 8: A Counterparty's Exposure Profile.

risk mitigants, such as netting agreements, margin agreements or option to early termination agreements, that have been entered into. The aggregation rule with and without netting were described above. The affect of margin on potential exposure is described below.

- *Calculate the counterparty's Exposure Profile.*

  The *Exposure Profile* of a counterparty is the current immediate exposure and the potential future exposure, calculated at a high confidence level (e.g. 99%), at a set of future dates. The Exposure Profile is defined in the context of the existing set of forwards and derivatives transacted with the counterparty, the risk mitigating legal agreements that have been entered into and the assumptions and methods underlying the long term simulation of changes in market factors. An example of a counterparty's exposure profile is shown in Figure 8. The exposure profile measures the potential exposure of the current set of transactions and assumes no additional transactions with the counterparty.

7.12 Effects of a margin agreement on exposure and risk

There are several ways of reducing or mitigating the potential exposure to a counterparty. One method, a netting agreement, was described above. Another risk mitigant is to require a counterparty to post some assets into a margin account to cover the potential cost of replacing the forwards and derivatives, should the counterparty default. Margin is a standard feature of exchange traded future and option contracts. It is also sometimes used in the over the counter (OTC) market between firms.

The simplest form of margin agreement would require the counterparty to post cash into a *margin account*, at the end of each day, equal to the end-of-day net market value of the contracts covered by the margin agreement. The amount of cash posted each day in the margin account is usually referred to as the *variation margin* (because it varies with the mark to market value of the portfolio).

In practice, actual margin agreements in the over-the-counter market may differ from the simple one just described: (a) the counterparty may have to post margin equal to the difference between the net market value of his contracts and a *threshold* amount. The threshold is the amount of 'naked' exposure the firm is willing to take with the counterparty; (b) the *frequency* of the variation margin call may be daily, weekly or even monthly; (c) the counterparty may be allowed to post an asset other than cash as collateral. When that occurs, the firm will usually ascertain a 'cash equivalent' of the asset posted as margin. The cash equivalent will be a fraction of the asset's current market value. The ratio of the cash equivalent of the asset to its current market value can be described as 1.0 minus a 'haircut'. The amount of haircut demanded will either be set by market convention or be based on an independent estimate of how much the asset could decrease in value during some specified period of risk. The period of risk used to assess the haircut will be based on the frequency at which the non-cash assets posted as margin are revalued (which can differ from the frequency at which the net market value of the derivatives are assessed) and the estimated time required to liquidate the non-cash assets. Let us call this period of risk the *margin-assets-liquidity period of risk*.

In summary, there are two types of revaluations that must be performed. The net marked to market value of the underlying derivatives determines the cash equivalent amount of margin that needs to be posted in the variation margin account. The amount of the required cash equivalent might be ascertained daily, weekly, etc. When a non-cash asset is posted as margin, its cash equivalent is ascertained as some fraction of its current market value. If the market value of the assets posted as margin decreases and if, as a consequence, the cash equivalent of the assets in the variation margin account is less than the required amount, the counterparty will be obligated to post additional assets in the margin account.

The intent of the margin agreement is to provide a cushion against a counterparty default. More specifically, if the counterparty were to default on any of its forward or derivatives or if the counterparty were to fail to post additional margin when required, the firm would terminate all forward and derivative contracts with the counterparty, seize any cash in the margin account, liquidate the non-cash assets in the margin account to generate additional cash and make itself whole by using all the cash it received from the margin account to replace the defaulted forwards and derivatives in the market. In short, the intent is that the cash in the margin account and the cash generated by selling non-cash assets will compensate the non-defaulting counterparty for some or all of its losses.

Let us assume that we have entered into the simplest form of margin agreement with a counterparty in which it posts cash, at the end of each day, equal to the end of day net market value of the contracts. In that context, what is the potential counterparty exposure?

The first question that must be asked concerns the legal certainty of the margin agreement. Before a firm reduces its measured pre-settlement credit exposure to a counterparty, it should have legal certainty that: (a) it can terminate its forwards and derivatives should the counterparty default on any obligation; (b) it has a perfected interest in the assets posted as margin; and (c) a court would not place a 'stay' on the assets in the margin account, should the counterparty default. If the court imposed a stay, the firm would be legally unable to liquidate the assets posted as margin until all claims on the defaulted counterparty were settled.

If the right legal context holds, the potential pre-settlement exposure in the simple example we are considering will be a function of how long it will take us to actually terminate the forward and derivative transactions with the counterparty and replace them with new transactions with another counterparty. Let us call this time interval the *margin period of risk* (which is different from the margin assets liquidity period of risk described above). Our potential exposure to the counterparty is the amount by which the net market value of the portfolio could increase during the margin period of risk.

The margin period of risk is the sum of several components:

- Even if the counterparty were notified at the end of the day to post additional margin, several days could potentially go by before we could be certain that the counterparty had defaulted on its obligations to post margin. Consequently, even if the counterparty were required to post margin every day, it might take five or ten business days to terminate the contract with the counterparty.
- Since we have assumed we have cash margin, we will assume that once the forward and derivative contracts have been terminated we can immediate enter the market to replace them. However, depending on market liquidity it may take some amount of time, perhaps days or weeks, to actually find another counterparty *willing to enter into the replacement forward or derivative.* This example differentiates between the time required to liquidate the assets posted as margin (the 'margin assets liquidity period of risk) and the time required to replace the defaulted forward or derivatives.

In the above example, let us assume that the margin period of risk is ten days. The exposure profile under such a margin agreement will have a shape

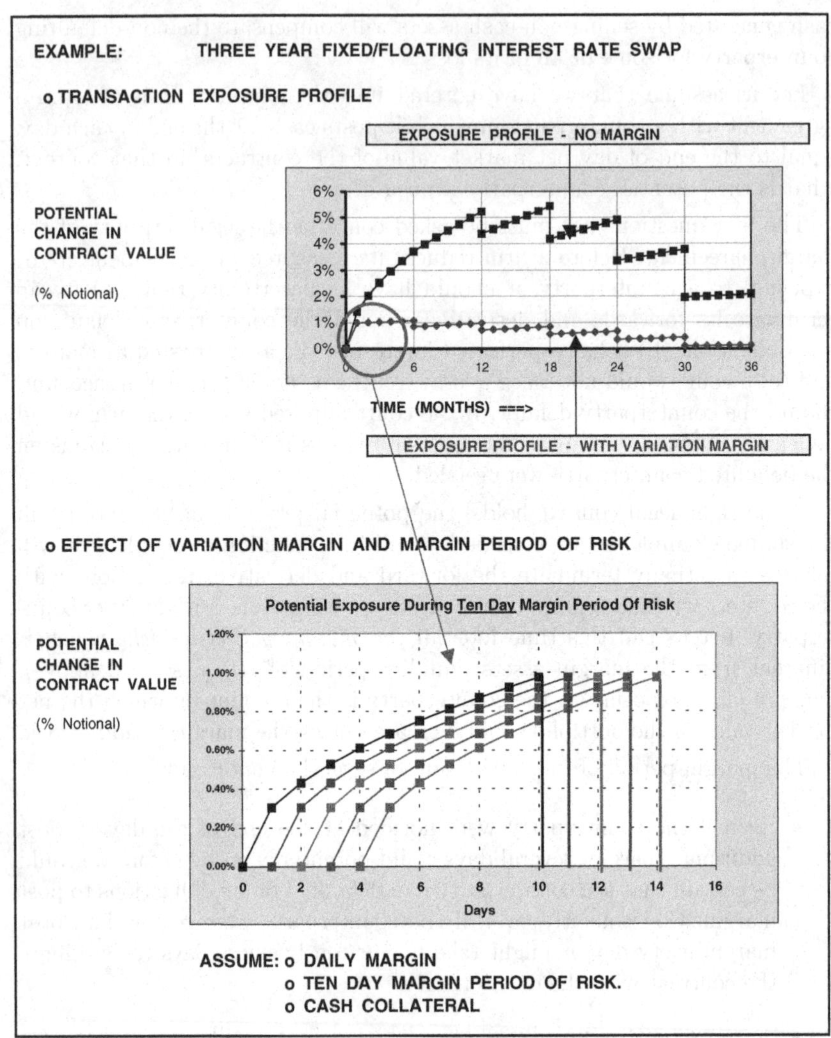

Figure 9: Single Contract.

of overlapping sawtooths, as shown in Figure 9. Even if the margin period of risk is ten days, there is a continual possibility that the counterparty might default anytime over the life of the transaction. Consequently there is a set of overlapping ten-day-margin-periods of risk over the life of the transaction. A margin agreement *reduces the magnitude* of potential exposure but not the tenor of exposure. There is potential exposure over the full life of the transaction (or portfolio).

In a stressful market, for which liquidity has dried up, the simple analysis could fall short:

- If a non-cash asset, even a Treasury security, is posted as margin, it might decrease more in value than anticipated by its haircut. Consequently the firm may receive less cash from selling the asset posted as margin than it had expected to.
- If market liquidity dries up, the margin period of risk will increase because it will take a longer amount of time to replace the contracts with the defaulted counterparty. Associated with a longer margin period of risk is the potential for a greater cost of replacing the contracts with the defaulted counterparty.

To mitigate the exposure it has to the counterparty during the margin period of risk, a firm might ask the counterparty to post additional margin. This additional margin is sometimes referred to as *'initial' margin.* Its function is to protect the firm against the incremental exposure during the margin period of risk. Under stressful conditions the initial margin may not be adequate to cover the incremental exposure during the margin period of risk.

7.13 Economic capital for pre-settlement risk – general principles

As explained above, the economic capital for pre-settlement credit risk is derived from the probability distribution of potential loss, as illustrated in Figure 5. The loss distribution will depend on the time horizon over which the potential loss is calculated (e.g. one year or life of portfolio) and on the definition of economic loss.

An essential difference between the potential loss distribution due to lending risk and the potential loss distribution due to pre-settlement counterparty risk is the uncertain exposure of the latter. If the future state of market rates could be known with certainty the credit exposure arising from pre-settlement risk would be known with certainty. Under that condition, we would have certainty about the magnitude and value of all future cash flows between each counterparty and our firm. Pre-settlement risk would consequently be equivalent to lending risk because, under the assumed condition of omniscience about future market rates, the magnitude and value of the future cash flows arising from forwards and derivatives could be represented by a set of spot and forward deposits and loans with fixed cash flows.

Of course we do not know what the future state of markets will be. However, we see that for each simulated path the market could take over time, we can construct an equivalent portfolio of fixed rate spot and forward deposits and loans with each counterparty. Consequently, whatever our definition of credit loss is, for each simulated path of market rates we can calculate a potential loss distribution for forwards and derivatives using the same methods developed to measure the potential loss distribution of loans.

The final total loss distribution due to counterparty risk will be the weighted sum of the loss distributions calculated for each simulated path of market rates.

7.14 Economic capital for pre-settlement risk – simple example

The simplest calculation of economic capital for pre-settlement risk would be that calculated by defining credit loss as only attributable to a default by the counterparty. Under that definition the loss distribution could be calculated in several steps:

- *Simulate thousands of paths of changes in market factors over time.* This step is identical to the first step in calculating a counterparty exposure profile on a portfolio basis.

- *For each simulated path of the market, calculate the potential exposure to each counterparty, at many future dates.* This is identical to the second and third step in the calculation of a counterparty's exposure profile on a portfolio basis, as described above.

- *For each simulated path of the market, calculate potential loss by simulating counterparty defaults, at many future dates.* In more detail, for each simulated path of the market we can generate thousands of simulations of potential defaults of counterparties over time. Each simulation of potential default would be like that done for lending risk: at each future date of the simulation we would randomly make a draw to determine how many counterparties were simulated to default and make another draw to ascertain which counterparties were simulated to default. For each defaulted counterparty a loss could occur if our firm had a simulated positive exposure to the counterparty at that point in time. If a default was simulated and the counterparty had a positive exposure we would make another simulation of the loss in the event of default. The result of thousands of simulations of defaults and recoveries, for each simulated path of the market, would be a probability distribution of potential loss. Note that for a given path of the market, not every counterparty simulated to default will have a positive exposure.

- *Repeat the simulation of potential defaults and recoveries for each stimulated path of market rates.* By taking into account many potential paths of market rates we introduce another stochastic element into the calculation of the final loss distribution due to counterparty exposure.

Figure 10: Market Risk vs. Pre-Settlement Exposure

	Market Risk	Pre-Settlement Exposure
Unit of Risk Analysis:	• Internal Organization	• External Customer
Risk:	• fall in Economic Value	• Default when Positive Value
Time Horizon	• Very short – usually overnight – thus, static portfolio	• Very long – usually life of portfolio – thus, changing portfolio
Legal Issues	• Not Relevant	• Critically Important e.g. Netting Marginning Option to early termination

- *Calculate the final loss distribution by appropriately aggregating the potential loss distribution for each simulated path of the market.*

A more complex calculation would be based on the assumption that an economic loss due to pre-settlement risk could occur even without default, simply if the counterparty's credit rating deteriorated or if the market spreads widened – the spreads used to discount the expected future cash flow of each counterparty back to present value (see Footnote 4).

The above process assumed there was zero correlation between the potential future state of market rates and the probability of a counterparty defaulting. In general that is a reasonable thing to assume for most counterparties because of the enormous difficulty in realistically assessing a non-zero correlation. However under some circumstances the correlation between changes in market rates (potential exposure) and counterparty default are clearly non-zero. A sophisticated methodology would incorporate the correlation of default and exposure where feasible.

8 Comparing and contrasting market risk and pre-settlement risk

The essential differences between market and pre-settlement credit risk are summarized in Figure 10.

The three most material differences between the measurement of market and counterparty pre-settlement credit risk are:

1. Unit of Risk Analysis Market Risk begins at the transaction level. Transactions are usually organized into portfolios, which are managed by traders, who work on trading desks, which are part of larger business organizations. In short, virtually any trading business is organized into some hierarchical structure and market risk can be measured and limited at any level of that organizational hierarchy.

More broadly, the three key dimensions for organizing, measuring, monitoring and limiting market risk are: (a) the business unit (e.g. trader, trading desk, trading business, etc.) taking the risk, for any particular level of the corporate hierarchy; (b) the type of market factors (e.g. three-month US$ LIBOR; spot US$/Japanese Yen exchange rate; etc.) for which the position has sensitivity; and (c) the product (i.e. the type of financial instrument) which is generating the market factor sensitivity. Within this three-dimensional space, for any particular subset of the organizational unit, type(s) of market factor and type(s) of financial instrument, factor sensitivities can be measured, for specified changes in the underlying market factors, and compared to limits; VAR can be measured and compared to limits and Stress Tests can be defined, measured and compared to limits. Normally VAR is only measured and monitored at a higher level of the corporate hierarchy. Factor sensitivity, in contrast, is the perspective most commonly used on the trading desk to measure, monitor and limit risk.

In contrast to market risk, the primary unit of organization for pre-settlement risk is the counterparty, not the trader or trading desk. Many risk mitigant agreements (such as an ISDA Master Netting agreement) are agreed to and legally enforceable across trading desks and booking centers (i.e. across the categories from which market risk is usually measurement and limited). For example, a netting agreement between two financial firms might include all fixed income derivatives, equity derivatives and commodity derivatives that the two counterparties have transacted, whether traded out of their New York or London trading desks.

More generally, the precision of the measurement of the potential Pre-Settlement Exposure will increase as more of the transactions with the counterparty are included in the calculation.

2. Time Horizon Market risk is usually measured over a very short time horizon. In fact, as explained above, market risk is usually measured under the assumption of a static portfolio. In contrast, the measurement of pre-settlement exposure requires that we simulate changes in market factors over time; simulate changes in contractual cash flows over time, for each scenario of changes in market rates, and finally simulate changes in each obligor's risk rating, and state of default over time.

3. Legally Enforceable Risk Mitigants Having legally enforceable risk mitigants can have a material effect on the magnitude of the credit exposure and credit risk to a counterparty. In contrast, the legal enforceability of risk mitigants is irrelevant to the measurement of market risk. Market risk exists simply from owning securities, holding positions in currencies or commodities and having entered into the various types of derivative contracts.

Value at Risk Analysis of a Leveraged Swap[1]

Sanjay Srivastava

Abstract
In March 1994, Procter and Gamble Inc. charged $157m against pre-tax earnings, representing the losses on two interest rate swaps. One risk measure designed to warn against the potential of large losses is Value at Risk (VaR). In this article, I conduct a Value at Risk analysis of one of the swap contracts. The VaR analysis is based on a one-factor Heath–Jarrow–Morton model of the term structure. The calculated VaR is approximately seven times the value of the contract. A complementary measure of risk (the 'conditional expected loss') is about ten times the value of the contract. An interesting by-product that emerges is that the one-factor model captured the yield curve evolution during that time rather well.

1 Introduction

In this article, I study the riskiness of a leveraged interest rate swap contract. The contract, initially worth about $6.65m, experienced an extreme change in value over a short period of time in 1993–1994, leading to a loss of over $100M. While large losses in financial markets have a long and significant history, there has rarely been a period of time like the mid 1990s when a string of losses occurred in a variety of financial markets. Since then, considerable effort has been devoted to the development of risk measures to warn against the potential of such losses. One such measure is Value at Risk (VaR). In this article, I conduct a Value at Risk analysis of the contract. I am specifically interested in understanding whether VaR would have provided a warning that losses of the magnitude experienced were possible.

The specific contract[2] is the swap agreement executed between Proctor and Gamble (P&G) and Banker's Trust (BT) in November 1993. The contract was terminated in March 1994 with a loss of approximately $100M. Briefly[3], P&G paid the floating rate on the 5-year, semi-annual swap. This floating rate was based on 30-day commercial paper and a spread (if positive). The spread was to be set on May 4, 1994. The magnitude of the spread depended on the yield

[1]First published in *Journal of Risk,* **1** (2), 87–101 (1999). Reproduced with permission of Risk Publications.

[2]This was only one of the major interest rate based losses experienced by various firms and municipalities during the 1994–1995 period. See Jorion (1996) for a survey.

[3]Details of the agreement are presented in the next section.

on the 5-year constant maturity Treasury and the price of a particular (30-year) Treasury bond. Once set, the spread would apply to the remaining term of the contract, and so represented a one-time bet on interest rates. Thus, P&G had sold an interest rate option to BT. As it turned out, interest rates moved so as to make the spread very large, resulting in the loss.

In light of this and other losses, attention has been focused on quantifying the losses that are possible on leveraged contracts such as this one as well as on portfolios of assets. One such measure is Value at Risk, which has rapidly gained acceptance as both a risk measure and as a regulatory tool.[4]

VaR attempts to answer the following question: what is the most I expect to lose with a certain probability over a given horizon? Typically, the probability is set to 1% or 5%. VaR attempts to explain what is the dollar amount that could potentially be lost over the time horizon. Formally, it is related to the tail of the distribution of portfolio value changes at the horizon. If we look at the 5% VaR, then if $F_T(\ )$ is the distribution of changes in the portfolio values at horizon T, then VaR satisfies $F_T(\text{VaR}) = 5\%$. This tells you that the probability of a loss greater than the VaR is 5%, so you do not expect to lose more than the VaR with 95% probability.

For the interest rate option in the contract, I use a variation of the parametric[5] method of calculating VaR. I calculate the future value distribution using an interest rate model. This requires an assumption about the stochastic process governing term structure movements over time. I use a one-factor Heath–Jarrow–Morton type model, using publicly available data on the Friday before the start date of the contract. The estimation is based on historical volatilities of forward rates, calibrated ('shifted') so that the premium paid equals the calculated value of the contract. The model is implemented as a 'tree'.

The horizon chosen is six months, corresponding to the time period after which the spread was to be set. The term structure model provides the distribution of yield curves in six months. This generates the distribution of contract values six months ahead, permitting the calculation of the VaR. Note that since the future probability distribution is calculated using the risk-neutral probabilities, what is being calculated is the 'economic' VaR.[6]

The calculated VaR is approximately seven times the value of the contract. One criticism[7] of VaR is that it does not provide information about the expected loss if a large loss was to occur. For example, suppose losses that occur with probability less than 5% occur. What is the expected loss? A complementary measure of risk, the 'conditional expected loss', which is a variant of shortfall risk, provides this information. For the contract at hand, this figure is about ten times the value of the contract.

---

[4]See Jorion (1996).

[5]This underlies the Riskmetrics methodology of J.P. Morgan.

[6]See Ait-Sahalia and Lo (1997).

[7]Another is that it is not subadditive, as discussed in Artzner *et al.* (1997).

In summary, the analysis indicates that VaR would have provided us with an accurate warning about the risk embedded in the contract. An interesting by-product that emerges is that the one-factor model captured the yield curve evolution during that time rather well.

One aspect of this study is that it examines the VaR of a specific contract at a point in time. By contrast, most studies of VaR have focused on how well the measure can track losses on portfolios across time. The latter give us information on whether the assumptions on asset price distributions that underlie the computation of VaR are supported by historical data. Such information is clearly important for a variety of reasons. The emphasis here is not so much on historical accuracy as on the use of the measure to evaluate the risk of a specific contract.

The details of the contract are described in the next Section. In Section 3, I summarize the movements in interest rates that occurred, and how the contract lost money. Section 4 contains the VaR analysis.

2 The Details of the Contract

The Original Contract

Procter and Gamble (P&G) was one party to the contract, while Bankers Trust (BT) was the counter-party. The original swap contract had the following features.[8]

- The contract commenced on November 2, 1993;
- The notional principal was US $200M;
- The contract would reset semi-annually and last for 5 years;
- The spread would be set on May 4, 1994 and would then remain fixed for the remainder of the contract;
- Every six months, BT would pay P&G the fixed rate of 5.3% times the notional principal;
- On May 4, 1994, P&G would pay BT the average of the 30-day commercial paper rates minus 75bp times the notional principal;
- Every six months thereafter, P&G would pay BT the average of the 30-day commercial paper rates between reset dates plus the spread minus 75bp times the notional principal;

[8]See Smith (1997) for a summary of the contract features, and for an insightful analysis of how the contract could have been replicated or hedged using Treasury options. Details of the contract are contained in Case No. C–1-94–735 filed at the US District Court, Southern District of Ohio, Western Division.

- The spread would be determined on May 4, 1994 by the following formula:

$$\text{spread} = \max\left\{\frac{\frac{98.5}{5.78\%}C_5 - T_{30}}{100}\right\},$$

where

C_5 is the yield on the 5-year constant maturity Treasury, and

T_{30} is average of the bid/ask clean price of the 6.25% 8/15/2023 Treasury bond, which at the time was the benchmark Treasury bond.

Note that the spread was to be determined once and would then apply for the remainder of the contract.

As will become clear shortly, the contract was essentially a bet that the level of interest rates would remain low.

Modifications of the Contract

The terms of the contract were not carried out. While not relevant for the analysis in this article, it is interesting to note that the contract was modified in January of 1994. This was prior to the first increase in interest rates by the Federal Reserve Board (Fed) in February of 1994. In January, the date on which the spread would be determined was moved to May 19, 1994, two days after the scheduled meeting of the Fed, and the discount of 75bp was increased to 88bp. Presumably, the additional discount was compensation for the risk of an additional rise in interest rates at the May 17 Fed meeting. In March of that year, the contract was terminated, with a loss of about $100M.

The Swap and the Embedded Option

One way to think about the leveraged swap is to separate it into two parts. One part is the more standard swap while the second part is the option.

The 'standard' part of the swap is the 5.3% fixed versus the 30-day commercial paper rate. This is not quite a standard swap because the floating rate is the average of the 30-day commercial paper rate between reset dates rather than the commercial paper rate on the reset date.

On October 29, 1993 (the Friday before start date of the contract) the (continuously compounded) yield on the 5-year Treasury was 4.82%. It follows that the fixed rate for a 5-year, semi-annual swap would be close to this number.[9] The difference between 5.3% and 4.82% is 48bp. It seems reasonable to argue that a 48bp spread over the Treasury rate is an appropriate spread over the Treasury rate for a commercial paper based swap for an AA corporation and given that the commercial paper rates were 10 to 20bp higher than the Treasury yields.

[9]The actual swap rate would be based on semi-annual, rather than continuous, compounding.

If we take this view, then the 75bp discount is the premium paid to P&G in return for selling the option to BT. The analysis conducted in this article will proceed on this assumption. In fact, from now on, I will ignore the 'standard' part of the contract and focus solely on the option component.

The Analysis Date and Data

Unless otherwise specified, the analysis in article is conducted using data for October 29, 1993, which is the Friday of the week before the contract commenced. It seems reasonable to assume that an analysis of this type would have taken place the week prior to the start date of the contract.

All interest rate data used in this article is the weekly H15 data provided by the Federal Reserve Board. In particular, term structures are computed by 'bootstrapping' the yields reported on constant maturity Treasuries. The price data on the August 15, 2023 Treasury bond was obtained from Reuters.

The Value of the Contract

To value the option, note that P&G is paid 75bp on a \$200M notional principal over 10 semi-annual periods. This means that every six months for five years, P&G is paid $0.0075 \times \$200\text{M}/2 = \$750,000$. The (continuously compounded) zero-coupon yield curve on October 29, 1993 was:

Maturity	Rate
0.25	3.19488
0.5	3.36274
1	3.57107
2	3.94032
3	4.25543
5	4.82093
7	5.22319
10	5.49831
30	6.31597

Using linear interpolation for rates in between the indicated maturities, we find that the present value of the premiums paid to P&G for selling the option was \$6.65m.

3 The Nature of the Bet

On November[10] 2, 1993, the 5-year CMT yield was 5.02% while the clean price of the August 2023 Treasury bond was $102\frac{31}{64}$, corresponding to a yield to

[10]On our analysis date (October 29, 1994), the 5-year CMT yield was 4.82% while the Treasury (clean) price was 103.94, implying that the second term was –0.2180= –21.80%.

maturity of 6.0679%. This means that the second term in the spread formula was

$$\frac{\frac{98.5}{5.78} \times 5.02 - 102.578125}{100} = -0.170297 = 17.029\%$$

implying that the spread was zero.

If yields had remained unchanged between November 1993 and May 1994, the contract would have implied that for 10 semi-annual periods, P&G would be receiving 5.3% and paying the average of the 30-day commercial paper rates less 75bp.

In the four months before November 1993, the difference between short term commercial paper and the equivalent maturity Treasury ranged from 10 to 25bp. Data over the previous year indicate that this difference was significantly below 75bp. Consequently, if the spread had remained at zero on May 4, 1994, and the basis remained below 75bp, P&G (an AA firm at the time) would have guaranteed itself receiving 5.3% and paying a floating rate below the Treasury rate for five years.

To understand the nature of the bet, it is instructive to look at what would have to happen to rates for the contract to have lost money. The question being posed is: by how much must the term structure shift for the contract to lose money?

This is complicated a little by the fact that the spread depends on both an interest rate and a price:

$$\text{spread} = \max\left\{\frac{\frac{98.5}{5.78\%}C_5 - T_{30}}{100}\right\}.$$

A simple way to convert yield changes to price changes is to use the modified duration of a bond,

$$dP = -MD \times P \times dy.$$

On October 29, 1993, the August 2023 Treasury had a modified duration of 13.18719. Thus, the change in the spread due to a change in yields can be written as

$$\Delta_{\text{spread}} = \frac{\frac{98.5}{5.78\%}\Delta C_5 - 13.18719\Delta y_{30}}{100},$$

where y_{30} is the yield on the 30-year bond.

Since the second term in the spread equation was initially −0.170297, a parallel move in yields would make the spread equal to zero if both C_5 and y_{30} moved up by 70.9bp. Note that it takes a relatively small change in yields to move the second term in the spread equation from −17% to 0, which gives a good indication of the leverage built into the contract.

Note that an increase in yields of 70.9bp is still profitable, since it implies receiving 75bp and not paying out anything on the option. To get a clearer idea of what would be required for the contract to break even, we need to find

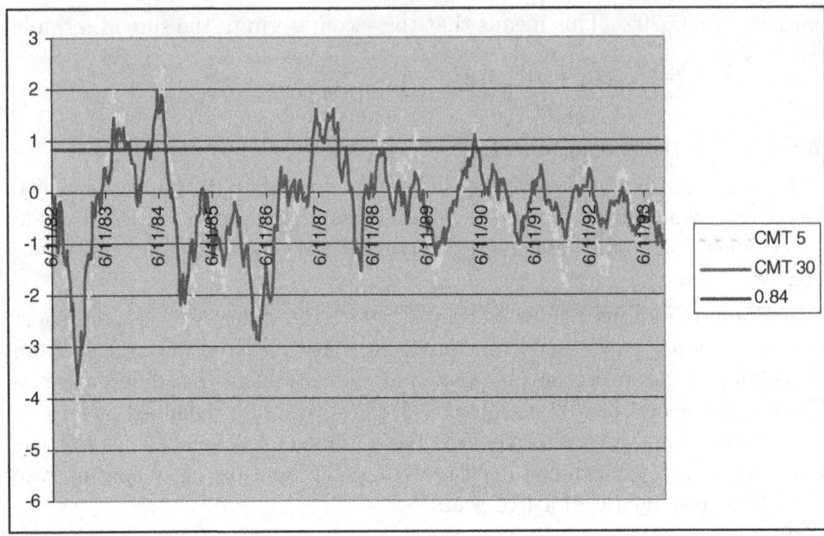

Figure 1: 6-Month Yield Differences

the value of the spread so that the present value of the 75bp discount equals the payout on the option given the spread. For the term structure prevailing on October 29, 1993, it turns out that an increase in yields of 84.3bp results in the contract breaking even.

One could ask how frequently the yield curve had shifted by more than 84bp over a 6-month period. Since 1982, it turns out that there has been a shift both the 5-year and 30-year of more than 84bp over 6 months in 63 out of 595 weeks from June 1992 to October 1993, which is a frequency of 10.59%. However, the last time this occurred was in May 1990. Figure 1 shows the 6-month yield differences for the two yields of interest.

Another way to think about the interest rate bet is to note that typically, the Fed raises interest rates by 25bp. This means that P&G could be betting that rates would not be increased more than three times in the period six months to May 4, 1994. Of course, the effect of monetary policy on the short end of the term structure does not mean that medium and long term rates cannot rise significantly. In fact, a significant steepening of the term structure could easily cause the spread to become significant.

Next, consider the sensitivity of the spread to changes in the level of the curve versus changes in the shape of the curve.[11] The spread depends positively on the 5-year CMT yield. Rewriting the constants that multiply the spread, we find that

$$\text{spread} = \max\{0, 17.0412C_5 - 0.01T_{30}\}.$$

[11]This is similar to Smith (1997), but conducted on a different date.

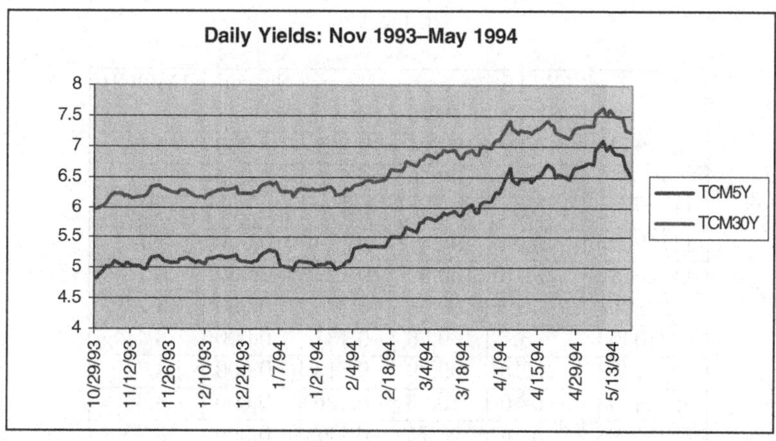

Figure 2: Treasury Yields

Using the modified duration formula, we find that

$$\Delta\text{spread} = \max\{0, 17.0412\Delta C_5 + 13.68279\Delta Y_{30}\},$$

where Y_{30} is the yield on the 30-year bond. It is evident that fundamentally, the bet is on the level of interest rates. A flattening of the curve (i.e., a fall in Y_{30} relative to C_5) and a steepening of the curve (i.e., i.e., a rise in Y_{30} relative to C_5) basically cancel each other. The fact that the option payoff depends primarily on the level of the curve provides some justification for using a one-factor model.

Finally, I note that if the spread changes by 1%, the implied payment is \$1m a period for 9 periods, so a 1% rise in yields leads to (undiscounted) future payments of \$9M. Thus, the future value of 1bp is \$90,000. This figure highlights the leverage embedded into the contract.

The *Ex Post* Behavior of Interest Rates and the Spread

Unfortunately for P&G, interest rates rose quite sharply between November 1993 and May 1994. Figure 2 shows the movement in the 5-year and 30-year constant maturity yields.

Weekly numerical values are given in Table 1. In summary, the 5-year CMT yield rose from 4.82% to 6.65% while the 30-year yield rose from 5.99% to 7.31%. The contract was renegotiated in January 1994, and the table shows that at that time, yields had risen by roughly 25bp. In March, when the contract was terminated, yields had risen by about 100bp from the beginning.

Table 1 also shows the weekly behavior of the spread. In January, the current value of the spread is still zero, but it rises to over 11% by the end of March, 1994. You can see that if P&G had not terminated the contract, the

Table 1

	CMT5	Clean Price	Term	Spread	1MOCP
10/29/93	4.82	103.94	-0.218	0.000	3.14
11/5/93	5.03	100.56	-0.148	0.000	3.15
11/12/93	5.04	101.44	-0.156	0.000	3.15
11/19/93	5.04	98.81	-0.129	0.000	3.14
11/26/93	5.13	100.09	-0.127	0.000	3.15
12/3/93	5.14	100.06	-0.125	0.000	3.27
12/10/93	5.1	100.66	-0.137	0.000	3.41
12/17/93	5.18	99.59	-0.113	0.000	3.34
12/24/93	5.16	100.5	-0.126	0.000	3.31
12/31/93	5.14	98.75	-0.112	0.000	3.35
1/7/94	5.21	100.28	-0.115	0.000	3.21
1/14/94	5.03	99.41	-0.137	0.000	3.12
1/21/94	5.06	99.56	-0.133	0.000	3.13
1/28/94	5.05	100.44	-0.144	0.000	3.11
2/4/94	5.14	98.69	-0.111	0.000	3.14
2/11/94	5.36	97.97	-0.066	0.000	3.41
2/18/94	5.4	95.19	-0.032	0.000	3.46
2/25/94	5.6	94.13	0.013	0.013	3.47
3/4/94	5.74	92.59	0.052	0.052	3.57
3/11/94	5.85	91.81	0.079	0.079	3.61
3/18/94	5.91	91.72	0.090	0.090	3.61
3/25/94	6	90.44	0.118	0.118	3.67
4/1/94	6.19	87.66	0.178	0.178	3.68
4/8/94	6.47	87.84	0.224	0.224	3.77
4/15/94	6.47	87.53	0.227	0.227	3.71
4/22/94	6.6	88.16	0.243	0.243	3.88
4/29/94	6.56	87.34	0.245	0.245	3.89
5/6/94	6.76	84.81	0.304	0.304	4.05
5/13/94	6.98	85.28	0.337	0.337	4.37
5/20/94	6.65	87.31	0.260	0.260	4.35

spread at the beginning of May would have been over 30%. This means that over the term of the contract, P&G would be receiving 5.3%, paying the (average) commercial paper rate less 75bp plus 30%. The 1-month commercial[12] paper rate changed from 3.14% to 4.05% on May 6. In broad terms, if the

[12]Note that the contract was based on the average commercial paper rate over the reset period, not the CP rate on a particular day.

contract had been carried out to its conclusion, P&G would have ended up paying approximately 30% plus CP minus 5.3% minus 0.75%. If we assume that commercial paper rates stayed fixed at 4.05%, then this implies a net payment of 28%; on the \$200M notional principal, this amounts to paying \$28M every six months for 9 periods[13], an undiscounted total payment of \$252M, \$217M if discounted at 5%, \$211M at 6%, and \$205M if discounted[14] at 7%.

4 Value at Risk

In this section, I use a one-factor term structure model to calculate the Value at Risk (VaR) of the contract. The model is the proportional volatility version of the Heath–Jarrow–Morton (1922) framework.

The Value at Risk of a contract is intended to measure the potential losses that can occur over a given time horizon. It is calculated as follows.

Let V be the current value of the contract (or a portfolio of assets). At a time horizon T, let $F_T(v)$ denote the distribution of values of the contract at the horizon, so

$$F_T(v) = \text{prob}(\tilde{V}_T \leq v).$$

For a confidence level α, let V_α be defined by

$$1 - \alpha = F_T(V_\alpha).$$

Therefore, with probability α, the portfolio value at the horizon will exceed V_α This means that losses in excess of V_α only occur with probability $1 - \alpha$. The Value at Risk is then defined as:

$$\text{VaR} = V - V_\alpha.$$

Typically, α is chosen to be 95% (e.g., the JP Morgan Riskmetrics methodology) or 99% (by most regulatory agencies). In this article, I use $\alpha = 95\%$. The interpretation of this number is that with 95% probability, the portfolio will not lose more than the VaR.

Clearly, what is critical here is the calculation of the future value distribution. There are two basic ways in which this is done. The first is based on a historical simulation, the second on an assumption about the distribution of future values. In the latter case, sometimes the distribution can be calculated analytically (for example, if we assume that all returns are normally distributed); otherwise, the value distribution is obtained by simulation.

[13]I have ignored the 75bp payment P&G would have received on May 4, 1994 in this calculation.

[14]P&G's debt at during this period was yielding about 7%.

In the case of the P&G swap, I will assume that that the stochastic evolution of interest rates is given by the following equation:

$$d(\log f(t,T)) = \mu(t,T)dt + \sigma(t,T)dz$$

where $f(t,T)$ is the instantaneous forward rate prevailing at time t for time T in the future, and μ and σ are known functions of t and T. I use the discrete time parameterization[15] described in Jarrow (1996, Chapter 12). The function μ is restricted by conditions of no arbitrage, and only the volatility function needs to be estimated.

The Horizon

Looking at the contract on October 29, 1993, a natural horizon date is 6 months. This when the spread will be determined, and the time period over which the bet has been placed.

The Model Implementation

The model was implemented as a (non-recombining) tree, using monthly time steps. This means that after 6 months, there are $2^6 = 64$ nodes, and so 64 term structures. In the implementation, every node is equally likely. The implementation calculates the full forward curve at every node. At each node, I calculated the 5-year CMT yield and the clean price[16] of the August 15, 2023 bond, and then the spread. Given the spread at a node, the payment made by P&G for the next 4.5 years is known. I then computed the term structure at each node, using this to discount the values of the payments. This yields the value of the contract at each node. Since each node is equally likely, we now have $F_T(v)$.

Notes on the Implementation and Choice of Model

Note that the size of the problem grows very quickly with the number of steps. A weekly estimation would involve 2^{26} nodes in six months, which is very large. An alternative is to use a short rate model, like the Black–Derman–Toy (1990) model, which has a recombining tree. A weekly estimation for such a model would produce 27, i.e. 26+1, yield curves after 6 months, but would require an enormous lattice. This is because after 6 months, we need the value of the 30-year Treasury bond, and this requires the lattice to extend out for the full 30 years. The total number of nodes in such a lattice would be about 300,000. In all this, the nodes do not need to be stored; however, not storing the information requires a lot of computing. By contrast, the HJM tree implemented here stores the entire forward curve at every node, so it is only necessary to go out 6 months, though at the cost of a non-recombining tree.

---

[15] This specification, like other 'log-normal' specifications, has the property that if the time steps are made very small, then forward rates explode. This was not a problem with the coarse time steps in our implementation.

[16] Assuming equally spaced coupon dates.

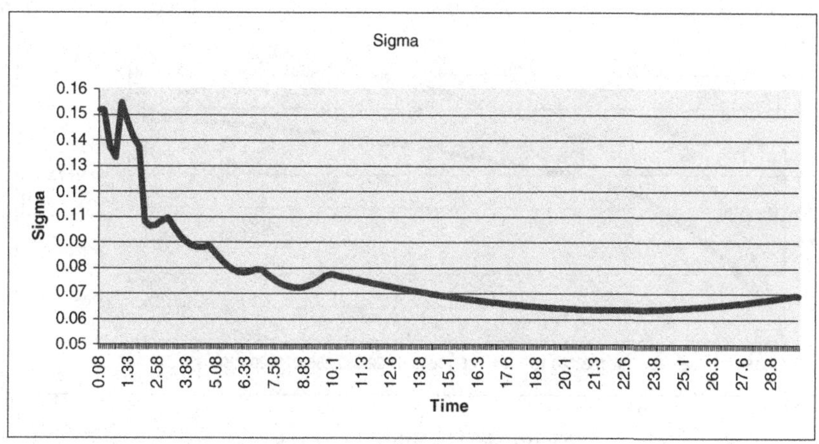

Figure 3: The Volatility Function

The Initial Term Structure and Forward Curve

The term structure and forward curve on October 29, 1993 are shown in Figure 3.

The Volatility Estimation

To implement the model, I need estimates of forward rate volatilities, the $\sigma(t, T)$. These we calculated as follows.

Step 1: bootstrap the CMT yields to produce continuously compounded term structures using linear interpolation, using data from January 8, 1982 to October 29, 1993.

Step 2: Calculate forward curves *monthly.*

Step 3: Interpolate the monthly forward curves so that all forward rates are at monthly time steps. Step 2 ensures that no overlapping data is used. This means that there are 360 forward rates (30 years times 12 months).

Step 4: Calculate the volatilities of the forward rates using the procedure described, e.g., in Jarrow (1996, Chapter 13) for the one-factor case.

The estimation produced the volatility function shown in Figure 4, where you can see the hump usually seen about 1 year out. I am not sure why there is an initial dip, though I believe this results from the interpolation.[17]

[17]Since the analysis is done monthly, the first forward rate is the 1-month spot rate. However, the shortest CMT has a three-month maturity, and so the initial rates were obtained using linear interpolation.

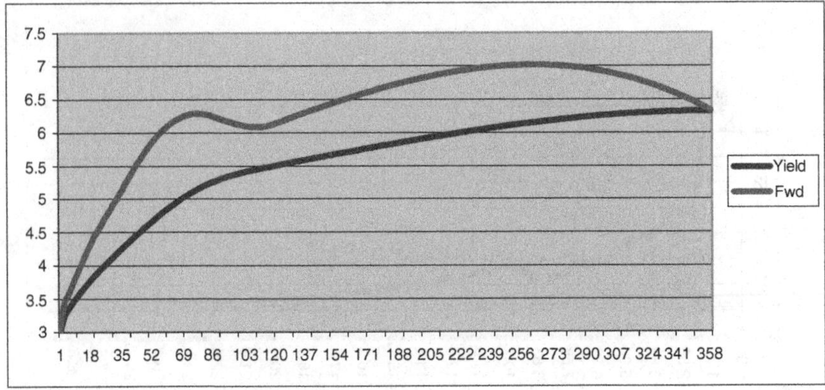

Figure 4: The Initial Curves

This forward curve produce an option value somewhat below the present value of the premium. I multiplied all calculated volatilities by 1.03535 to equate the option value to the premium paid.

The Option Delta

A simple way to see how the option value depends on interest rates is to calculate the delta of the contract. Since I am using a one-factor model, I calculated the change in option value due to a change in the one-month forward rate. This turns out to be 3418.91, which clearly indicates the bet on falling rates.

Sample Term Structures after 6 Months

These, as well as the term structure on October 29, 1993, are shown in Figure 5. Recall that the implemented tree has 64 terminal nodes, labeled 63 to 126. The higher the node number, the higher the curve. It can be seen that the curves retain their basic shape, which is an artifact of the one-factor model.

The Spread and Future Contract Values

Out of the 64 nodes, the spread was positive on 7 nodes, or 10.93%. The relevant information is shown in the following table.

Spread	Value	Probability	CMT5	T30 price
0.062232	49.07267	0.09375	0.059289	94.81414
0.062527	49.30414	0.07813	0.059298	94.79932
0.062829	49.54152	0.06250	0.059306	94.78411
0.06312	49.77079	0.04688	0.059315	94.76911
0.063425	50.01055	0.03125	0.059323	94.75354
0.063731	50.25117	0.01563	0.059332	94.73767
0.17427	136.0099	0.00000	0.063672	91.07946

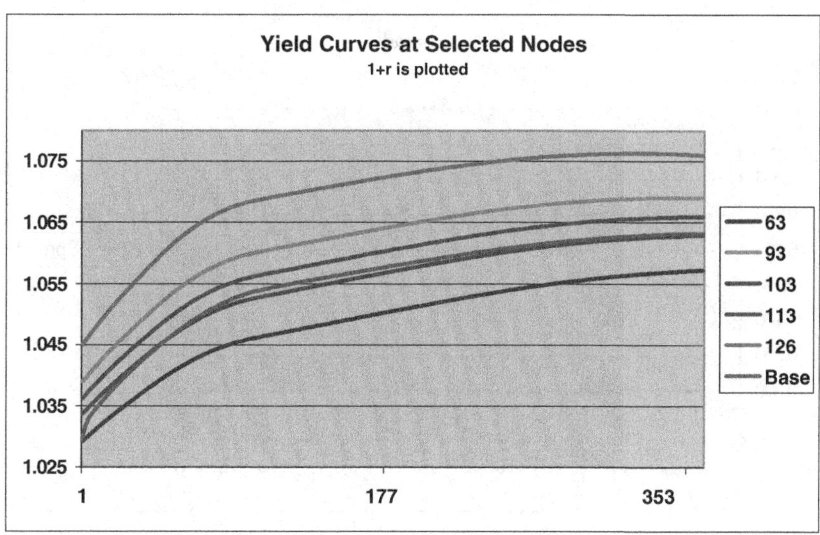

Figure 5: Sample Term Structures in 6 Months

The table shows that with 5% probability, the payment on the option would exceed \$49.5M, so the Value at Risk is \$42.85M. Recall that the value of the option is \$6.65M. so the VaR is about 7 times the value of the contract. The table also shows that in the worst case in the model, the 5-CMT yield is 6.37% while the clean price of the August 2023 Treasury is 91.08. On May 1, 1998, the actual CMT yield was 6.78% while T30 price was \$87.3. Given the coarseness of the estimation, this indicates to me that the model actually captured the possible paths of the term structure quite well.

The spread is shown in Figure 6 for the 64 terminal nodes of the tree; the level of interest rates rises with the node, so rates are rising from left to right. In Figure 7, I show the gains and losses at the different nodes. It is evident that the contract had fairly limited upside potential and allowed for the possibility of a very large loss.

Another risk measure, which complements VaR, is the conditional expected loss. This is defined as the expected loss conditional on being in the region that occurs with less than 5% probability. In our model, this turns out to be \$64.8m, about ten times the value of the contract.

5 Conclusion

In summary, it is clear that the Value at Risk of the contract clearly points out the potential loss from the contract. While the actual events in the months after October 29, 1993 were worse than the worst case in the model, and

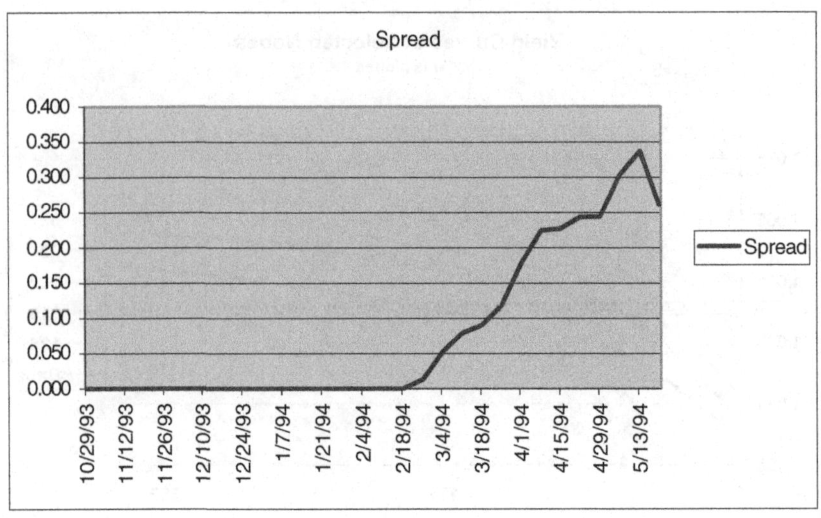

Figure 6: Behavior of the Spread

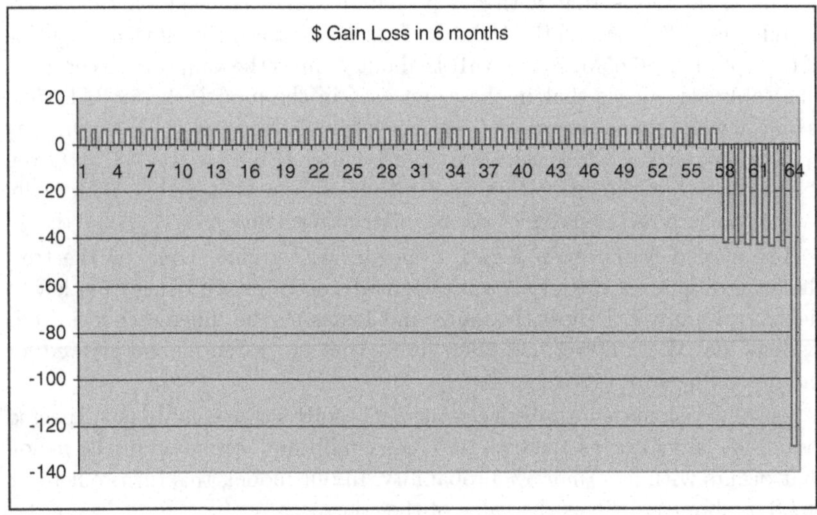

Figure 7: Gains and Losses at the Horizon

while it is unfortunate that this worst case was realized, it is evident that a VaR analysis of the type conducted here would have clearly indicated the risk inherent in the contract.

Acknowledgments

I thank Philippe Jorion, and participants at the Workshop on Risk Management at the Isaac Newton Institute for Mathematical Sciences, Cambridge University, and seminar participants at the University of Amsterdam, MIT, Northwestern University, and the University of Lausanne, for their comments.

References

Ait-Sahalia, Y. and A. Lo, 'Nonparametric Risk Management and Risk Aversion', National Bureau of Economic Research Working Paper 6130, August 1997.

P. Artzner, J.-M. Eber, F. Delbaen and D. Heath, 'A Characterization of Measures of Risk', *Risk Magazine*, November 1997.

Black, F., E. Derman and W. To, '"A One-Factor Model of Interest Rates and its Application to Treasury Bond Options', *Financial Analysts Journal* **46** (1990), 33–39.

Jarrow, R., *Modeling Fixed Income Securities and Interest Rate Options*, McGraw-Hill, New York, 1996.

Heath, D., R. Jarrow and A. Morton, 'Bond Pricing and the Term Structure of Interest Rates: a New Methodology for Contingent Claims Valuation', *Econometrica* **60** (1992), 77–105.

Jorion, P., *Value at Risk: the New Benchmark for Controlling Market Risk*, Irwin Professional Publishing, Chicago, 1996.

Smith, D.J., 'Aggressive Corporate Finance: a Closer Look at the Procter & Gamble–Bankers Trust Leveraged Swap', *Journal of Derivatives* **4** (1997), 67–79.

Stress Testing in a Value at Risk Framework[1]

Paul Kupiec

Abstract

This article proposes a methodology that can be used to parameterize stress test scenarios using the conditional probability distributions that are commonly used in VaR calculations. This new approach allows for a complete characterization of the value change distribution of a portfolio in a stress scenario. Statistical evidence demonstrates that the proposed loss exposure measure is substantially more accurate than the stress exposure measures that financial institutions commonly use.

The results also suggest, contrary to popular perception, that historical VaR risk factor covariances and the assumption of conditional normality can be used to construct reasonably accurate loss exposure estimates in many stressful market environments.

Introduction

Value at Risk, or VaR, is a commonly accepted methodology for measuring the loss magnitudes associated with rare 'tail' events in financial markets. On occasion, either to satisfy management interest or for regulatory compliance, it becomes necessary to quantify the magnitude of the losses that might accrue under events less likely than those analyzed in a standard VaR calculation.[2]

The procedures used to quantify potential loss exposures under such special circumstances are often called a 'stress test'. While this term is commonly used by risk managers and financial institution regulators, there is no accepted definition of what constitutes a stress test. There is even less published information about how best to accomplish a stress test of a trading portfolio or mark-to-market balance sheet.

Given the nature of institutions, it is impossible to document the specific details about how they actually conduct stress tests. Informal discussions, published research (Litterman [1996]), and the public comments of

[1]First published in *Journal of Derivatives*, Fall 1998, 7–23 (1998). Reproduced with permission of Institutional Investor Journals.

[2]For example, The qualitative standards of the Basle internal models approach [1995] for market risk capital requirements specify that a bank must conduct periodic stress tests in addition to its VaR calculations. The standards, however, are silent on how banks should conduct these stress tests.

senior bank officials suggest that many institutions' stress-testing programs are designed around procedures that attempt to generate an approximate full revaluation on the institution's portfolio (or its balance sheet) under selected scenarios that mimic a set of extremely large pricing factor movements that either actually have occurred in the past (historical stress test) or are judged to be remotely possible, given changes in a socioeconomic or political climate (prospective stress test).[3]

An alternative approach to stress testing, an approach not related to history or prospective events, uses an algorithm in an attempt to identify the set of movements in the constellation of pricing factors that will generate the largest losses for an institution or portfolio without regard to the potential for the occurrence of such movements.

While stress testing approaches differ among institutions, the output of all stress tests is an estimate of the loss that would be suffered by an institution or portfolio if a particular scenario were ever to be realized. How should one specify the set of pricing factor movements that are analyzed in a stress test? Are there some statistical relationships that can help in defining stress test parameters? Must the values of all pricing factor changes be specified, or is it appropriate to parameterize the stress-test only partially and set 'nuisance' pricing factor changes to zero?

This article outlines alternative procedures for constructing stress tests and analyzes alternative stress test performance measures for US dollar portfolios that include exposures to US, Asian, and European bond and stock markets. The procedures build on the so-called variance-covariance VaR framework popularized by JP Morgan's *RiskMetrics* [1996]. These stress testing procedures provide a way to construct stress test estimates that are consistent with historically observed volatility and correlation patterns.

While some would argue that stress testing should not use historical correlations or the assumption of normality because it is alleged that these assumptions are materially compromised in stress scenarios, this study suggests that when stress test statistics are appropriately constructed, the common statistical assumptions that underlie VaR provide a useful guide for estimating losses in many stress scenarios. At least for portfolios with exposures broadly distributed across risk factors, the standard VaR normality assumption and the use of historical volatility and correlation patterns do not seem to introduce an unacceptable level of bias and distortion in stress event losses measures.

Moreover, statistical evidence is presented to show that the standard approach for specifying stress test scenarios produces highly inaccurate exposure estimates. Thus, not only do historical volatilities and correlations provide a

[3]Thomas Labrecque, president of Chase Manhattan Corporation, has described the Chase stress testing process [1998]. The bank revisits several specific historical stress episodes including the 1987 stock market crash, the ERM crisis, the Mexican peso devaluation, and the interest rate reversal of 1994. In addition, the Chase staff prepares several prospective stress scenarios.

useful basis for constructing stress loss measures, but failure to incorporate such data in stress event loss measures will also, in most instances, lead to highly inaccurate exposure estimates.

The statistical evidence focuses on evaluating a stress-testing methodology that utilizes historical volatility and correlation data. It is, however, recognized that a user may at time wish to calculate results based upon alternative volatility and correlation assumptions. While it provides no statistical evidence to support the accuracy of such alterations, this study does discuss methodologies that can be used to specify stress test scenarios that incorporate user-selected volatility and correlation patterns that differ from the historical record. Once again, the proposed stress test methodology is integrated into the context of VaR methodology so the user can specify partial 'what-if' scenarios and use the VaR structure to 'fill in' the most likely values for the remaining factor values in the system.

1 A Review of the VaR Framework

The basic VaR framework begins with the assumption that position value changes can be modeled as if they are being 'driven' by changes in some set of underlying fundamental factors. Let $\mathbf{F}_t = [f_{1t}, f_{2t}, f_{3t}, \ldots, f_{Nt}]$ represent an N-dimensional vector of values for the underlying pricing factors on day t. For example, in the *RiskMetrics* approach, the individual pricing factors are the so-called vertices that are used in cash flow mapping procedures. In practice, at any point in time the pricing factor values (the cash flow vertices) are given by the values of USD exchange rates, estimates of foreign and domestic (USD) zero coupon bond prices (or yields), and the values of selected foreign and domestic stock market indices.

Let $\mathbf{x}_{1t}$ represent the total cash flow from all portfolio positions that are mapped on to pricing factor (vertex) $\mathbf{f}_{1t}$ and $\mathbf{X}_t = [\mathbf{x}_{1t}, \mathbf{x}_{2t}, \mathbf{x}_{3t}, \ldots, \mathbf{x}_{Nt}]$ represent the entire vector of these mapped cash flow values. Under the standard VaR assumptions, the day-to-day changes in the values of the individual asset pricing factors (cash flow vertices) evolve according to the relationship:

$$\log(\mathbf{f}_{i\,t+1}) - \log(\mathbf{f}_{it}) \approx \frac{\mathbf{f}_{i\,t+1} - \mathbf{f}_{it}}{\mathbf{f}_{it}} = \tilde{r}_{it}, \tag{1}$$

where $\tilde{r}_{it}$ is a normally distributed random variable.

Let $\tilde{\mathbf{R}}_t = [\tilde{r}_{1t}, \tilde{r}_{2t}, \tilde{r}_{3t}, \ldots, \tilde{r}_{Nt}]$ represent the $(N \times 1)$ random vector of the day-to-day pricing factor returns. By assumption,

$$\tilde{\mathbf{R}} \sim N(0_{\mathbf{N}}, \Sigma_t) \tag{2}$$

where $N(.)$ indicates the multivariate normal density function, $0_{\mathbf{N}}$ represents an $(N \times 1)$ vector of zeros (the mean return) and represents Σ_t an $(N \times N)$

covariance matrix. VaR estimation assumes that the change in the portfolio's value can be written in terms of the vertex return vector as,

$$\Delta\tilde{\mathbf{V}}_t = \mathbf{X}_t\tilde{\mathbf{R}}_t. \tag{3}$$

Under these assumptions, the 1-day change in the market value of a portfolio is approximately equivalent to a normally distributed random variable with a mean of 0, and a variance of $\mathbf{X}_t\Sigma_t\mathbf{X}_t^{\mathrm{T}}$ where the superscript T indicates the transpose. Accordingly, its 95% VaR estimate, VaR(95), is given by

$$\mathrm{VaR}(95) = -1.65\sqrt{\mathbf{X}_t\Sigma_t\mathbf{X}_t^{\mathrm{T}}}. \tag{4}$$

Unlike a routine daily VaR calculation, which estimates the potential losses that may accrue with a given maximum likelihood under a historical estimate of the complete factor price-change probability distribution, a stress test measures the loss that could be experienced if a set of pricing factors take on a specific set of exogenously specified values. Typically, the specific pricing factor settings that are the basis of a single stress test represent a highly unlikely event when measured against the historical conditional pricing factor return density function (i.e. according to equation (2)).

Specification of such a stress test requires as input the magnitudes of the specified pricing factor changes. For example, a stress test might specify that portfolio positions be revalued under an 'inflation shock' scenario in which US equity indices fall substantially, long-term US dollar interest rates rise markedly, and the dollar simultaneously depreciates against a basket of currencies. While a stress test generally specifies the movements that are to be associated with key pricing factors of interest, a stress scenario may be agnostic as to the magnitudes of the pricing factor changes that are appropriate for the remaining pricing factors in the system.

What values should be specified for the changes accorded to any risk factors that play only supporting roles in a stress scenario? The most common practice is to set to zero the changes on risk factors that do not play a 'starring role' in a stress test scenario.[4]

While such a practice may be common, it is not the only way to specify stress test inputs. Indeed, it will be argued that the daily VaR methodology can be adapted to provide a framework that is useful both for completing the specification of stress test scenarios and for generating a more informative set of risk exposure statistics. Moreover, the statistical evidence will show that the standard approach for specifying stress test scenarios produces highly inaccurate exposure estimates when it is applied to estimate the potential losses that would have been realized in many historical stress situations.

---

[4] This is the approach taken in the scenario analysis utilities of all commercial risk management software programs I am familiar with.

2 Stress Testing in the Context of VaR

Consider the VaR representation of the portfolio value-change given in equation (3). It is useful to partition the pricing factor return matrix into the set of k factors that will play 'starring roles' in a stress test scenario, $\tilde{\mathbf{R}}_{2t}$, a $(k \times 1)$ vector, and all other pricing factors, $\tilde{\mathbf{R}}_{1t}$, an $(N-k \times 1)$ vector. If the covariance matrix is partitioned accordingly, the pricing factor conditional return distribution can be written

$$\begin{bmatrix} \tilde{\mathbf{R}}_{1t} \\ \tilde{\mathbf{R}}_{2t} \end{bmatrix} \sim N\left(\begin{bmatrix} \mu_{1t} \\ \mu_{2t} \end{bmatrix}, \begin{bmatrix} \Sigma_{11t} & \Sigma_{12t} \\ \Sigma_{21t} & \Sigma_{22t} \end{bmatrix} \right), \tag{5}$$

where the mean return vectors μ_{1t} and μ_{2t} are both set to zero under the standard VaR assumptions. The portfolio weighting vector, $\mathbf{X}_t$ can also be partitioned to conform with the stress test partition of the factor return matrix, $\tilde{R}_t$. Incorporating this partitioned matrix, the equation describing the portfolio value change (equation (3)) can be written:

$$\Delta\tilde{\mathbf{V}}_t = \begin{bmatrix} \mathbf{X}_{1t} & \mathbf{X}_{2t} \end{bmatrix} \begin{bmatrix} \tilde{\mathbf{R}}_{1t} \\ \tilde{\mathbf{R}}_{2t} \end{bmatrix}. \tag{6}$$

Let the pricing factors that play key roles in the stress test scenario experience a specific set of returns,

$$\tilde{\mathbf{R}}_{2t} = \mathbf{R}_2 = [r_1, r_2, r_3, \ldots, r_k]. \tag{7}$$

The magnitudes of the R_2 returns in $\mathbf{R}_2$ are set according to management judgement based on analysis that accounts for economic factors and information beyond that which is reflected in the historical pricing factor data alone. Given the specific set of pricing factor returns specified in the stress test, the portfolio's value-change equation becomes:

$$\Delta\tilde{\mathbf{V}} = \mathbf{X}_{2t}\mathbf{R}_2 + \mathbf{X}_{1t}\tilde{\mathbf{R}}_{1t}. \tag{8}$$

The most common approach to stress testing would set $\tilde{\mathbf{R}}_{1t} = 0_{N-k}$ and thereby estimate potential losses as $\mathbf{X}_{2t}\mathbf{R}_2$. This method of parameterizing a stress test simply revalues the instruments that are priced off of the factors that play starring roles in the stress test, and ignores the influence of potential pricing factor changes that accrue to other factor. The risk exposures that arise from the remaining risk factors can be accounted for using the portfolio value change equation (8) and the properties of the multivariate normal distribution that underlie the standard VaR methodology. We consider alternative methods that can be used to parameterize stress test scenarios and calculate loss exposures. The alternatives differ according to the detail specified in the stress test scenario.

3 Stress Tests Scenarios based upon Historical Covariances

An alternative approach for constructing stress test loss exposure estimates is to assume that the historical covariance matrix, Σ_t, is unaltered in the stress test. Using Σ_t, the assumption that the risk factor returns have a multivariate normal distribution function enables one to calculate the probability distribution of the remaining risk factors (those in $\tilde{\mathbf{R}}_{1t}$) conditional on the fact that the risk factors in $\tilde{\mathbf{R}}_{2t}$ take on the specific values specified exogenously in the stress test. Conditional on these values, the remaining factors are distributed as,

$$\tilde{\mathbf{R}}_{1t}|_{\tilde{\mathbf{R}}=\mathbf{R}_2} \sim N[\mu_c, \Sigma_c], \tag{9}$$

where

$$\mu_c + \Sigma_{12}\Sigma_{22}^{-1}\mathbf{R}_2 \tag{10}$$

and

$$\Sigma_c = \Sigma_{11} - \left(\Sigma_{12}\Sigma_{22}^{-1}\Sigma_{21}\right). \tag{11}$$

Using the conditional probability density function to define the characteristics of the remaining risk factors in the stress test scenario, the stress-scenario change in portfolio value is a normally distributed random variable with an expected value given by:

$$\mathrm{E}(\Delta\tilde{\mathbf{V}}_t) = \mathbf{X}_{2t}\mathbf{R}_2 + \mathbf{X}_{1t}\mu_c \tag{12}$$

and a standard deviation of $\sqrt{\mathbf{X}_{1t}\Sigma_c\mathbf{X}_{1t}^{\mathrm{T}}}$. Once the joint distributional properties of the pricing factor are accounted for, it is clear that the stress test scenario generates an implied distribution for the portfolio value change, not just a single loss estimate. The output data generated by a stress test can be further distilled by using a summary measure of the potential stress scenario loss distribution to characterize the portfolio's risk exposure. The stress test loss measure analogue to VaR would suggest the use of a loss exposure measure equal to a left-hand critical tail value of the portfolio's value-change probability distribution. For example, using a 5% left-hand tail critical value threshold would generate a so-called 95% stress scenario VaR value [StressVaR(95)]. Such a loss measure would be calculated according to the expression:

$$\mathrm{StressVaR}(95) = \mathbf{X}_{2t}\mathbf{R}_2 + \mathbf{X}_{1t}\Sigma_{12}\Sigma_{22}^{-1}\mathbf{R}_2 - 1.65\sqrt{\mathbf{X}_{1t}(\Sigma_{11} - \Sigma_{12}\Sigma_{22}^{-1}\Sigma_{21})\mathbf{X}_{1t}^{\mathrm{T}}}. \tag{13}$$

Alternatively, a risk manager could decide to ignore the losses that might be generated by the unexpected variation in the pricing factors in $\tilde{\mathbf{R}}_{1t}$, and use the expected value of the stress test portfolio value change distribution to measure the loss in the stress scenario. This measure, the expected stress test loss [E(StressVar)] is given by:

$$\mathrm{E}[\mathrm{StressVaR}] = \mathbf{X}_{2t}\mathbf{R}_2 + \mathbf{X}_{1t}\Sigma_{12}\Sigma_{22}^{-1}\mathbf{R}_2. \tag{14}$$

Regardless of the metric selected to summarize the stress test portfolio value-change distribution, when a stress test is integrated into the VaR framework, provided that $\Sigma_{12} \neq 0$, stress tests will produce estimates that differ from those that would be produced by the common approach of assuming that $\tilde{\mathbf{R}}_{1t} = 0_{N-k}$.

4 Accuracy of Stress Tests that use Historical Distributions

While there may be limitations associated with using the normality and historical volatility and correlation assumptions that underlie the daily VaR calculations to estimate stress event loss exposures, in many cases these distributional assumptions do provide a useful basis for constructing stress event exposure estimates.

Data

The accuracy of the alternative techniques for estimating stress event loss exposures are compared using *RiskMetrics* data over the period, April 1, 1993 to Jan 13, 1998. Stress scenario calculations are simulated for portfolios with exposures to equity index and zero coupon interest rate risk factors from Asian, European, and US markets.

The accuracy of alternative stress test loss estimators are compared for portfolios with linear exposures to the five-year AUD zero coupon bond, ten- and twenty-year GBP zero-coupon bonds, the twenty-year JPY zero-coupon bond, five-, ten-, twenty-, and thirty-year US zero-coupon bond prices, and equity index exposures to markets in Australia, Switzerland, Germany, the UK, Japan, the Philippines, Thailand, and the US. When these exposures are combined in a USD-based portfolio, including the relevant foreign exchange risk factors, the portfolio has exposures to (i.e., cash flows are mapped onto) twenty-six separate risk factors.

Identifying Stress Events

The analysis of alternative stress-testing procedures requires a sample of market stress events. For purposes of this study, market stress events are identified using an empirical criterion. The sample of stress events examined are the set of events in which an individual interest rate factor (zero-coupon bond price) equity index or foreign exchange rate risk factor experiences a decline in value that exceeds 3.5 unconditional standard deviations of its sample log return distribution.[5] Declines of this order of magnitude are substantially larger than the 1.65 (or 2.33) standard deviation moves that underlie 95% (or 99%) VaR calculations.

Using this criterion, the sample period contains 102 stress events. These stress events are not independent as extreme moves occur simultaneously in many markets.

For example, on October 27, 1997, four risk factors posted declines in excess of 3.5 (return) standard deviations. While it is possible to condition a stress loss estimate on multiple extreme market moves, it is perhaps a more realistic model of actual stress test formulation to condition on a single extreme risk factor move and project the response of other risk factors in the stress scenario using the stress testing procedures described earlier.[6]

Stress Test Scenarios

Suppose management correctly anticipates the potential for an extreme move in a single risk factor. While the accuracy of management's forecast of the magnitude of the risk factor's move is clearly important to the ultimate quality of the stress test exposure estimates, it will add realism and generate little controversy to assume a relatively naive forecast of a stress event.

For this analysis, it is assumed that management conducts stress tests under the assumption that the risk factor will decline in value by an amount equal to 3.5 times its daily VaR return volatility estimate. As the VaR approach taken in *RiskMetrics* uses a time-varying volatility estimate, the magnitude of a 3.5 standard deviation return will vary over time.[7] This procedure for specifying a stress test scenario is a transparent extension of the daily VaR calculation.

Initially, the properties of alternative stress loss exposure estimates are analyzed for a portfolio that is 'long' all risk factors. In this benchmark case, the interest rate and equity risk factor weights are selected so that, measured in US dollars, the cash flow mappings on all interest rate and equity factors are roughly comparable in magnitude.

---

[5] In the Thai bhat foreign exchange market, there were six instances that qualify as events under this definition but are excluded from the analysis. Although they were events as measured by the sample unconditional standard deviation, these FX moves were small when compared with the time-varying conditional VaR standard deviation on the event day. Exclusion of these events does not alter the results, as all proposed stress measures performed well on these excluded event days.

[6] The difference between conditioning on a single risk factor move and multiple risk factor movements is, from the perspective of predicting the movements in the non-stressed factors, the difference between using univariate and multivariate regressions.

[7] In this analysis, the VaR covariance matrix is calculated using 250 days of return data and a decay factor of 0.97. The production version of *RiskMetrics* uses 550 days of data and a decay factor of 0.94. The qualitative results of this study are not sensitive to the decay factor and the sample size assumptions. No claim is made regarding the optimality of these parameter settings.

For example, the US equity exposure is long the S&P500 index. Depending on the interest rate or non-US dollar equity index in question and the date of the event, multiple zero-coupon bonds positions or multiples of the non-dollar equity indices must be held to generate US dollar investments of a size comparable to an investment in the S&P500 index. While investment proportions are set to achieve a portfolio that has reasonable balance among risk factor exposures, only integer multiples of risk factors are selected, so the portfolio is not equally weighted.[8]

Stress events are grouped according to the time zone in which the stress market correction occurs: Asia, Europe, or the US. The time zone differentiation distinguishes the temporal ordering of the risk factor return data. The *RiskMetrics* data assumes the day begins in Asia and ends in the US. If an event is initiated in an Asian market, the standard *RiskMetrics* return day is an appropriate ordering for calculating inter-market correlations. If, alternatively, an event is initiated in the US, then the event will be reflected in returns on the following day in Asia and Europe.

In the analysis, the return data is adjusted so that the return day begins in the time zone of the stress event. Thus, the Asian stress tests assume that the event begins in Asia and the day unfolds according to the standard *RiskMetrics* temporal assumptions. European event estimates assume the event begins in Europe and is reflected in Asian markets on the following day. Similarly, the US stress event estimates assume that the event begins in the US and is reflected in Asian and European markets on the following day.

While this assumption may seem innocuous, the results that follow will show that the temporal ordering of risk factor returns can be an important determinant of the 'accuracy' of stress test loss estimates.

Results

Tables 1, 2, and 3 report, respectively, the individual results for stress events in the Asian, European, and US time zones. For each stress event, the tables report: the event market factor and date; the total value of the portfolio in US dollars; the US dollar value of the portfolio's position in the stressed risk factor; the portfolio's 95% VaR, the predicted stress loss based on the current stress testing method ($\mathbf{X}_{2t}\mathbf{R}_2$); the actual change in the portfolio's value; the actual number of standard deviations of the stress move; the expected stress value change E(StressVaR); and the 95% stress test VaR estimate, StressVar(95).

While these tables document the event-specific performance of the alternative stress exposure measures, the general tendencies demonstrated by the analysis are more easily seen if the data is analyzed graphically. Figures 1

[8]The choice of integer weights is completely arbitrary.

Table 1: Stress Test Results for Asian Time Zone Events

Stress market	Stress currency	Event date	Total portfolio value USD	Stress position value USD	Portfolio 95% VaR	Current method predicted value change	Actual stress event portfolio value change	Actual magnitude of stress event	Expected portfolio value change new measure	95% stress VaR measure
15 yr zero	AUD	19 Apr 94	53407	2029	336	−113	−36	−3.52	−25	−361
15 yr zero	AUD	20 Jun 94	52965	1721	291	−108	−275	−3.19	−311	−563
15 yr zero	AUD	12 Sep 94	53702	1652	267	−89	−48	−4.34	−73	−338
15 yr zero	AUD	6 Mar 95	54073	1581	219	−75	46	−3.60	−57	−274
15 yr zero	AUD	30 June 95	59168	1783	388	−85	15	−3.76	−272	−638
15 yr zero	AUD	21 Feb 90	61232	2065	292	−98	124	−4.47	−79	−358
15 yr zero	AUD	11 Mar 96	59993	1990	436	−127	−231	−4.00	208	−632
15 yr zero	AUD	6 Dec 96	65086	2620	369	−98	−835	−4.76	−501	−785
15 yr zero	AUD	3 Jan 97	63944	2472	395	−109	68	−3.95	−86	−479
Equity	AUD	27 Jun 94	52474	1470	352	−38	369	−4.06	−79	−429
Equity	AUD	11 Mar 96	59993	1742	436	−47	−231	−4.58	−295	−708
Equity	AUD	6 Dec 96	05086	1903	369	−42	−835	−4.63	−472	−766
Equity	AUD	27 Oct 97	69078	1797	415	−60	−731	−3.48	−493	−836
Equity	AUD	28 Oct 97	68347	1713	389	−94	164	−4.59	−196	−574
Equity	JPY	23 Jan 95	52663	3798	237	−142	−616	−5.23	−407	−547
Equity	JPY	3 Apr 95	55532	3735	288	−182	−33	−3.38	14	−273
Equity	JPY	19 Nov 97	69406	2660	422	−188	−91	−2.62	−482	−838
Equity	JPY	25 Nov 97	69678	2661	448	−201	−423	−2.74	−555	−918
Equity	JPY	19 Dec 97	70196	2518	325	−168	−318	−2.76	−375	−648
20 yr zero	JPY	22 Mar 94	54619	3124	384	−84	285	−2.89	31	−353
20 yr zero	JPY	14 Jul 95	61089	5577	408	−158	−562	−3.65	−470	−813
20 yr zero	JPY	2 Aug 95	59523	5229	338	−138	120	−3.18	−273	−586
Equity	PHP	13 Jan 95	52865	3239	166	−143	10	−4.49	−76	−238
Equity	PHP	10 Apr 97	67122	2289	333	−167	−226	−4.46	−216	−533
Equity	PHP	23 Oct 97	69602	1665	383	−93	−651	−3.12	−189	−561
Equity	PHP	28 Oct 97	69347	1579	389	−102	163	−3.43	−59	−447
Equity	PHP	11 Dec 97	69994	1679	344	−9.5	−262	3.06	−194	−517
Equity	PHP	8 Jan 98	70270	1156	324	−65	−140	−3.26	−155	−471
Equity	PHP	9 Jan 98	70130	1106	330	−83	−359	−3.97	−259	−566
Equity	THB	7 Oct 96	63084	1185	283	−69	−120	−3.63	−7	−290
Equity	THB	18 Nov 96	65220	1181	268	−63	47	−3.83	98	−362
Equity	THB	28 Oct 97	68347	382	389	−23	163	−3.56	55	−332
FX	AUD	9 Nov 95	60193	3504	232	−52	−118	−4.14	−47	−279
FX	AUD	3 Dec 96	66288	4655	290	−84	−552	−4.74	−405	−623
FX	AUD	23 May 97	65872	4350	401	−67	283	−4.31	162	−233
FX	AUD	22 Oct 97	69857	4644	345	−98	−255	−4.74	−151	−489
FX	JPY	26 Apr 95	56992	9278	245	−301	−208	3.17	−305	−503
FX	JPY	2 Aug 95	59523	8936	339	−202	120	−4.10	5	334
FX	JPY	15 Aug 95	58333	8391	329	−236	−476	−4.07	−284	−585
FX	PHP	21 Sep 94	53350	3409	244	−59	−161	−5.00	−22	−265
FX	PHP	6 Oct 94	52705	3467	228	−94	−39	−3.66	−44	−271
FX	PHP	10 Nov 94	53825	3818	258	−110	−382	−2.97	−173	−417
FX	PHP	21 Mar 95	54493	2768	250	−74	−197	−3.04	−13	−203
FX	PHP	11 Jul 97	6998	2849	382	−215	13	−5.50	3	−379
FX	PHP	14 Aug 97	68068	2646	373	−120	9	−2.20	1	−372
FX	PHP	3 Sep 97	66873	1903	312	−79	26	−3.66	27	−285
FX	PHP	10 Sep 97	67541	2140	324	−87	−513	−2.17	−147	−464
FX	PHP	17 Sep 97	67965	1964	363	−80	214	−2.57	−111	−470
FX	PHP	14 Oct 97	70285	1888	372	−85	−10	−2.91	85	−285
FX	PHP	18 Nov 97	69544	1665	434	−55	−138	−2.60	108	−323
FX	PHP	11 Dec 97	69994	1679	344	−46	−262	−3.70	−63	−400
FX	PHP	12 Dec 97	69732	1550	328	−58	38	−4.06	−20	−348
FX	PHP	15 Dec 97	69770	1466	324	−70	−221	−3.68	−96	−417
FX	PHP	2 Jan 98	70074	1414	304	−53	113	−2.56	−33	−337
FX	PHP	5 Jan 98	70187	1379	335	−57	575	−2.70	137	−192
FX	PHP	7 Jan 98	70762	1320	334	−66	−209	−3.40	39	−295
FX	THB	11 July 97	69998	664	382	−33	13	−2.08	222	−146
FX	THB	3 Sep 97	66873	439	312	−23	26	−4.44	14	−299
FX	THB	11 Dec 97	69994	275	344	−10	−262	−2.76	−15	−359
FX	THB	12 Dec 97	697.12	250	328	−10	38	−2.79	4	324
FX	THB	15 Dec 97	69770	245	324	−13	−221	−3.86	−95	−406
FX	THB	5 Jan 99	70187	235	335	−11	575	−2.53	68	−265

and 2 provide a useful visual summary of the results of the individual stress tests.

Figure 1 plots, for each event, the actual portfolio change in value, the common measure of stress test loss exposure, and the proposed expected loss exposure measure, E(StressVaR). While the events are not a true time series,

Table 2: Stress Test Results for European Time Zone Events

Stress market	Stress currency	Event date	Total portfolio value USD	Stress position value USD	Portfolio 95% VaR	Current method predicted value change	Actual stress event portfolio value change	Actual magnitude of stress event	Expected portfolio value change new measure	95% stress VaR measure
Equity	CHF	22 Aug 97	67827	3739	410	−155	−390	−2.87	−512	−843
Equity	CHF	28 Aug 97	66956	3613	366	−160	−110	−2.80	−389	706
Equity	CHF	28 Oct 97	68091	3804	471	−177	601	−3.44	−368	−805
Equity	DEM	20 Jun 94	52725	1258	326	−40	−244	−4.36	−240	546
Equity	DEM	6 Dec 96	64647	1972	357	−59	−299	−4.53	−36	−393
Equity	DEM	22 Aug 97	67827	2309	410	−98	−390	−3.25	−451	−801
Equity	DEM	23 Oct 97	69403	2342	404	−110	−410	−3.45	−436	783
Equity	DEM	27 Oct 97	69093	2278	477	−120	−1002	−2.83	−652	1016
Equity	DEM	28 Oct 97	68091	2204	471	−157	601	−3.94	−183	648
Equity	GBP	23 Oct 97	69403	8418	404	−243	−410	−3.70	−572	−872
Equity	GBP	27 Oct 97	69093	8122	477	−256	−1002	−2.89	−757	o73
Equity	GBP	19 Dec 97	70110	8583	337	−289	−293	−2.98	−364	−654
10 yr zero	GBP	24 Mar 94	54911	6919	455	−207	−673	−3.04	−779	1047
10 yr zero	GBP	18 Apr 94	54805	6773	448	−210	−880	−3.16	−667	986
10 yr zero	GBP	31 May 94	53461	6327	365	−187	−261	−3.03	−567	−81,5
10 yr zero	GBP	30 Jun 94	52983	6505	432	−237	−619	−2.44	−766	−1002
10 yr zero	GBP	27 Jul 94	53556	6504	362	−189	−426	−3.18	−612	831
10 yr zero	GBP	9 Sep 94	54228	6444	347	−186	−761	−3.28	−582	−794
10 yr zero	GBP	29 Dec 94	53503	6576	207	−141	−162	−4.84	−190	377
10 yr zero	GBP	9 Jun 95	59817	7142	373	−159	−276	−4.01	−461	−764
FX	CHF	30 May 95	.55141	2197	297	−66	−172	−3.46	−210	490
FX	CHF	15 Aug 95	58376	2353	335	−56	−478	−4.15	−360	−649
FX	CHF	7 Apr 97	62481	3123	254	−61	32	−4.41	116	132
FX	DEM	30 Mar 95	55141	1393	297	−35	−172	−3.30	−206	−487
FX	DEM	15 Aug 95	58376	1540	335	−34	−478	−4.39	−341	−635
FX	GBP	10 Mar 95	53759	14519	290	−310	−26	−3.80	−235	503
FX	GBP	3 Dec 96	66253	18536	287	−247	−473	−4.60	−364	−594
FX	GBP	28 Jul 97	70284	20665	423	−324	−456	−4.16	−437	806

Table 3: Stress Test Results for US Time Zone Events

Stress market	Stress currency	Event date	Total portfolio value USD	Stress position value USD	Portfolio 95% VaR	Current method predicted value change	Actual stress event portfolio value change	Actual magnitude of stress event	Expected portfolio value change new measure	95% stress VaR measure
Equity	USD	8 Mar 96	61245	6537	365	−169	−855	−4.17	−593	−827
Equity	USD	11 Apr 97	62373	7583	261	−203	−252	−3.56	−305	−523
Equity	USD	15 Aug 97	68233	9248	299	−246	−380	−3.41	−351	−600
Equity	USD	27 Oct 97	68933	9416	352	−459	−374	−4.93	−368	−674
Equity	USD	9 Jan 98	70313	9561	282	−290	−331	−3.43	−206	−471
5 yr zero	USD	6 May 94	53328	7138	289	−87	−450	−3.47	−322	−566
5 yr zero	USD	20 Feb 96	62251	7655	250	−75	−646	−4.86	−440	−580
5 yr zero	USD	8 Mar 96	61245	7490	365	−111	−855	−4.52	−667	−852
5 yr zero	USD	5 Apr 96	60881	7343	284	−96	−268	−3.62	−410	−618
5 yr zero	USD	5 Jul 96	61137	7205	237	−83	−460	−4.48	−354	−522
30 yr zero	USD	8 Mar 96	61245	2569	365	−100	−955	−4.24	−577	−820
30 yr zero	USD	5 Jul 96	61137	2401	237	−85	−460	−4.61	−390	−539

they are plotted as a time series in order to illuminate the character of the results. Figure 2 incorporates a similar set of plots including the StressVaR(95) loss exposure estimate in place of the expected loss measure.

The test results clearly show that the common approach taken to estimate potential stress test losses, $\mathbf{X}_{2t}\mathbf{R}_2$, provides on average a very poor measure of stress event potential loss exposure. The loss exposure estimates it generates are frequently exceeded, and loss estimates are often only a fraction of actual losses experienced.

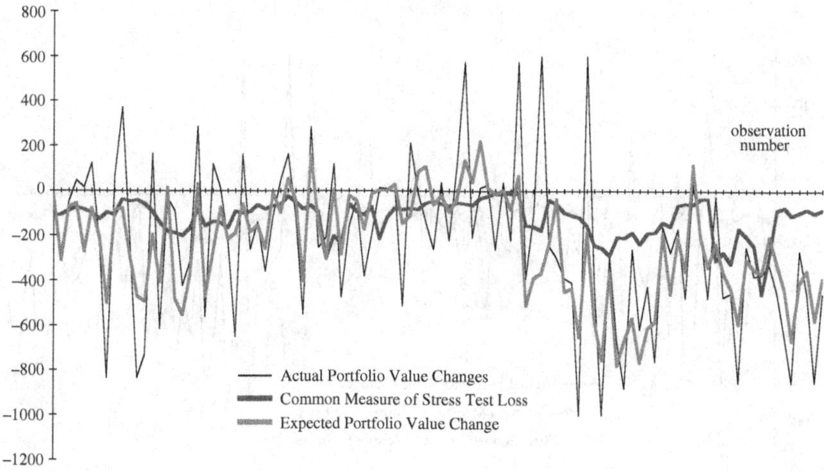

Figure 1: A Comparison of the Common Stress Test Loss Measure to the Expected Loss Measure

The plots in Figure 1 illustrate that the proposed expected loss exposure measure, E(StressVaR), improves upon the current method of calculating stress exposures, but again this measure is also frequently violated by actual stress event losses. While the incorporation of the conditional expected values of the non-stressed risk factors improves the accuracy of stress loss exposure estimates, ignoring the uncertainty in the non-stressed risk factor return distribution clearly leads to an underestimate of the potential losses that may be engendered in market stress situations.

In contrast to the alternative stress event measures, the data plotted in Figure 2 show that the StressVaR(95) exposure estimate places a useful lower bound on actual portfolio losses in all 102 stress events examined. The StressVaR(95) exposure measure is itself violated in six events, but three of these violations are immaterial in magnitude. Although there is some overlap in the sample and the 102 stress events are not all independent, using the sample proportion of failures test (PF test) outlined in Kupiec [1995], it is not possible at the 5% level of the test to reject the hypothesis that the StressVaR(95) statistic is unbiased.[9]

These results provide relatively strong evidence that (1) highlight the significant weaknesses of current stress testing practice, and (2) recommend the use of stress test exposure estimates that are consistent with conditional VaR distributional assumptions.

[9]Exhibit 4 in Kupiec [1995] shows that, given six observed violations of a loss exposure estimate, it is impossible to reject the null hypothesis that the loss exposure estimate provides true 95% coverage unless there are fewer than fifty independent observations.

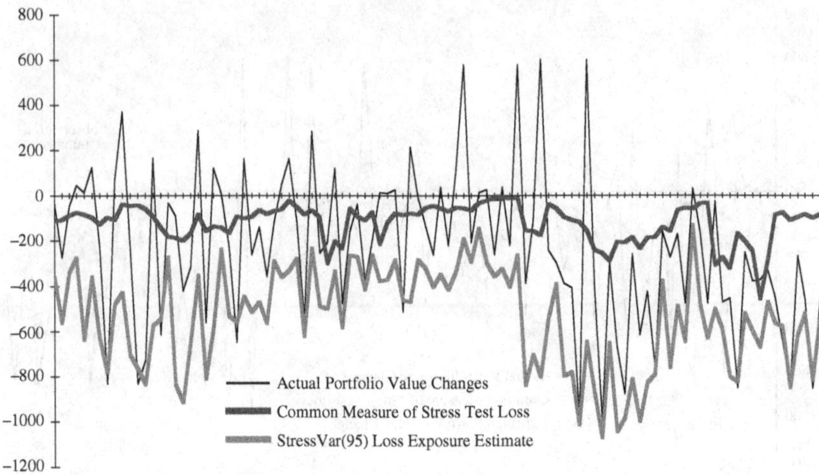

Figure 2: A Comparison of the Common Stress Test Loss Measure to the Stress VaR (95) Loss Exposure Estimate

The results summarized in Figures 1 and 2 are not definitive. It is possible that the results for the sample portfolio of long positions may not be representative of the stress test results that would prevail for a general portfolio with a mixture of long and short positions.

To address this potential issue, the 102 stress events are analyzed using a methodology that is unaffected by the choice of portfolio weights. For each event, an independent random investment of between –100 USD and +100 USD (from a uniform distribution) is selected for each interest rate and equity risk factor. The resulting portfolio is stress tested and potential loss exposures are measured using the current method, the StressVaR(95), and the StressVar(99) estimates. The actual portfolio gain or loss is compared with the alternative stress test loss exposure estimate and violations are tabulated. These calculations are repeated 1000 times for each event and so each event produces a distribution of outcomes generated from the distribution of randomized portfolio weights. Tables 4, 5, and 6 report the results of the analysis. The results document, with rare exceptions, the overwhelming inadequacy of the current approach for calculating stress event loss exposures.

If the objective of a stress test exercise is to generate an accurate lower bound on potential portfolio losses, the results in these tables suggest that the proposed StessVaR loss exposure measures provide a practical alternative approach for measuring stress event exposures. Consistent with expectations, the StressVar(99) measure is the most conservative exposure estimate and provides, except in rare occurrences, a useful lower bound regardless of portfolio composition.

Table 4: Stress Test Results for Randomized Portfolios — Asian Time Zone Events

Stress market	Stress currency	Event date	Number of portfolios for which actual losses exceed loss predicted by current method	Number of portfolios for which actual losses exceed 95% stress VaR loss estimate	Number of portfolios for which actual losses exceed 99% stress VaR loss estimate
15 yr zero	AUD	18 Apr 94	965	0	0
15 yr zero	AUD	17 Jun 94	996	0	0
15 yr zero	AUD	9 Sep 94	910	1	0
15 yr zero	AUD	3 Mar 95	322	9	0
15 yr zero	AUD	29 Jun 9S	577	94	1
15 yr zero	AUD	20 Feb 90	436	120	35
15 yr zero	AUD	8 Mar 96	996	9	0
15 yr zero	AUD	5 Dec 96	1000	9	0
15 yr zero	AUD	2 Jan 97	668	10	0
Equity	AUD	24 Jun 94	673	126	0
Equity	AUD	8 Mar 96	879	0	0
Equity	AUD	5 Dec 90	986	0	0
Equity	AUD	8 Mar 90	879	0	0
Equity	AUD	24 Oct 97	1000	0	0
Equity	AUD	25 Oct 97	976	4	0
Equity	JPY	20 Jan 95	1000	0	0
Equity	JPY	31 Mar 95	760	0	0
Equity	JPY	18 Nov 97	848	207	32
Equity	JPY	24 Nov 97	983	58	3
Equity	JPY	19 Dec 97	988	0	0
FX	JPY	15 Aug 95	924	2	0
FX	PHP	21 Sep 94	921	0	0
FX	PHP	6 Oct 94	733	0	0
FX	PHP	10 Nov 94	904	0	0
FX	PHP	21 Mar 95	893	0	0
FX	PHP	11 Jul 97	792	0	0
FX	PHP	14 Aug 97	854	0	0
FX	PHP	3 Sep 97	290	65	0
FX	PHP	10 Sep 97	997	0	0
FX	PHP	17 Sep 97	482	73	0
FX	PHP	14 Oct 97	660	7	0
FX	PHP	18 Nov 97	377	0	0
20 yr zero	JPY	19 Apr 94	175	11	0
20 yr zero	JPY	20 Jun 94	887	0	0
20 yr zero	JPY	12 Sep 94	332	312	72
Equity	PHP	6 Mar 95	719	2	0
Equity	PHP	30 Jun 95	995	0	0
Equity	PHP	21 Feb 96	997	0	0
Equity	PHP	11 Mar 96	979	1	0
Equity	PHP	6 Dec 96	1000	0	0
Equity	PHP	3 Jan 97	948	27	1
Equity	PHP	27 Jun 94	998	0	0
Equity	THB	11 Mar 96	197	116	0
Equity	THB	6 Dec 96	484	31	0
Equity	THB	11 Mar 96	197	116	0
Equity	THB	27 Oct 97	980	0	0
FX	AUD	9 Nov 95	376	4	0
FX	AUD	3 Dec 96	924	0	0
FX	AUD	23 Mar 97	29	0	0
FX	AUD	22 Oct 97	621	1	0
FX	AUD	26 Apr 95	695	1	0
FX	AUD	2 Aug 95	242	15	0
FX	PHP	11 Dec 97	1000	0	0
FX	PHP	12 Dec 97	960	0	0
FX	PHP	15 Dec 97	972	0	0
FX	PHP	2 Jan 98	339	292	26
FX	PHP	5 Jan 98	738	59	11
FX	PHP	7 Jan 98	1000	0	0
FX	THB	11 July 97	936	0	0
FX	THB	3 Sep 97	122	1	0
FX	THB	11 Dec 97	1000	0	0
FX	THB	12 Dec 97	920	18	1
FX	THB	15 Dec 97	951	0	0
FX	THB	5 Jan 98	729	193	50

Table 5: Stress Test Results for Randomized Portfolios — European Time Zone Events

Stress market	Stress currency	Event date	Number of portfolios for which actual losses exceed loss predicted by current method	Number of portfolios for which actual losses exceed 95% stress VaR loss estimate	Number of portfolios for which actual losses exceed 99% stress VaR loss estimate
Equity	CHF	22 Aug 97	977	42	7
Equity	CHF	28 Aug 97	966	58	5
Equity	CHF	28 Oct 97	29	994	907
Equity	DEM	20 Jun 94	935	0	0
Equity	DEM	6 Dec 90	610	0	0
Equity	DEM	22 Aug 97	967	4	0
Equity	DEM	23 Oct 97	994	0	0
Equity	DEM	27 Oct 97	1000	1	0
Equity	DEM	28 Oct 97	5	971	666
Equity	GBP	23 Oct 97	993	1	0
Equity	GBP	27 Oct 97	1000	0	0
Equity	GBP	19 Dec 97	998	0	0
10 yr zero	GBP	24 Mar 94	600	290	41
10 yr zero	GBP	18 Apr 94	743	8	0
10 yr zero	GBP	31 May 94	161	731	348
10 yr zero	GBP	30 Jun 94	839	47	10
10 yr zero	GBP	27 Jul 94	645	121	1
10 yr zero	GBP	9 Sep 94	680	65	13
10 yr zero	GBP	29 Dec 94	459	0	0
10 yr zero	GBP	9 Jun 95	624	1	0
FX	CHF	30 May 95	622	50	12
FX	CHF	15 Aug 95	717	1	0
FX	CHF	7 Apr 97	444	0	0
FX	DEM	30 Mar 95	633	55	14
FX	DEM	15 Aug 95	667	0	0
FX	GBP	10 Mar 95	490	7	0
FX	GBP	3 Dec 96	899	0	0
FX	GBP	28 Jul 97	483	0	0

Anomalies

Of the 102 events examined for the random portfolios in Tables 4 through 7, there are three events for which the proposed stress test exposure measures perform remarkably poorly. These events are the DEM and CHF equity events on October 27, 1997, and the GBP interest rate shock that occurred on May 31, 1994.

The poor performance of the newly proposed stress test measures can be understood if these events are examined in the reports of the financial press.

Table 6: Stress Test Results — US Time Zone Events

Stress market	Stress currency	Event date	Number of portfolios for which actual losses exceed loss predicted by current method	Number of portfolios for which actual losses exceed 95% stress VaR loss estimate	Number of portfolios for which actual losses exceed 99% stress VaR loss estimate
Equity	USD	8 May 96	477	111	21
Equity	USD	11 Apr 97	277	0	0
Equity	USD	15 Aug 97	983	0	0
Equity	USD	27 Oct 97	946	0	0
Equity	USD	9 Jan 98	998	0	0
5 yr zero	USD	6 May 94	835	0	0
5 yr zero	USD	20 Feb 96	542	15	1
5 yr zero	USD	8 Mar 96	551	52	11
5 yr zero	USD	3 Apr 96	553	0	0
5 yr zero	USD	5 Jul 96	542	72	19
30 yr zero	USD	8 Mar 96	543	34	6
30 yr zero	USD	5 Jul 96	534	90	18

Table 7: European Events, October 28, 1997. Trading Day Begins in New York

Stress market	Stress currency	Event date	Number of portfolios for which actual losses exceed loss predicted by current method	Number of portfolios for which actual losses exceed 95% stress VaR loss estimate	Number of portfolios for which actual losses exceed 99% stress VaR loss estimate
Equity	CHF	28 Oct 97	995	2	0
Equity	DEM	28 Oct 97	997	0	0

The May 1994 interest rate move is notable in that the *Financial Times* reported that all European bond markets suffered large losses in the face of inflation worries, while European stock markets were largely unaffected by the bond market selloffs. Such an episode would appear to be a classic 'market decoupling' event, when the historical correlation between bond and stock markets would be of little use in predicting stock index returns conditional on a bond market stress event.

If markets decouple, historical conditional correlation estimates are biased, and their use may impart a significant source of error in a stress test loss exposure estimate. Improvements in loss exposure estimates can, at least in theory, be achieved by anticipating correlation and volatility changes and altering historical conditional correlations appropriately. Alternatively, it is possible that loss exposure accuracy may be improved by ignoring historical relationships entirely, as in the current stress testing standard.

While market decoupling can cause substantial bias in the new proposed stress-testing measures, it is notable that the results generated using these 102 stress events suggest that significant decoupling is not pervasive characteristic of the stress events examined.

Consider now the stress events of October 27, 1997. The equity market stress events of October 27, 1997 actually began on October 26 in the US market. The Dow index fell 550 points during the afternoon of October 26, after European markets had closed. In apparent sympathy with these losses, markets in Europe posted significant losses when they opened on October 27.

As the trading day continued, the US market ultimately recovered a significant portion of its prior-day losses. While the European markets were tracking the US recovery, markets on the continent closed too early to fully participate in the gains recorded in US markets. London markets, however, benefited from extra trading time and consequently more fully reflected the strong gains that were recorded in the US.

Thus, while October 27 is recorded as a stress event in both the DEM and the CHF equity markets, the event began in the US. If correlations were estimated using return data that began in the US, it is possible that the contagion in the European markets could have been more accurately anticipated by the stress exposure measures.

The results in Table 7 suggest that this is indeed the case. Table 7 reports the randomized portfolio results for these two equity events using return data that assumes that the trading day began in the US. The poor results reported in Table 5 apparently do not owe to market decoupling, but rather are a consequence of measuring VaR covariances using the wrong temporal risk factor return ordering.

5 Stress Testing with Volatility and Correlation Shocks

Notwithstanding the results that suggest that effective stress exposure estimates can be constructed using historical probability distribution estimates, it is possible to construct stress test scenarios in which a specific set of pricing factor returns are shocked; a potentially different set of pricing factor return volatilities are set to reflect forecasts that differ from those represented in a historical covariance matrix estimate, Σ_t; and still a third set of historical correlations are altered in a stress scenario. Such a stress test scenario would arise for example under an 'inflation shock' scenario if, in addition to the US stock, bond, and FX rate shocks, it is assumed that the volatilities of long-maturity US dollar zero-coupon bonds increased markedly while short-maturity volatilities remain unchanged. In order to estimate the potential losses under these types of stress test scenarios, it is necessary to re-write the VaR portfolio value change equation in terms of factor return correlations. Let ρ_{ijt} represent the correlation on date t between pricing factor i and j:

$$\rho_{ijt} = \frac{\mathrm{Cov}(r_{it}, r_{jt})}{\sigma_{it}\sigma_{jt}}, \tag{15}$$

where σ_{kt} is the standard deviation of the return to pricing factor k on day t. Recall that correlations are symmetric, $\rho_{ijt} = \rho_{jit}$, bounded in value between −1 and +1, and the correlation of a factor return with itself is 1, i.e. $\rho_{iit} = 1$ for all i. Let Ω_t represent the correlation matrix among risk factor returns:

$$\Omega_i = \begin{bmatrix} 1 & \rho_{12t} & \rho_{13t} & \cdots & \rho_{1Nt} \\ \rho_{21t} & 1 & \rho_{23t} \cdots & \rho_{2Nt} & \\ \cdots & \cdots & \cdots & \cdots & \cdots \\ \rho_{N1t} & \rho_{N2t} & \rho_{N3t} & \cdots & 1 \end{bmatrix}. \tag{16}$$

Define D_t to be an $N \times N$ diagonal matrix with the corresponding factor return standard deviations as individual elements. Using these definitions, the VaR covariance matrix can be written in terms of pricing factor return correlations:

$$\Sigma_t = D_t \Omega_t D_t. \tag{17}$$

Stress test scenarios that 'shock' pricing factor returns, return volatilities, and correlations using historical correlations can be constructed using this decomposition.

Stressing Volatilities

Consider a stress test scenario that specifies only specific pricing factor and volatility shocks. The volatility shocks can be easily accommodated using the covariance matrix decomposition given in expression (17). Construct an $N \times N$ diagonal matrix Δ in which the elements are equal to the increments by which historical standard deviations differ from those desired in the stress test scenario.

While the elements in Δ are implicitly restricted by the requirement that variances must be positive, this restriction is generally not binding, as volatilities are typically increased in stress scenarios. If a factor's standard deviation (volatility) is not to be altered in the stress test, the diagonal element in Δ that corresponds with that factor should be set to 0. Using these modified pricing factor return volatilities, the stress test covariance matrix that should be used in stress test calculations is given by, Σ':

$$\Sigma' = (D_t + \Delta)\Omega_t(D_t + \Delta). \tag{18}$$

Using this modified covariance matrix, we can complete the stress test calculations. That is, the pricing factor return vector, weighting matrix, and the 'stressed' covariance matrix can be partitioned so that the factors that take on exogenous factors in the stress test simulation appear in the $\mathbf{R}_2$ partition and all other factors are accounted for in $\tilde{\mathbf{R}}_{1t}$. Under these conditions, the relevant stress test statistics are given by equations (13) and (14) after the pricing factor return covariance matrix partitions are replaced by appropriate partitions of the 'stressed' return covariance matrix, Σ'.

Shocking Volatilities and Correlations

The most general set of conditions that can be accommodated in the VaR-based stress exposure measure allows for specific exogenous shocks to individual pricing factor returns and return volatilities, and to the correlations among pricing factor returns. The specification of these types of stress scenarios becomes more involved because changes in the correlation matrix must satisfy certain restrictions in order to preserve mathematical properties that are required by any multivariate normal VaR methodology.[10]

To parameterize the most general set of stress scenarios, first consider the scenario changes that relate to pricing factor correlations. Assume that the stress scenario requires that the correlations among g different factors are to be stressed from their historical values. The pricing factors can be re-ordered so that the g pricing factors which are to experience correlation shocks are written in the first g rows of the pricing factor matrix.

Partition the pricing factor return vector into two elements, $\tilde{\mathbf{R}}_{at}$, the $g \times 1$ vector of pricing factor returns that will experience correlation shocks in the stress scenarios, and $\tilde{\mathbf{R}}_{bt}$, the $(N-g) \times 1$ vector of remaining pricing factor returns. Using this partition, the pricing factor return distribution can be written:

$$\begin{bmatrix} \tilde{\mathbf{R}}_{at} \\ \tilde{\mathbf{R}}_{bt} \end{bmatrix} \sim N\left(\begin{bmatrix} 0_g \\ 0_{N-g} \end{bmatrix}, \begin{bmatrix} D_{at} & 0 \\ 0 & D_{bt} \end{bmatrix} \times \begin{bmatrix} \Omega_{aat} & \Omega_{abt} \\ \Omega_{bat} & \Omega_{bbt} \end{bmatrix} \begin{bmatrix} D_{at} & 0 \\ 0 & D_{bt} \end{bmatrix} \right), \tag{19}$$

where D_{at} is a $g \times g$ diagonal matrix with elements equal to the standard deviations of the g factors whose correlations are to be stressed, and D_{bt} is

[10]See Finger [1997] for a discussion.

the $(N-g) \times (N-g)$ diagonal matrix of the remaining factors' standard deviations.

Consider the partitioned correlation matrix, Ω_t:

$$\Omega = \begin{bmatrix} \Omega_{aat} & \Omega_{abt} \\ \Omega_{bat} & \Omega_{bbt} \end{bmatrix}. \tag{20}$$

By design, the leading submatrix, Ω_{aat}, is the $g \times g$ correlation matrix that contains all the pairwise correlations that are to augmented in the stress test. Note that all pairwise correlations in this matrix need not be altered in the stress scenario.

Let Ω'_{aat} be the $g \times g$ matrix of stressed pricing factor correlations that are specified in the scenario. Because Ω'_{aat} is a correlation matrix, it should itself be symmetric, positive (semi-) definite, with diagonal elements equal to unity and all other elements less than or equal to one in absolute value. While Ω'_{aat} must satisfy these conditions to be a properly specified stressed correlation matrix, these conditions alone are not sufficient to ensure that the entire stressed factor correlation matrix, Ω'_t:

$$\Omega'_t = \begin{bmatrix} \Omega'_{aat} & \Omega_{abt} \\ \Omega_{bat} & \Omega_{bbt} \end{bmatrix} \tag{21}$$

is positive (semi-) definite as is required in a multivariate normal VaR framework. The matrix Ω'_t is a properly specified correlation matrix if all its eigenvalues are positive.[11]

Suppose for a moment that, Ω'_{aat} , the leading $g \times g$ partition of the stressed correlation matrix, is a well-specified correlation matrix. This implies that, in additions to the conditions stated above, all g of the eigenvalues of this leading submatrix are positive.

Even if Ω'_{aat} is well-specified as a stand-alone correlation matrix, it could turn out that the entire 'stressed' correlation matrix, Ω'_t, may not be well-specified. That is, altering one 'block' of the correlation matrix, even if the block itself is altered in a manner consistent with a lower dimensional multivariate normal distribution, may destroy the statistical properties required of the larger multivariate normal correlation matrix in which it is embedded.

A concrete example of such a violation is presented in the aforementioned article by Finger [1997]. If such a case occurs, Ω'_{aat} must be altered so that Ω'_t is a positive definite matrix.

If a proposed set of stress correlations violate the properties required of a covariance matrix, there are many avenues that can be taken to alter the scenario so that it becomes well-specified. One approach is to alter the stressed

[11] The eigenvalues, λ, are the solutions to the equation: $(\Omega_i - \lambda \mathbf{I}_{N\times N})\mathbf{x}$, $\forall \mathbf{x} \in \mathbb{R}^N$. If all the eigenvalues are positive, the matrix is positive-definite, and no pricing factor is redundant. Strictly speaking, the correlation matrix need only be positive semi-definite for normality to hold (some eigenvalues may be 0). This will be true if some pricing factors are linear combinations of other included factors.

correlation values by trial and error until the overall correlation matrix has the appropriate properties. Another alternative is to apply a systematic adjustment to selected correlation groups using the technique proposed by Finger [1997]. Still another approach is to use the originally proposed stressed correlations and so-called shrinkage techniques to adjust the overall correlation matrix until it has the appropriate mathematical properties.

The shrinkage technique is intuitively simple but may require a moderate amount of computation. Assume that the stress test measurement process begins with a well specified risk factor return correlation matrix; that is, Ω_t is symmetric and positive definite. A well-specified stress test correlation matrix can be constructed using an algorithm. Let $c \in (0, 1)$. Construct the matrix

$$\widehat{\Omega}'_t = c\Omega_t + (1 - c)\Omega'_t. \tag{22}$$

Begin by choosing a value of c very close to 0. Evaluate equation (22) and use a numerical algorithm to calculate the eigenvalues of $\widehat{\Omega}'_t$. If all the eigenvalues are positive, reduce the magnitude of c slightly and repeat the calculations. Alternatively, if some of the eigenvalues are negative or zero, increase the magnitude of c and repeat the calculations.

The object of the algorithm is to generate a correlation matrix Ω with strictly positive eigenvalues using the smallest possible value of c.[12] This algorithm merely adjusts the desired stressed correlations toward their historical correlations until the overall correlation matrix has the appropriate mathematical properties. By using the smallest possible value of c, the required mathematical properties of the correlation matrix are preserved and the stressed correlation components are disturbed as little as is possible from their desired stress test settings. Once the correlation matrix is modified to account for the correlation components of the stress test, volatilities and factor return stress can be incorporated as described above.

6 Issues Regarding Full Repricing

The stress-testing techniques and statistical examples outlined in this analysis are based on the variance-covariance approach for calculating VaR measures. It is well known that such an approach is inaccurate when a portfolio contains significant option-like exposures. When a portfolio has significant optionality, maximum potential portfolio losses may not occur at the extreme risk factor movements that are implicitly analyzed in the variance-covariance measurement framework. Monte Carlo simulation approaches that incorporate full or

[12]While positive eigenvalues are a sufficient condition in theory, in practice, if eigenvalues are close to 0, the correlation matrix may not be numerically stable. It may be more appropriate to select the smallest value of c that is consistent with an acceptable condition value for the matrix. For a discussion of the condition property of a matrix, see Belsley [1980, chapter 3].

pseudo-full revaluation of positions are an alternative methodology that can be used to accurately assess option-like risk exposures.

The specific stress event risk measures derived in this study are appropriate only for portfolios with predominately linear risk exposures. The results relating to the use of a conditional stress test distribution are, however, more general.

The historical stress test results suggest that stress test loss estimates can be constructed for portfolios with significant optionality by full revaluing instruments in a Monte Carlo simulation that uses the conditional normal stress scenario distribution characterized in equations (9), (10), and (11). In many market stress environments, the assumption of normality and the conditional time-varying covariance structure typical in VaR analysis can provide a useful characterization of the conditional probability distribution of the value changes of a broadly diversified portfolio.

7 Practical Issues Related to Scenario Specification

When applying the *StressVaR* methodology, it may be tempting to incorporate a view on multiple factors in the set of exogenous shocks included in a stress scenario. Practical considerations, however, suggest some cautions to consider when choosing which factors to shock.

While the procedures outlined can, in theory, accommodate many separate factor shocks in a stress scenario, as a practical matter, it is important for the covariance matrix of shocked factors, Σ_{22} (or Σ'_{22}) be well-conditioned. That is, the covariance matrix of the shocked factor partition, $\mathbf{R}_2$, must be of full rank and have a stable inverse.[13]

In practical applications, matrix singularity problems may be encountered if, for example, one shocks the set of all long-dated (or short-dated) zero-coupon bond price factors that represent a term structure system, as these set of factors typically have close to perfect correlations in many samples. While this problem may be addressed (in a brute force manner) using a generalized inverse, a simpler solution is to avoid the problem by specifying scenarios that include an exogenous shock for only one factor from each highly correlated group of factors. When factors are highly correlated, one factor can be shocked and the remaining factor values can be set automatically through the conditional mean calculation used in the stress test procedures.

A special case may arise if the stress test scenario is specifically designed to include a set of uncharacteristically divergent shocks to a set of risk factors whose movements are normally very highly correlated. An example of such

[13]The author is indebted to Peter Zangari for identifying this issue.

a scenario might be one in which a 'hump' at an intermediate maturity is added to an otherwise flat segment of the yield curve.

This scenario may, for example, involve a negative shock to the ten-year zero-coupon bond price and a positive shock to the twenty-year bond price. Over many sample periods, movements in these bond prices are very close to perfectly positively correlated, so the Σ_{22} matrix may be ill-conditioned.

In such a scenario it is perhaps most appropriate to reduce (i.e., stress) the pairwise correlations of the bond factors in concert with the specified set of zero-coupon bond price shocks. By including the correlation stress, the conditioning properties of Σ'_{22} can be improved and thereby facilitate calculation of the stress test conditional means and covariances.

8 Concluding Remarks

This new methodology can be used to completely parameterize stress test scenarios using conditional probability distributions that are easily derived from the inputs to daily VaR calculations. The methodology allows for stress scenarios that shock risk factors, risk factor volatilities, and risk factor correlations, and can easily be extended to measure options risk using full-revaluation Monte Carlo simulation. The approach allows for a complete characterization of the value-change probability distribution of a portfolio in a stress scenario.

Statistical evidence is presented that suggests that the new measure of stress event loss exposure is substantially more accurate than the stress exposure estimates that are commonly used in financial institutions. Although conditional covariances may not be stable in all market stress environments, the results also suggest, contrary to popular perception, that historical VaR model risk factor covariances and the assumption of conditional normality do provide information that can be used to construct reasonably accurate loss exposure estimates in many stressful market environments.

While market experiences have convinced many practioneers that stress events are often accompanied by shifts in factor correlation structures, it remains an open (and difficult) question whether correlations undergo systematic changes in stress environments, and whether these changes can be identified, anticipated, and incorporated effectively into stress test measures.

References

Belsley, D.A. 1980 *Conditioning Diagnostics: Collinearity and Weak Data Regression.* New York: John Wiley & Sons.

Finger, Christopher 1997. 'A methodology for stress correlations.' *RiskMetrics Monitor*, fourth quarter, 1997.

Kupiec, Paul 1995. 'Techniques for verifying the accuracy of risk management models.' *Journal of Derivatives* **3**, No. 2, 73–84.

Labrecque, Thomas 1998. Comments at *Financial Services at the Crossroads: Capital Regulation in the 21st Century.* Federal Reserve Bank of New York Conference, February, 1998.

Litterman, Robert 1996. 'Hot spots and hedges.' *Journal of Portfolio Management*, December, 1996.

RiskMetrics Technical Document. New York: J.P. Morgan, 1996.

The Supervisory Treatment of Market Risks. Bank for International Settlements, Basel, Switzerland, 1995.

Dynamic Portfolio Replication Using Stochastic Programming

M.A.H. Dempster and G.W.P. Thompson

Abstract

In this article we consider the problem of tracking the value of a 'target' portfolio of European options with a range of maturities within a one year planning horizon using dynamic replicating strategies involving a small subset of the options. In defining a dynamic replicating strategy we only allow rebalancing decision points for the replicating portfolio at the payout dates of the options in the target, but for one application we measure the tracking error between the value of the two portfolios daily. The target portfolio value has a Bermudan path-dependency at these decision points and it is likely that a carefully chosen dynamic strategy will out-perform simpler static or quasi-static strategies. Here we construct trading strategies by solving appropriate stochastic programming formulations of two principal tracking problems: portfolio compression for risk management calculations and dynamic replicating strategies for simplified replicating portfolios which may be used for hedging or actual target portfolio simplification. We demonstrate the superior performance of dynamic strategies relative to both more static strategies and delta hedging in a number of numerical tests.

1 Introduction

In this article we construct periodically rebalanced dynamic trading strategies for a portfolio containing a small number of tradable instruments which tracks the value of a large 'target' portfolio daily over a long period of time. A successful solution to this 'tracking' problem is useful in a number of practical financial applications. One such is the investment problem of index-tracking, where the target portfolio consists of the constituent assets of some equity or bond index such as the FTSE-100 or the EMBI+ (see e.g. Worzel, Vassiadou-Zeniou & Zenios, 1994) but here we will address a more complex problem involving nonlinear instruments from a risk management perspective. For risk managers the resolution of the tracking problem can be seen as a way of reducing extensive and complex daily *value at risk* (VaR) calculations for the target portfolio to the simpler task of evaluating the VaR of the tracking portfolio—we refer to this application as *portfolio compression*. In an investment bank the target portfolio would typically be very large. The approach can also be useful as a hedging tool—which extends Black–Scholes delta hedging replication for a single option to a typical large *portfolio* of options or other derivatives of various maturities by finding a *dynamic replication strategy* for the target portfolio. A practical application might involve using a collection of liquid instruments to track the value of a much less liquid target.

Here we analyze instances of the daily tracking problem involving a specified target portfolio of 144 European options on the S&P 500 index of different strikes and maturities *within* a one year planning horizon, using the technique of *dynamic stochastic programming* (DSP). Thus over the horizon considered in our experiments the values of both the target and dynamic tracking portfolios will exhibit a Bermudan path-dependency at rebalance points of the latter. Gondzio, Kouwenberg & Vorst (1998) have studied the use of DSP techniques to implement the Black–Scholes dynamic hedging strategy for a *single* European option—a simple special case of the tracking problem studied here. The present paper is to our knowledge the first application of DSP to a realistically large portfolio containing instruments which mature within the planning horizon.

Over the last few years leading edge risk management practice has evolved from current mark-to-market to one period forward VaR and mark-to-future techniques (Jamshidian & Zhu 1997, Chisti 1999). When such static methodologies are applied over long horizons to target portfolios containing instruments with maturities within the horizon, they take no account of changes in portfolio composition due to instruments maturing—for replication of such portfolios dynamic trading strategies are required which may be found optimally using dynamic stochastic programming techniques. DSPs are a form of stochastic dynamic programming problem—but solved by mathematical programming techniques—which allow very large numbers of decisions and high dimensional data processes at a smaller number of natural decision points or *stages* (such as option maturity dates) than the fine timestep typically considered in traditional dynamic programming problems. For practical purposes these latter are restricted by the large number of timesteps to only a few state and decision variables—Bellman's curse of dimensionality. DSPs are multi-stage stochastic programming problems for which the term *dynamic* signifies that the underlying uncertainties are being modelled as evolving in *continuous* time and the corresponding scenarios approximating the data process paths are to be simulated with a much finer timestep (here daily) than the (multi-day) interval between decision points. In this paper we demonstrate that the use of such an approach—which we term *dynamic portfolio replication*—provides a better solution to the daily tracking problem with respect to two definitions of tracking error than other approaches such as static hedging. We also confirm in the context of risk-management the general view in the literature (*cf.* Dempster *et al.*, 2000) regarding other DSP problems that the scenario trees required for such an approach reduce tracking error when initial branching is high.

The paper is organized as follows. In Section 2 we describe the DSP approach more fully and review the relevant recent literature. Section 3 discusses in detail the process of constructing dynamic trading strategies using DSP. We describe our particular problem and how the DSP approach is applied in this situation in Section 4. In Section 5 we report a number of numerical tests to compare the daily tracking performance of our dynamic trading strategies with simpler hedges including the portfolio delta hedge. Section 6 concludes and discusses current research directions.

2 The dynamic stochastic programming approach

The DSP approach to optimization of a continuous state vector stochastic system is as follows (see e.g. Dempster (1980), Birge & Louveaux (1997) and Wets & Ziemba (1999)). First fix a sequence of times at which decisions will be made (with the current moment the first decision time). Now replace the law of the continuous time paths of the continuous state variables with a sampled 'scenario tree'; this has a single node representing the current moment from which a number of branches extend, representing possible discrete time transitions which the state variables may follow from the current moment to the second decision time. Each of these branches ends at a node which itself has further branches, representing discrete time continuations of the paths of the state variables from the second decision time to the third. This process is repeated until every decision time is represented by a collection of nodes in a tree (see Figure 1). Once this *scenario tree* is constructed, we re-phrase our stochastic optimization problem as an *equivalent deterministic* optimization problem (typically a linear programming problem) by associating a set of decision variables with each node in the tree, and by expressing the objective and constraints of the problem in terms of these decision variables and the values of the state variables on the nodes. The drawback of this approach—which extends both classical decision tree analysis involving a finite number of possible decisions at each node and stochastic dynamic programming as noted above—is that the scenario tree (and hence the size of the optimization problem) grows exponentially as the number of decision times (stages) increases, often necessitating parallel computation and sequential sampling schemes (Dempster & Thompson, 1999).

Although dating from the 1950s, the practical use of SP in financial problems is still fairly new since the associated optimization problems as noted above can become extremely large. Most stochastic programming problems in the current financial literature are of two types: portfolio insurance and asset liability management. A *portfolio insurance* problem also specifies a 'target portfolio' and asks the analyst to find a dynamic portfolio which when added to the target portfolio reduces the losses of the combined portfolio but retains

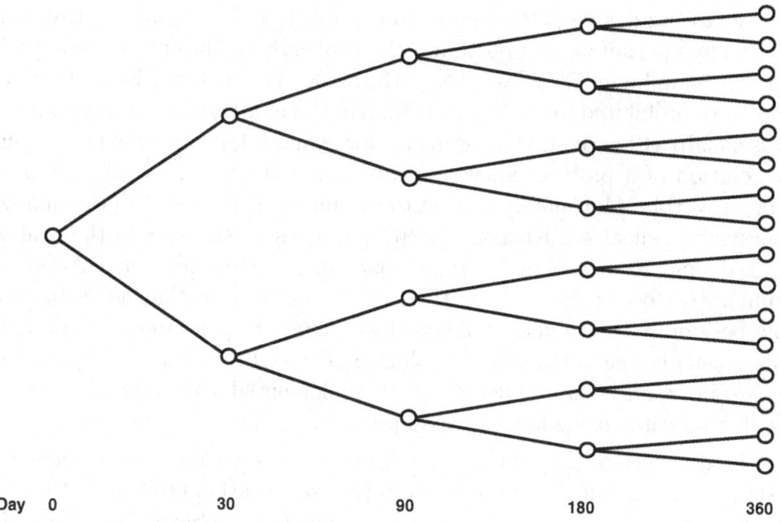

Figure 1: A binary branching scenario tree whose paths are generated by a data process simulator with daily timestep.

as much as possible of the profits of the original. *Asset liability management* (ALM) problems consider a sequence of liabilities (which may change over time stochastically) and try to maximize some overall measure of profit after the liabilities have been paid by allocating investment capital between a set of asset classes such as stocks, bonds and cash (see e.g. Mulvey & Ziemba, 1998). In both cases the objective function does not usually penalize over-performance, in contrast to our *tracking problem* where we penalize both under- *and* over-performance relative to the target. This has subtle implications for the construction of the scenario tree.

As in Figure 1, the simplest scenario trees have the same number of branches at each node, and use random sampling to construct the state variable paths associated with each branch. A more sophisticated approach is to try to fit moments of a theoretical conditional distribution of the state variables at decision times (Kouwenberg, 1998). In either case, obviously the more scenarios used to construct the stochastic programming problem the closer the approximation will be to the intractable stochastic optimization problem defined by the continuous time and state (vector) data process assumed to underly the situation being modelled. (In general weak convergence of the law defined by the scenario tree to the law of the underlying stochastic data process observed only at decision points can be established under reasonable conditions as the branching at each node tends to infinity.)

A problem can arise if the scenario tree contains an *arbitrage*: a trading opportunity within the deterministic equivalent of the DSP which can be seen to generate a positive profit from zero capital. If the tree and constraints of the problem permit such opportunities, it is likely that the optimization problem will be unbounded unless over-performance is penalized. Even if infinite profits are prohibited by other constraints in the model such as position limits (as is usually the case in ALM problems for example), it can be expected that the solution of a problem allowing arbitrages will perform poorly, although in practice this is intimately related to the methods used and to the number of scenarios generated. Klaassen (1997) discusses these issues in the context of ALM, and Kouwenberg & Vorst (1998) in portfolio insurance. A feature common to most of the scenario trees in the literature is that nodes in early time periods have a higher number of branches than in later time periods. This is usually due to the way the solution of the DSP problem is expected to be used in practice—initial decisions are implemented and the whole problem is rolled forward to the next decision point.

A final aspect of the DSP approach on which we should comment is the 'testing' of the quality of the solution to the optimization problem in the context of the original continuous time continuous state problem using simulated paths of the state variables. As noted above it is frequently assumed that after the optimization problem is solved, the optimal decision variables at the node representing the current moment will be implemented for some period of time, after which a new scenario tree will be constructed, and a new optimization problem solved. This process is repeated for as long as is required. A suitable testing procedure given this approach is to construct a large number of test scenarios involving the underlying stochastic processes (interest rates, stock prices, etc.), and for each test scenario and decision point, to construct a scenario tree and solve an optimization problem. Golub *et al.* (1997) and Fleten, Høyland & Wallace (1998) perform this type of test. It has the drawback that the scenario trees cannot be very large or the optimization problems involved will take too long to solve. A typical problem using this scheme might thus have weekly decisions, and generate scenario trees extending over three, four or five weeks. Gondzio & Kouwenberg (1999) give an example of an ALM problem where the scenario tree has 6 decision times and a branching of 13. Solving a single instance of this problem even with state-of-the-art software and hardware takes over 3 hours and improvements to such times are only possible by utilizing sophisticated parallel algorithms and hardware techniques. Another easier to implement testing procedure is to use each test scenario to construct a path through the scenario tree which, at each stage chooses the branch which is 'closest' to the path of the test scenario over the appropriate time interval and then uses the optimal decision variables at the nodes in the scenario tree, together with the genuine statespace variables from the test scenario (Dempster & Ireland, 1988). This is the approach we will adopt for the portfolio compression application which needs fast computation. Gondzio, Kouwenberg & Vorst (1998) use a similar approach, choosing test scenarios from a fine-grained tree which was previously 'aggregated' to form the scenario tree used in the optimization problem. Another approach is to generate test scenarios by picking a random path through a previously generated scenario tree or lattice, but this is likely to give rather optimistic answers, especially for trees if later time periods have very few branches or in the case of arbitrage-free lattices, since this property is destroyed by sampling. Gondzio *et al.* (1998) describe one way of constructing smaller arbitrage-free trees starting from an initial (but possibly very large) arbitrage-free tree.

3 Constructing dynamic replicating strategies using DSP

In this section we describe in detail how we apply the DSP approach to tracking problems. Recall that a tracking problem defines a target portfolio, whose value is to be 'tracked' and asks us to find a self-financing tracking strategy using a prescribed set of trading instruments. We assume that the value of the target and tradables at time t are functions of the path over $[0, T]$ of some observable *state process* $\{\boldsymbol{S}(t) \in \mathcal{S} : 0 \leq t \leq T\}$, where boldface is used to denote random entities throughout.

Tracking problems

As mentioned above, we will assume a number of discrete *decision points* $0 = t(1) \leq t(2) \leq \cdots \leq t(T) \leq T$ in time and consider the data of our problem to be observations of the state process at these discrete time points which for simplicity are represented by integer labels $t = 1, 2, \ldots, T$.

We consider a multistage version of a static stochastic optimization problem whose deterministic equivalent was studied in Dembo & Rosen (1999). Given a discrete time process $\{\boldsymbol{\tau}_t \in \mathbb{R} : t = 1, \ldots, T\}$ of *values* of a *target portfolio* the stochastic dynamic *tracking problem* we study is given by

$$\text{(SPL1)} \qquad \inf_{\boldsymbol{x}} \frac{1}{T} \sum_{t=1}^{T} \mathbb{E}\left[\boldsymbol{y}_t^+ + \boldsymbol{y}_t^-\right] \tag{1}$$

such that

$$\begin{aligned}
p_1' x_1 &\leq p_1' x_0 & & \qquad (2)\\
\boldsymbol{p}_t'[\boldsymbol{x}_t - \boldsymbol{x}_{t-1}] &= 0 & \text{a.s. } t = 2, \ldots, T & \qquad (3)\\
\boldsymbol{p}_t'\boldsymbol{x}_t - \boldsymbol{y}_t^+ + \boldsymbol{y}_t^- &= \boldsymbol{\tau}_t & \text{a.s. } t = 1, \ldots, T & \qquad (4)\\
\boldsymbol{x}_T - \boldsymbol{x}_{T-1} &= 0 & \text{a.s.} & \qquad (5)\\
\boldsymbol{y}_t^+ \geq 0, \quad \boldsymbol{y}_t^- &\geq 0 & \text{a.s. } t = 1, \ldots, T. & \qquad (6)
\end{aligned}$$

Here the fundamental discrete time *decision process* $\boldsymbol{x}$ is a set of portfolio positions (long and short) in the securities of the problem and the corresponding *price process* $\boldsymbol{p}$ is used to value these positions at each decision point. At time point 1, prices p_1 and initial *endowment* $p_1'x_0$ are known, and an initial position x_1 must be taken. Subsequently all decisions are state dependent and hence stochastic; so that the tracking portfolio will have an *almost surely* (a.s.)—i.e. with probability one—upside ($\boldsymbol{y}_t^+ \geq 0$) or downside ($\boldsymbol{y}_t^- \geq 0$) *tracking error* (6) in value terms. The objective (1) is to minimize the *average absolute* (or L_1 norm) tracking error subject to *budget* (2), *self-financing* (3) and *tracking error* (4) constraints which must hold a.s.. The stochastic tracking portfolio $\boldsymbol{x}_{T-1}$ set at the penultimate decision point must be held (5) over the period to the horizon at T. Here $\mathbb{E}$ denotes expectation and prime denotes transpose. Many variations on the problem are possible and several will be used in this paper.

First if (6) is replaced by

$$\rho \geq \boldsymbol{y}_t^+ \geq 0, \quad \rho \geq \boldsymbol{y}_t^- \geq 0 \quad \text{a.s. } t = 1, \dots, T \tag{7}$$

and the objective (1) is replaced by

$$\inf_{\boldsymbol{x}} \rho \tag{8}$$

we obtain the problem (SPLINF) whose solution minimizes the *worst case* (or L_∞ norm) tracking error a.s. over both decision points and scenarios.

For either (SPL1) or (SPLINF) applied to the portfolio compression problem – since all positions in the tracking portfolio are *virtual*, i.e. for computational purposes only – we ignore transaction costs and allow an arbitrarily large initial endowment $p_1'x_0$. In other applications however an optimal dynamic replicating strategy must be implemented and hence proportional transaction costs are incurred. This results in a slightly more complex model involving a real budget constraint and *buy* and *sell* decisions $\boldsymbol{x}_t^+$ and $\boldsymbol{x}_t^-$ respectively. The L_1 variant of this model becomes

$$\text{(SPL1}'\text{)} \qquad \inf_{\boldsymbol{x},\boldsymbol{x}^+,\boldsymbol{x}^-} \frac{1}{T} \sum_{t=1}^{T} \mathbb{E}\left[\boldsymbol{y}_t^+ + \boldsymbol{y}_t^-\right] \tag{9}$$

such that

$$p_1'x_1 \leq p_1'x_0 \tag{10}$$

$$\boldsymbol{x}_{t-1} + \boldsymbol{x}_t^+ - \boldsymbol{x}_t^- = \boldsymbol{x}_t \qquad \text{a.s. } t = 1, \dots, T \tag{11}$$

$$\boldsymbol{p}_t'(\boldsymbol{x}_t - \boldsymbol{x}_{t-1}) + \boldsymbol{p}_t'K(\boldsymbol{x}_t^+ + \boldsymbol{x}_t^-) = 0 \qquad \text{a.s. } t = 2, \dots, T \tag{12}$$

$$\boldsymbol{p}_t'\boldsymbol{x}_t - \boldsymbol{y}_t^+ + \boldsymbol{y}_t^- = \boldsymbol{\tau}_t \qquad \text{a.s. } t = 1, \dots, T \tag{13}$$

$$\boldsymbol{x}_t^+ \geq 0, \quad \boldsymbol{x}_t^- \geq 0, \quad \boldsymbol{y}_t^+ \geq 0, \quad \boldsymbol{y}_t^- \geq 0 \qquad \text{a.s. } t = 1, \dots, T. \tag{14}$$

Here the new constraint (11) expresses position *inventory balance* over time, and the self-financing constraint (12) now involves a diagonal matrix $K := \text{diag}(\kappa_1, \ldots, \kappa_I)$ of (two-way) proportional *transaction costs.*

To construct our dynamic replication strategy, we will approximate the law of the process $\boldsymbol{S}$ with a tree, solve an appropriate optimization problem on this tree, and then interpret the solution as a dynamic trading strategy for the tracking portfolio.

Scenario trees

To define the optimization problem used to construct our dynamic replication strategy, we first make precise the notion of a 'scenario tree' and how it approximates the law of the state process $\boldsymbol{S}$.

A *scenario tree* is a tree with nodes in a set $\mathcal{N}$, with one node, $r \in \mathcal{N}$, designated the *root node.* For $n \in \mathcal{N}, n \neq r$, we let $P(n)$ denote the *predecessor* of n: the unique node adjacent to n on the unique path from n to r, and refer to those nodes with a common predecessor as the *successors* of their predecessor. We define $P(r) = r$.

With each node $n \in \mathcal{N}$ we assume that there is given a (*decision*) *time* $t(n)$ and a *state* $s(n) \in \mathcal{S}$ satisfying: $t(r) = 1$ (or the current real time 0), $s(r)$ is the observed current state and $t(P(n)) < t(n)$ for all $n \neq r$. We will also assume that if $P(n) = P(n')$ then $t(n) = t(n')$. For each node n, we link $s(P(n))$ and $s(n)$ with an arc representing a path segment of a suitable discrete time approximation of $\boldsymbol{S}$ over the number of time points between the decision times corresponding to $P(n)$ and n respectively (see Figure 1). Thus we can associate a path of arcs $\{s(u) : u = t(1), \ldots, t(T)\}$ with each sequence of nodes $n_1, n_2, n_3, \ldots, n_T$ for which (with a slight abuse of notation) $t(1) := t(n_1) < t(n_2) < \cdots < t(n_T) := T$ and $P(n_{i+1}) = n_i$ for $i = 1, 2, \ldots, T-1$. In order to implement holding tracking portfolios over the period between the last two decision points we must distinguish *leaf nodes* $\ell \in \mathcal{L} \subset \mathcal{N}$ with $t(\ell) = T$. We assume that we are given a set of strictly positive real numbers $\{\pi(n) : n \in \mathcal{N}\}$ such that $\pi(r) = 1$ and if n has successors, then $\pi(n) = \sum_{n':n=P(n')} \pi(n')$ for all $n \in \mathcal{N}$. In this situation the conditional probability of $s(n')$ being realized after $s(n)$ is $\pi(n')/\pi(n)$.

To show how to interpret a scenario tree as a discrete time stochastic process approximating the law of $\boldsymbol{S}$, it suffices to show how to generate a single sample path. Starting at the root node, select one of its successors n at random with conditional probability proportional to $\pi(n)$; then the arc from $s(r)$ to $s(n)$ represents a sample path for $\boldsymbol{S}$ over $[t(r), t(n)]$. Now move to node n and select one of its successors $\tilde{n}$ with conditional probability proportional to $\pi(\tilde{n})$, and so on.

By identifying a node n with the path from the root node to n, and thence with a path $\{s(u) : u = t(1), \ldots, t(T)\}$ in $\mathcal{S}$, we can speak of the *value* of the target and the *price* of the tradable instruments at n. We denote these quantities $\tau(n)$ and $p(n)$ respectively, where $p(n) \in \mathbb{R}^I$ is a vector giving the prices of the I tradable instruments. We denote by T the *depth* of the tree: the number of nodes on a path of maximal length starting at the root node.

The optimization problem

We are now in a position to modify the deterministic equivalent of the static optimization problem introduced in Dembo & Rosen (1999) to state a recursive version of the dynamic deterministic equivalent of (1) applicable to our tree setting. Using the notation above, we consider the problem:

$$\text{(L1)} \qquad \underset{x(n):n\in\mathcal{N}}{\text{minimize}} \quad \frac{1}{T}\sum_{n\in\mathcal{N}} \pi(n)[y^+(n) + y^-(n)] \tag{15}$$

$$\begin{aligned} \text{subject to,} \qquad & p(r)'x(r) \le M \\ & p(n)'[x(n) - x(P(n))] = 0 \\ & p(n)'x(n) - y^+(n) + y^-(n) = \tau(n) \\ & x(\ell) - x(P(\ell)) = 0 \\ & y^+(n) \ge 0, \quad y^-(n) \ge 0, \quad \text{for all } n \in \mathcal{N},\ \ell \in \mathcal{L}, \end{aligned}$$

which is the problem of minimizing the expected total tracking error subject to an initial budget of M, and also the worst-case version:

$$\text{(LINF)} \qquad \underset{x(n):n\in\mathcal{N}}{\text{minimize}} \quad \rho \tag{16}$$

$$\begin{aligned} \text{subject to,} \qquad & p(r)'x(r) \le M \\ & p(n)'(x(n) - x(P(n))) = 0 \\ & p(n)'x(n) - y^+(n) + y^-(n) = \tau(n) \\ & y^+(n) \le \rho, \quad y^-(n) \le \rho \\ & x(\ell) - x(P(\ell)) = 0 \\ & y^+(n) \ge 0, \quad y^-(n) \ge 0. \end{aligned}$$

In both cases, the variables $x(n) \in \mathbb{R}^N$ are interpreted as the holdings in the N tradable instruments to be used if the state process follows the path from the root node to n. The second constraint is the self-financing constraint at node n and the third defines $y^{\pm}(n)$ as the upside and downside tracking errors at n. Note that since $P(r) = r$, these constraints also make sense when $n = r$. Observe that at leaf nodes without successors the variables $x(\ell)$ must be identical to those at their predecessors $x(P(\ell))$.

We impose additional constraints in both forms of the problem. First, we restrict the holding in each tradable instrument to the interval

[−100,000,000, 100,000,000] (to ensure that the solution is finite), and secondly if any tradable instrument has value within 5×10^{-5} of zero at node n, or at every successor of n, then we impose $x(n) = 0$. This second condition stops the optimal strategy from holding large positions which appear to have low or zero tracking error at the nodes of the tree on the paths corresponding to the law implied by the scenario tree, but which have very high daily tracking error under the discrete time approximation to the true law in between decision points. We also impose proportional transaction costs in the deterministic equivalent of (9), which we now state for our particular problem.

The transaction cost model

The deterministic equivalent form of (9)–(14) to be used in our particular problem has objective the same as that of the previous problem (L1). Now however we must distinguish between the different components of the decision variables x: the vector $x(n)$ has components denominating the holdings in the underlying, cash and the options respectively. The values of the tradable instruments at node n are given by the vector $p(n)$. The 'depth' of the scenario tree is denoted by T. The predecessor node of node n is $P(n)$ and r denotes the root node. There are two transaction costs: κ_{snp} is the cost associated with purchases/sales of the index, while κ_{opt} applies to options transactions. The initial portfolio holdings are given by the vector h. Thus we have

$$
\text{(L1}'\text{)} \qquad \underset{x(n):n\in\mathcal{N}}{\text{minimize}} \quad \frac{1}{T}\sum_{n\in\mathcal{N}} \pi(n)[y^+(n) + y^-(n)] \qquad (17)
$$

$$
\begin{aligned}
\text{subject to,} \quad & h + x^+(r) - x^-(r) = x(r) \\
& x(P(n)) + x^+(n) - x^-(n) = x(n) \\
p(r)'(x(r) - h) = & -\kappa_{\text{snp}} p_0(r)(x_0^+(r) + x_0^-(r)) - \kappa_{\text{opt}} \sum_{i=2}^{5} p_i(r)(x_i^+(r) + x_i^-(r)) \\
p(n)'(x(n) - x(P(n))) = & -\kappa_{\text{snp}} p_0(n)(x_0^+(n) + x_0^-(n)) \\
& -\kappa_{\text{opt}} \sum_{i=2}^{5} p_i(n)(x_i^+(n) + x_i^-(n)) \\
& p(n)'x(n) - y^+(n) + y^-(n) = \tau(n)
\end{aligned}
$$

$$
x^+(n) \geq 0, \quad x^-(n) \geq 0, \quad y^+(n) \geq 0, \quad y^-(n) \geq 0, \quad \text{for all } n \in \mathcal{N},\, n \neq r.
$$

We also impose the constraint that options cannot be held once they have expired.

Turning the solution into a trading strategy

Once one of the above optimization problems has been solved, the problem arises of interpreting the solution as a replicating trading strategy for the tracking portfolio. We will evaluate the tracking error of our replicating portfolio by simulating many independent realizations of the underlying (daily) discrete time data process and then valuing both target and tracking portfolios at each timestep. We can use the portfolio associated with the root node of an optimal solution up to the first branch time, but then we must decide how to re-balance. If the value of the data process at that time were exactly equal to the value of this process at one of the successors of the root node, we would just use the optimal portfolio associated with that node, but unfortunately this is unlikely to be the case.

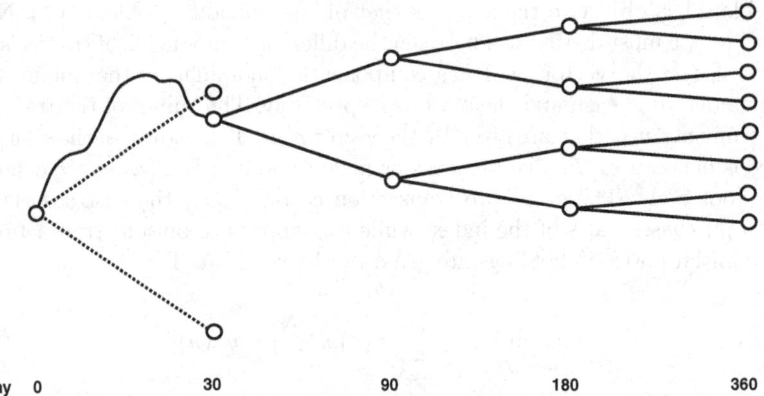

Figure 2: Graphical illustration of the full dynamic replication strategy showing part of the initial tree and the tree used at the second decision point.

In the context of dynamic portfolio replication, we will re-solve the stochastic programming problem by sampling a new scenario tree with the actual value of the process at the second decision point assigned to the root, and then take the calculated optimal portfolio at the new root node (see Figure 2). This would be far too time consuming in portfolio compression applications.

In this latter case we will use a simple procedure based on having a metric $m(\cdot,\cdot)$ on $\mathcal{S}$ for measuring the distance from a simulated realization of $\boldsymbol{S}$ to the nearest node. For example, if the state distribution of the process $\boldsymbol{S}$ at t were Gaussian with covariance matrix V, then the choice of the *Malanobis metric* $m(s,s') = (s-s')^T V^{-1}(s-s')$ would be very natural. Denoting the solution to the optimization problem by x^*, the *portfolio compression trading strategy* is defined as follows (see Figure 3): let $n = r$ initially, let t' be the (common) time at the successors of node n, and use the portfolio $x^*(n)$ over $[t(n), t']$. At time t' consider each of the successors of n which them-

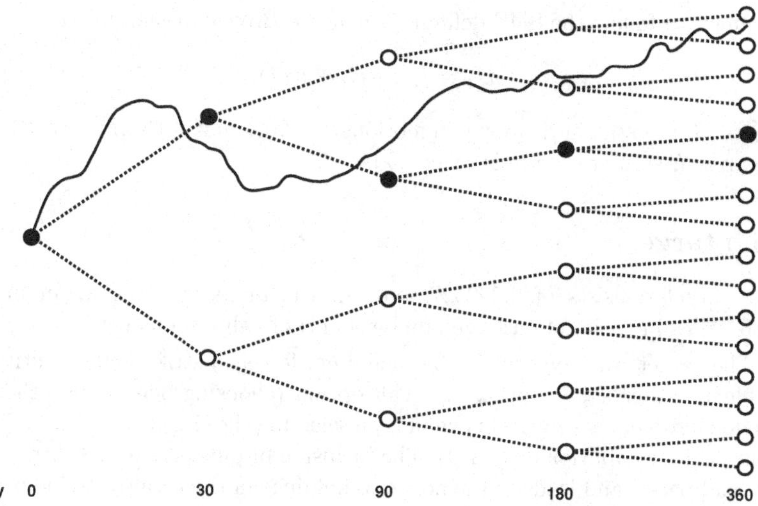

Figure 3: Graphical illustration of the compression trading strategy (black nodes).

selves have successors and find the successor $\tilde{n}$ which minimizes $m(S(t'), s(\tilde{n}))$ where $S(t')$ is the observed value of the state process at time t', and implement $x^*(\tilde{n})$. Then replace n with $\tilde{n}$ and repeat from the start. Once a node n is reached which has no successors, use the portfolio $x^*(n)$ for all future time. If this strategy turns out to be non-self-financing at any times of decision points, any 'slack' is to be absorbed by investing/borrowing using one of the tradables, chosen arbitrarily (preferably one with a low volatility). Thus this method is based on constructing a dynamic trading strategy from the solution of a single *fixed* DSP with sufficiently many scenarios (and hence tree nodes) to give a finely resolved number of alternative portfolios at each decision point.

Other methods are also worth exploring, such as interpolating between optimal portfolios on paths through the scenario tree which are close to the observed path or fitting a parameterized trading rule to the optimal solution, but we will not consider them here.

4 The options problem

We consider a portfolio of standard European options on a stock index $\boldsymbol{I}_t$ with a wide range of strikes and maturities—each of which is taken as a decision point—and try to find a dynamic trading strategy which uses cash, the underlying index and possibly a small subset of the options. The index is assumed to follow the SDE defining *geometric Brownian motion, viz.*

$$d\boldsymbol{I}_t = \boldsymbol{I}_t(\sigma\, d\boldsymbol{B}_t + \mu\, dt), \tag{18}$$

where $\boldsymbol{B}_t$ is a standard Brownian motion, $\sigma \approx 20\%$, $\mu \approx 10\%$ and $I_0 = \$1275$. The risk-free interest rate is taken to be 5%.

The target

The target consists of 0.299527 units of the underlying index, an initial −\$8.74774 in cash, and a large number (144) of call options (see Table 1).

The target was constructed by including, for each strike and maturity in Table 1, either a put option or a call option (choosing one or the other at random with equal probability) and then selecting the size of the position from a uniform distribution on $[-1, 1]$. The inclusion of puts is accomplished using put-call parity and leads to the non-zero holdings in the underlying index and the cash account.

Figure 4 shows the highly nonlinear payouts of this portfolio as its options mature.

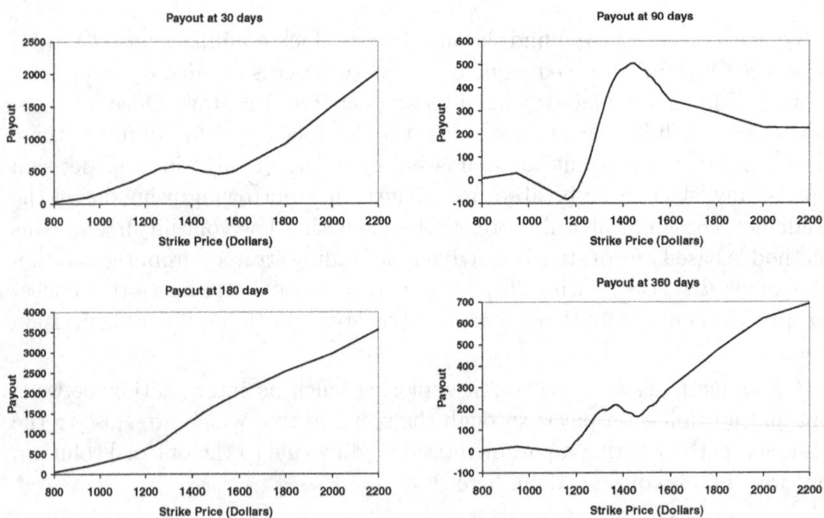

Figure 4: Target payouts at the four maturities.

The tradable instruments

We will assume that the set of tradable instruments consists of the underlying index, cash, and perhaps also a single option for each of the four maturities.

Strike	Maturity (days) 30	90	180	360
750	0.499804	0.15194	0.998693	0.113471
950	0.74093	−0.700248	0.391195	−0.24525
1150	0.571599	0.302505	0.972411	0.579243
1165	−0.262467	0.909999	0.309761	0.648487
1180	0.49019	0.571087	0.36354	0.456401
1195	−0.292544	0.142507	−0.336683	−0.475631
1210	−0.479556	0.925102	−0.681272	−0.0208652
1225	0.55358	−0.126283	0.205513	0.699144
1240	0.151576	0.846815	0.962714	−0.276082
1255	−0.833992	0.53772	−0.0719259	0.209064
1270	−0.780106	0.181939	0.689277	−0.118359
1285	−0.218627	0.284286	0.473013	−0.567698
1300	0.918133	−0.861963	−0.0463901	−0.387226
1315	−0.673621	0.548627	−0.00481466	−0.820417
1330	−0.479278	−0.757467	−0.111879	0.867178
1345	−0.128155	−0.850984	−0.336303	0.237092
1360	−0.744766	−0.270088	0.16514	−0.773939
1375	0.881284	−0.610828	0.349776	−0.85157
1390	−0.691578	−0.676415	−0.528584	−0.531592
1405	−0.690527	0.101874	0.916708	0.869105
1420	0.753697	0.0554727	−0.85114	0.148518
1435	−0.472188	−0.572643	−0.23437	−0.758292
1450	0.540183	−0.997261	0.433229	0.997215
1465	0.542085	−0.045124	−0.957911	0.727558
1480	−0.562721	0.524317	0.257921	0.459888
1495	−0.176574	−0.970444	0.0958814	−0.0225336
1510	0.859913	0.689543	−0.214078	−0.775679
1525	0.374881	−0.817983	−0.383831	0.398134
1540	0.216602	0.577868	0.222183	0.765986
1555	0.265243	−0.801022	−0.678699	−0.579844
1570	0.264488	0.494393	0.60596	0.37497
1585	0.921523	0.0242848	0.435875	−0.471075
1600	−0.106634	0.923895	−0.716491	−0.0146881
1800	0.905168	−0.0522193	−0.514874	−0.10789
2000	−0.203584	0.30069	0.742959	−0.406423
2200	0.910171	0.171869	0.508814	0.499529

Table 1: The static target portfolio's option holdings.

The strikes of these options are chosen to be close to the expected price of the index at that maturity, and are given in Table 2. We set the proportional cost for index transactions κ_{snp} to be 1% and the corresponding cost for options κ_{opt} transactions to be 2.5%.

The scenario tree

To generate the scenario tree we will use either *random sampling* of the discretization of (18) with daily timestep or the following simple *discretiza-*

Maturity	30	90	180	360
Strike	1285	1300	1330	1405

Table 2: The options available to replicating portfolios.

tion procedure at decision points: if $\tilde{n}$ is the kth successor of node n, $k = 0, 1, \ldots, K-1$, set

$$I(\tilde{n}) := I(n) \exp\left(\Phi^{-1}\left(\tfrac{1+2k}{2K}\right)\sqrt{\Delta t}\,\sigma + \left(\mu - \tfrac{1}{2}\sigma^2\right)\Delta t\right), \tag{19}$$

where $\Delta t := t(\tilde{n}) - t(n)$ is the time difference between n and its successors, Φ is the normal distribution function and $I(n)$ denotes the value of the index at node n. The probability of a particular successor $\tilde{n}$ of n is taken to be $1/K$. The aim here is to set up a uniform grid of $2K$ levels and use the inverse Gaussian distribution function to create the corresponding grid for the logarithm of the index values. In both cases, the depth of the tree – i.e. the number of decision points – is T.

5 Numerical Experiments

We consider two types of experiment illustrating our two intended applications: portfolio compression, where we replicate the target with a trading strategy which can be simulated very quickly, and dynamic portfolio replication, in which we model future trading decisions and optimize the current trading decision taking the effects of future decisions into account.

In both cases we will consider the effects of allowing dynamic strategies to use just cash and the underlying index as tradable instruments versus allowing additionally trading in the four options described in Section 4.

As a 'benchmark' we will compare our dynamic replication strategies to a static simple strategy: which uses a simple scenario tree branching only after the first decision point and optimizes a *single* trading decision which is used at every node of the tree (see Figure 5). A second alternative to the full dynamic replication strategy is a strategy which is allowed to rebalance at each decision point of a test scenario according to a static simple strategy, but *assumes* at each such point that the implemented portfolio of the static simple solution will not subsequently change. We term such a strategy *quasi-static* (see Figure 6).

A further natural benchmark is to construct on each test scenario the portfolio *delta hedge* which trades only in cash and the underlying to rebalance at each decision point and holds the new portfolio to the next decision point.

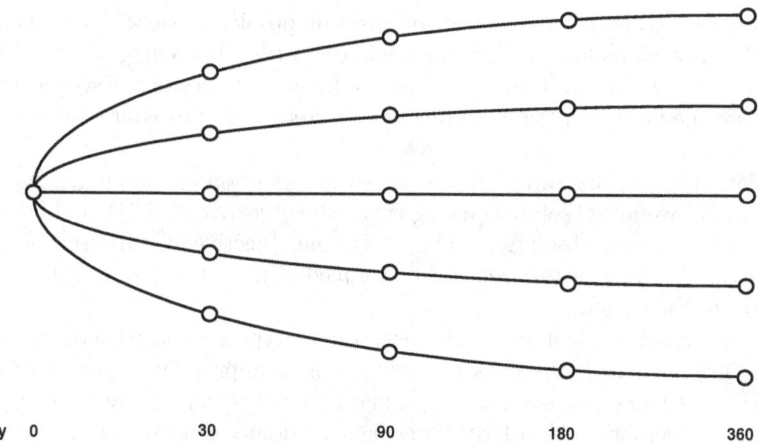

Figure 5: Graphical illustration of the static simple trading strategy in which a common portfolio is used at all nodes of the scenario tree.

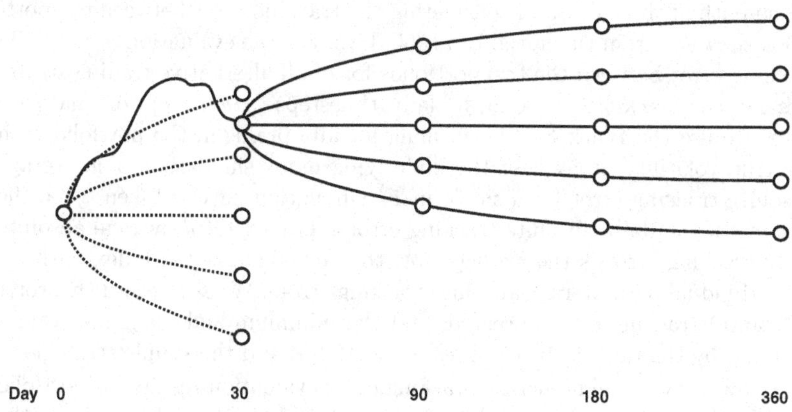

Figure 6: Graphical illustration of the quasi-static simple trading strategy which uses a static simple trading strategy from the realized simulation path at the second and subsequent decision points.

Portfolio compression

Our first experiments compress the target by solving the first two optimization problems presented in Section 3, with either L_1 or L_∞ objectives, and interpreting the solutions as *dynamic* trading strategies using the nearest node technique described in Section 3. For this purpose we ignore the budget constraint and could even ignore the self-financing constraint, except that we would like to know whether we can use the nearest node method of Section 3 to construct a dynamic trading strategy, so we must impose it. We consider two alternative scenario trees for the optimization problems: one with a branching of five at each node (625 scenarios), and one with a branching of ten (10,000 scenarios). To generate the value for the index at nodes in the scenario tree we use the inverse distribution function discretization procedure described in Section 4.

We also consider two *static* simple strategies obtained by minimizing the expected average absolute tracking error (the objective of (L1)), and by minimizing the worst absolute tracking error (the objective of (LINF)) over a set of 1000 scenarios for the index path, sampled at time 0 and the four maturity dates in the target.

Our experimental design with each strategy type is as noted above to solve the DSP versions of the tracking problem for an appropriately generated scenario tree whose nodes are at option maturity dates and then test the quality of these decisions with a large number of randomly generated *test scenarios* simulating the index with a daily timestep up to the last option payout date at 360 days ($T = 5$). The numerical results evaluating the tracking performance of the various approaches were obtained by running 10,000 test simulations of the path of the index and considering the tracking error between the portfolios *each day* from the initial time 360 days. For the evaluation of the daily tracking error between the two portfolios for of all alternative trading strategies for the tracking portfolio at daily timesteps *between* option maturity dates we use the Black–Scholes formula for all options in the portfolio with the true volatility utilised for the underlying index simulation. The average absolute tracking error for a single index simulation path is taken to be the average of the daily absolute tracking errors. The *expected* average absolute daily tracking error is the average over the 10,000 simulated index paths of the individual path average absolute tracking errors. We also record the worst (absolute) tracking error observed, and the minimum and maximum values attained by the target, the compressed portfolios and the simple strategies.

A summary of the numerical tracking error evaluation results for portfolio compression is presented in Tables 3 to 6. The 'objective' column shows the optimal value for the initial optimization problem. This value is usually very different from the expected average absolute daily tracking error obtained when the solution is implemented as a trading strategy, since the simulated paths of the index are highly unlikely to pass through the nodes in the scenario tree. All optimization problem CPU total solution times shown are for an Athlon 650 Mhz PC with 256 MB memory.

Note that allowing options trading improves the simple strategies but is detrimental to the dynamic strategies when the nearest node strategy is used. This suggests that using the nearest node technique of Section 3 to interpret the solution of the initial optimization problem as a trading strategy is questionable with non-linear instruments unless the scenario tree is very large (well in excess of the 10,000 scenarios used here) to give a much more finely resolved strategy.

Method	Objective	Expected average absolute daily tracking error	Optimization CPU time (s)
static simple, L1	209.8	301.0 (1.52)	13
static simple, LINF	1542.0	500.0 (2.60)	17
5-branch dynamic, L1	49.8	97.9 (0.96)	1
5-branch dynamic, LINF	218.4	112.0 (0.89)	1
10-branch dynamic, L1	57.4	90.5 (0.91)	73
10-branch dynamic, LINF	301.0	117.0 (0.72)	126

Table 3: Portfolio compression evaluation results: strategies using only cash and the index as tradable instruments. One standard error in the estimate of the expected average tracking error is indicated in brackets.

Method	Objective	Expected average absolute daily tracking error	Optimization CPU time (s)
static simple, L1	139.4	187.0 (1.5)	98
static simple, LINF	774.0	330.0 (1.0)	59
5-branch dynamic, L1	8.2	387.0 (22.2)	1
5-branch dynamic, LINF	62.3	441.0 (22.4)	2
10-branch dynamic, L1	16.1	209.0 (22.2)	117
10-branch dynamic, LINF	112.0	237.6 (22.4)	1729

Table 4: Portfolio compression evaluation results: using cash, the index and four options as tradable instruments. One standard error in the estimate of the expected average tracking error is indicated in brackets.

Method	Expected average absolute daily tracking error	Expected worst absolute daily tracking error	Test scenario CPU times (s)
Static simple, L1	187.0	494.0	0.0228
Static simple, LINF	330.0	496.6	0.0228
Dynamic L1	90.5	200.3	0.0004
Dynamic LINF	117.0	218.3	0.0004
Delta hedge	93.3	170.2	0.0009

Table 5: Portfolio compression results: A comparision of the best static and dynamic hedges with delta hedging.

Method	Cash	Underlying	30-day	60-day	180-day	360-day
static simple, L1	−3172.5	4.21	0.73	1.78	3.37	−3.51
static simple, LINF	518.1	1.00	1.36	1.65	3.36	−0.13
10-branch dynamic, L1	−5960.9	6.43	N/A	N/A	N/A	N/A
10-branch dynamic, LINF	−5193.8	5.78	N/A	N/A	N/A	N/A

Table 6: Portfolio compression results: Optimal initial portfolio holdings using only cash and the underlying in the dynamic strategies. Reported are the best static hedges (i.e. those from Table 4) and the best dynamic hedges (i.e. those from Table 3).

The dynamic strategies using just cash and the underlying index not suprisingly have a significantly lower expected average absolute daily tracking error than any of the other strategies except the delta hedge, particularly for the versions minimizing the (L1) objective. The frictionless delta hedge strategy has the next best performance to the 10-branch dynamic L_1 strategy but takes over twice as long to evaluate on a test scenario due to Black–Scholes option evaluations at decision points (Table 5). Thus for portfolio compression used in Monte-Carlo VaR calculations it would be significantly inferior.

Figures 7 and 8, showing respectively the density functions of the average absolute daily tracking error and the worst tracking error for the 10-branch dynamic and simple strategies with each objective, demonstrate that although the dynamic worst case (L_∞) trading strategy better controls the upper tails of both the average and worst case absolute daily tracking error distributions, the average (L_1) trading strategy appears better overall with respect to both criteria.

It is also of interest to see if the compressed portfolios still track the target in extreme market conditions. Figure 9 shows the distribution functions of the *minimum* value of the target and of the four tracking strategies as before, while Figure 10 shows the distribution functions of the corresponding *maxima*. These figures show that the dynamic strategies are also better at estimating both the lower tail of the minimum value of the target and the upper tail of the target value's maximum.

Dynamic portfolio replication

The second set of experiments use dynamic stochastic programming iteratively to construct a dynamic replication strategy for the target using a prescribed set of tradable instruments. For each test scenario at each of the four maturity dates we construct a new scenario tree, re-solve the stochastic programming problem and re-balance to the portfolio associated with the initial node of this problem (*cf.* Figure 2).

For this dynamic replication strategy we assume a proportional transaction cost of 1% on trades involving the underlying index and 2.5% on trades

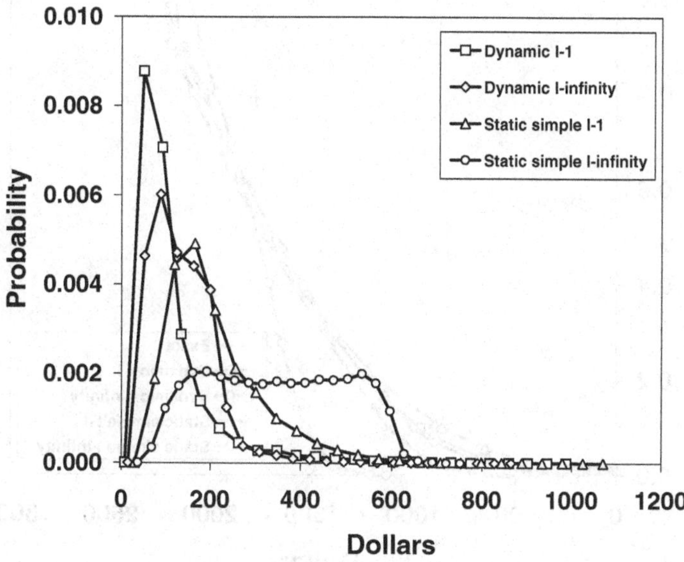

Figure 7: Density functions of the average absolute daily tracking error for two compression techniques and two static simple strategies.

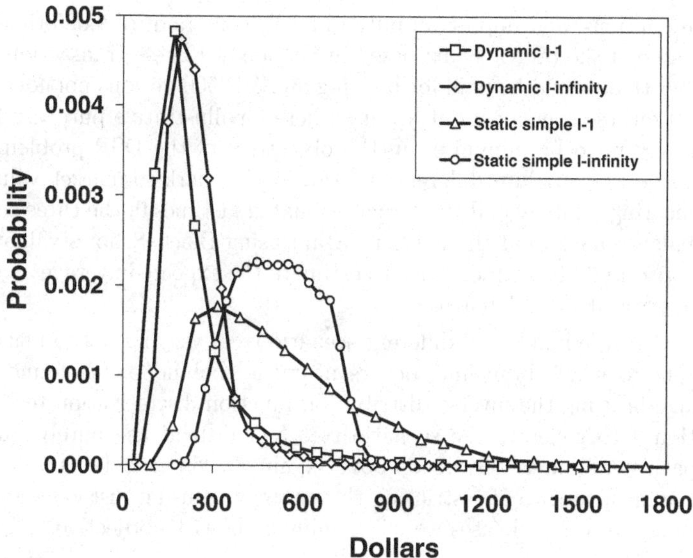

Figure 8: Density functions of the worst absolute daily tracking error for two compression techniques and two static simple strategies.

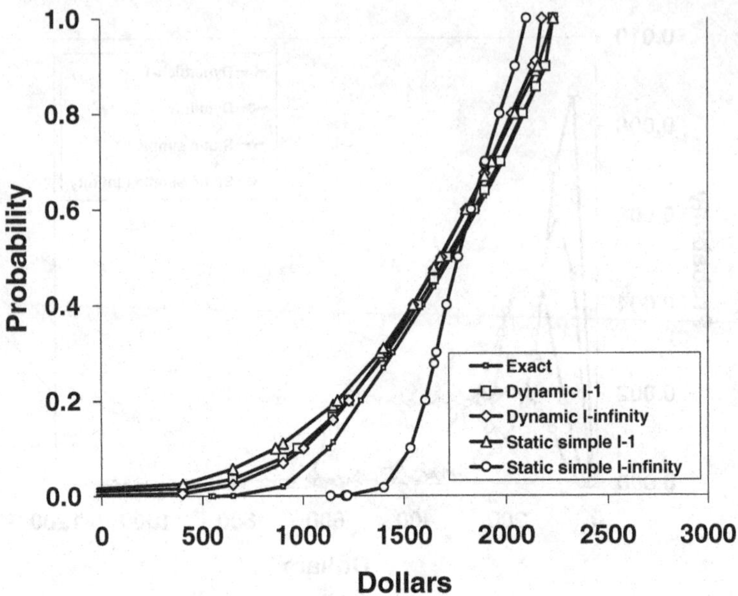

Figure 9: Distribution functions of the minimum value of the target and of several compressed portfolios.

involving options, and impose an initial budget constraint of 1.025 times the initial value of the target as discussed in Section 4. (These transaction costs are higher than would be paid for hedging an S&P 500 options portfolio with S&P 500 futures contracts, but are used here for illustrative purposes.) The tracking errors to be minimised in the objectives of the DSP problems are taken to be the absolute differences between the mark-to-market values of the replicating strategy and the target evaluated at times 0, the three successive rebalance dates and the horizon (again using Black–Scholes valuation). Here we use 1,000 test simulations to estimate the expected average absolute tracking error at these 5 dates.

We try a large number of different scenario trees varying both in size and the extent to which branching occurs near the root node. We found little advantage in using the inverse distribution function discretization technique of Section 4 to generate the scenario tree and instead use purely random index path sampling as described above. Again we will consider the effect on the dynamic strategies of restricting the strategy to using just cash and the underlying index. In all cases we will minimize the (L1′) objective.

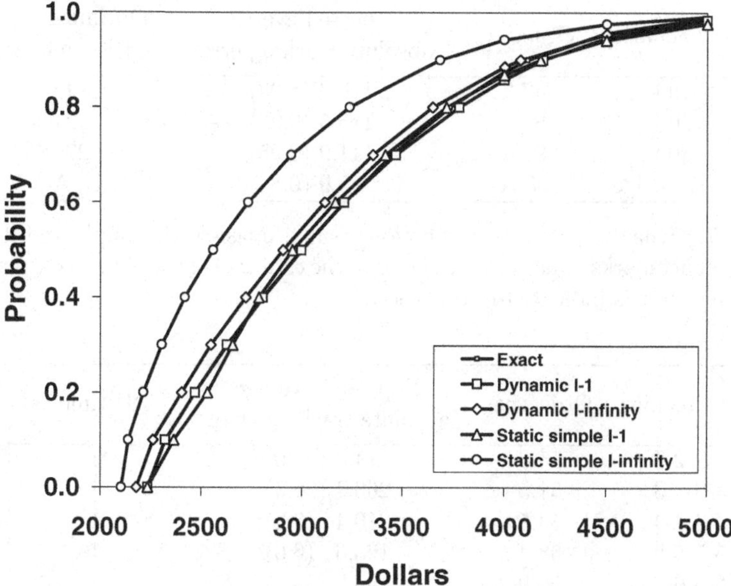

Figure 10: Distribution functions of the maximum value of the target and of several compressed portfolios.

Benchmark results

Our benchmark replication strategies will be quasi-static simple strategies based on respectively 100, 200 and 300 random scenarios for the path of the index from its initial value and the delta hedge rebalanced at decision points *with* transactions costs. We allow the quasi-static strategies to trade using cash, the underlying index and the four tradable options. Note that the simple strategies are quasi-static in the sense that the optimization problem solved at each step *assumes* no further re-balancing, but is in fact allowed to rebalance the portfolio according to the root node solution of the static simple problem corresponding to the current timestep. As the problem is re-solved at each timestep, the positions in the portfolio will generally change.

A summary of the numerical results for the benchmark strategies is given in Table 7. As the number of scenarios increases, although the objective of the optimization problem used in the quasi-static strategeies increases, the expected average absolute tracking error of the resulting strategy decreases (as we would expect) and is comparable to that of the delta hedge. Similar remarks apply to the expected *worst* absolute tracking error.

No. Scenarios	Objective	Expected average absolute tracking error	Optimization CPU time (s)
100	87.2	110.9 (8.03)	4
200	91.1	115.5 (0.97)	12
300	92.4	111.9 (0.95)	27
delta hedge	N/A	115.9 (0.89)	N/A

Table 7: Dynamic portfolio replication results: quasi-static simple and delta hedge benchmarks. One standard error in the estimate of the expected average tracking error is indicated in brackets.

Branching	Objective	Expected average absolute tracking error	CPU time (s)
2-2-2-2	15.6	414.8 (27.0)	1
3-3-3-3	51.5	260.4 (20.3)	1
4-4-4-4	43.5	210.4 (7.4)	1
5-5-5-5	58.4	163.6 (6.0)	1
6-6-6-6	50.0	145.5 (3.7)	2
7-7-7-7	54.3	142.4 (3.6)	8

Table 8: Dynamic portfolio replication results: dynamic strategy with varying tree sizes. Only cash and the underlying index are available as trading instruments. One standard error in the estimate of the expected average tracking error is indicated in brackets.

Experiments varying tree size

For dynamic portfolio replication we consider scenarios trees which have the same branching factor at each node, and we gradually increase the branching factor expecting that larger trees should lead to better replicating trading strategies.

The results are presented in Tables 8 and 9. Again we see that allowing the dynamic strategy to trade in options is detrimental, although the difference decreases as the branching-factor increases. None of the dynamic strategies based on balanced scenario trees with equal branching at each node and trading in only cash and the underlying index beat the simple benchmarks. However we will see that by varying the branching structure over decision points these results can be improved.

Branching	Objective	Expected average absolute tracking error	time (s)
2-2-2-2	0.0	987.0 (98.9)	1
3-3-3-3	23.6	703.2 (74.9)	1
4-4-4-4	1.5	694.2 (83.8)	1
5-5-5-5	21.3	433.7 (29.0)	4
6-6-6-6	18.6	312.5 (37.0)	9
7-7-7-7	26.0	188.5 (16.0)	61

Table 9: Dynamic portfolio replication results: dynamic strategy with different branching structures. Cash, the underlying index and four options are available as trading instruments. One standard error in the estimate of the expected average tracking error is indicated in brackets.

Experiments varying initial branching

Since with the dynamic replication strategy, although we re-solve the optimization problem at each timestep using a new scenario tree, we ultimately only use the optimal portfolio associated with the root node appropriate to the timestep. Given this procedure, it seems sensible to make scenario trees branch more near the root node of each successive DSP problem. Here we keep the total number of nodes in the tree roughly constant (in an attempt to keep the optimization time constant across experiments) and gradually increase the branching at the root node of each successive DSP problem.

The results are summarized in Tables 10 and 11. As the initial branching increases, both the objective and the expected average tracking error tend to decrease, but with some sampling fluctuation due to the randomly generated scenario trees. This random decrease is more prominent when the dynamic trading strategy is allowed to use options (Table 11), when the final strategies in the table (in which almost all branching occurs at the root node), give a *significant improvement* over the best benchmark strategy in less than one third the computing time (*cf.* Table 7). Initial positions chosen by the various strategies are set out in Table 12.

Finally, a comparison of the distributions of the average and worst absolute tracking errors (measured at the initial time and the four subsequent decision points) for the quasi-static simple benchmarks and the dynamic replication strategies based on the tree with highest branching in the initial time stage, both with and without the availability of trading options, is shown in Figures 11 and 12. Again this shows that the best dynamic replication strategies continue to track the target in extreme market conditions.

Branching	Objective	Expected average absolute tracking error	CPU time (s)
6-6-6-6	50.0	145.5 (3.7)	2
8-6-5-4	43.6	137.3 (3.3)	3
9-7-5-3	40.5	134.0 (3.9)	5
13-8-4-2	47.5	135.8 (3.1)	3
21-9-3-1	38.8	122.4 (2.6)	2
22-8-3-1	30.1	119.7 (2.6)	2
31-8-2-1	34.1	118.6 (2.4)	4
43-9-1-1	25.9	122.8 (2.5)	7
55-7-1-1	28.9	120.6 (2.5)	4
75-5-1-1	24.7	119.3 (2.5)	5
120-3-1-1	23.0	115.7 (2.0)	6
300-1-1-1	17.1	116.5 (2.3)	5

Table 10: Dynamic portfolio replication results: dynamic strategy with different branching structures. Only cash and the underlying index are available as trading instruments. One standard error in the estimate of the expected average tracking error is indicated in brackets.

Branching	Objective	Expected average absolute tracking error	CPU time (s)
6-6-6-6	18.7	312.5 (37.0)	9
8-6-5-4	9.3	261.5 (29.7)	5
9-7-5-3	7.1	215.9 (17.2)	7
13-8-4-2	19.9	132.5 (5.3)	8
21-9-3-1	16.9	133.4 (14.3)	10
22-8-3-1	12.2	107.2 (6.0)	7
31-8-2-1	17.9	116.7 (4.5)	8
43-9-1-1	10.2	103.9 (7.2)	6
55-7-1-1	12.2	132.1 (34.4)	73
75-5-1-1	10.2	94.8 (2.0)	6
120-3-1-1	8.1	96.1 (2.6)	7
300-1-1-1	3.6	93.0 (1.8)	8

Table 11: Dynamic portfolio replication results: dynamic strategy with different branching structures. Cash, the underlying index and four options are available as trading instruments. One standard error in the estimate of the expected average tracking error is indicated in brackets. The anomalous runtime is due to chance degeneracy of the deterministic equivalent linear programme.

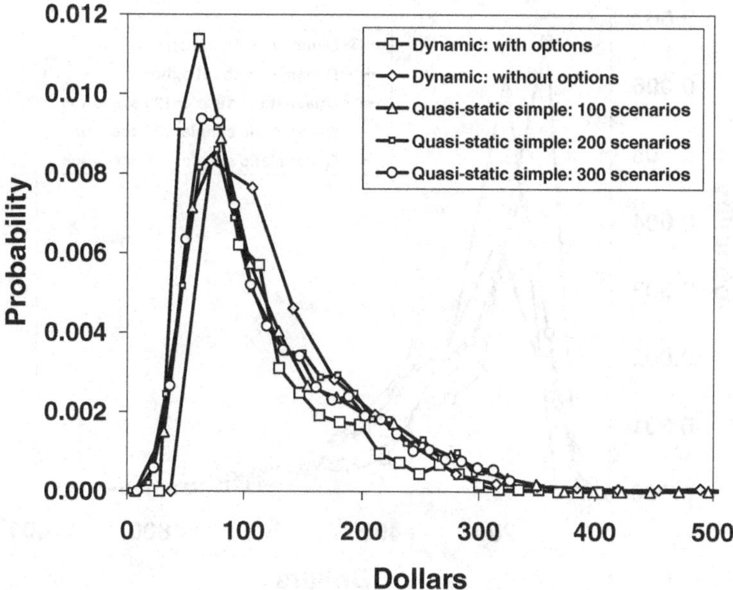

Figure 11: Dynamic portfolio replication results: density functions of the average tracking error for two dynamic hedging techniques and three benchmark strategies.

Method	Cash	Underlying	30-day	60-day	180-day	360-day
quasi-static simple, 100 scenarios	−329.2	1.73	0.27	2.97	3.85	0.00
quasi-static simple, 200 scenarios	−171.4	1.62	0.00	2.97	3.65	0.00
quasi-static simple, 300 scenarios	−387.6	1.80	0.00	2.41	3.95	0.00
dynamic with options	−1426.8	2.16	−4.23	0.00	0.00	13.45
dynamic w/o options	−5194.0	5.81	N/A	N/A	N/A	N/A

Table 12: Dynamic portfolio replication results: Optimal initial portfolio holdings.

6 Conclusions

For both applications considered in this paper—portfolio compression and dynamic portfolio replication—the solutions from a stochastic programming approach are superior to optimized 'quasi-static' approaches and a delta hedge. For compression however, is not clear how the solution of the stochastic programming problem should be interpreted as a trading strategy; the method

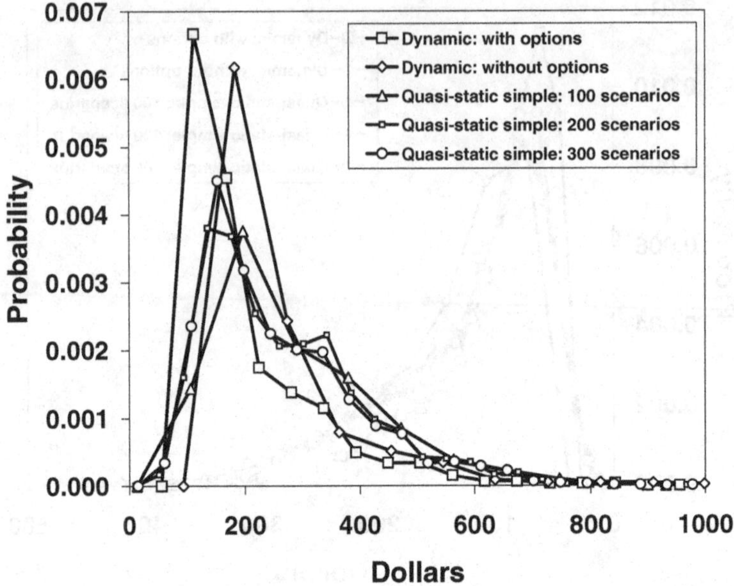

Figure 12: Dynamic portfolio replication results: density functions of the worst absolute tracking error for two dynamic hedging techniques and three benchmark strategies.

used here is not robust enough to allow the strategy to trade in options as well as cash and the underlying.

From Figures 9 and 10 we see that the extreme values taken by the target portfolio over the planning horizon have medians close to 1750 and 3000 respectively, indicating that the expected average absolute tracking error achieved by the best dynamic hedge (90.5) is approximately 3%–5% of the target value.

In the dynamic portfolio replication problem, very different trees should be used from those effective for portfolio compression: almost all the branching should occur at the root node. In this case the method is robust enough to allow the dynamic strategy to trade options in order to reduce tracking error and produce an overall significantly best strategy in a running time three times faster than the best quasi-static benchmark alternative.

In a companion paper we will describe the extension of these ideas to a swaption portfolio where, unfortunately, further ingenuity is required.

Acknowledgements

The authors wish to thank members of the Centre for Financial Research and of Algorithmics Incorporated's Research Department, in particular David Saunders, for helpful comments and suggestions which materially improved earlier drafts.

References

Birge, J.R. & Louveaux, F. (1997), *Introduction to Stochastic Programming*, Springer.

Dembo, R. & Rosen, D. (1999), 'The practice of portfolio replication', *Ann. Op. Res.* **85**, 267–284.

Chisti, A. (1999), 'Simulation of fixed-income portfolios using grids', *Algo Res. Quart.* **2**(2), 41–50.

Dempster, M.A.H., ed. (1980), *Stochastic Programming*, Academic Press.

Dempster, M.A.H. & Ireland, A.M. (1988), 'A financial decision support system'. In *Mathematical Models for Decision Support Systems*, G. Mitra, ed., Springer, 415–440.

Dempster, M.A.H. & Thompson, R.T. (1999), 'EVPI-based importance sampling solution procedures for multi-stage stochastic linear programmes on parallel MIMD architectures', *Ann. Op. Res.* **90**, 161–184.

Dempster, M.A.H., Hicks-Pedron, N., Medova, E.A., Scott, J.E. & Sembos, A. (2000), 'Planning logistics operations in the oil industry', *J. Op. Res. Soc.* **51**, 1271–1288.

Fleten, S., Høyland, K. & Wallace, S. (1998), 'The performance of stochastic dynamic and fix mix portfolio models', Working Paper, Norwegian University of Science and Technology.

Golub, B., Holmer, M., Polman, L. & Zenios, S. (1997), 'Stochastic programming models for money management', *Euro. J. Op. Res.* **85**, 282–296.

Gondzio, J. & Kouwenberg, R. (1999), 'High performance computing for asset liability management', Working Paper, Erasmus University.

Gondzio, J., Kouwenberg, R. & Vorst, T. (1998), 'Hedging options under transaction costs and stochastic volatility', Working Paper.

Jamshidian, F. & Zhu, Y. (1997), 'Scenario simulation: theory and methodology', *Finance and Stochastics* **1**, 43–67.

Klaassen, P. (1997), 'Discretized reality and spurious profits in stochastic programming models for asset/liability management', Working Paper, Vrije Universiteit, Amsterdam.

Kouwenberg, R. (1998), 'Scenario generation and stochastic programming models for asset liability management', Working Paper, Econometric Institute, Erasmus University, Rotterdam.

Kouwenberg, R. & Vorst, T. (1998), 'Dynamic portfolio insurance: A stochastic programming approach', Working Paper, Department of Finance, Erasmus University, Rotterdam.

Wets, R.J.-B. & Ziemba, W.T., eds (1999), 'Stochastic Programming: State of the Art, 1998', *Ann. Op. Res.* **85**.

Worzel, K.J., Vassiadou-Zeniou, C. & Zenios, S.A. (1994), 'Integrated simulation and optimization models for tracking indices of fixed-income securities', *Op. Res.* **42**(2), 223–233.

Ziemba, W.T. & Mulvey, J., eds (1998), *Worldwide Asset Liability Modelling*, Cambridge University Press.

Credit and Interest Rate Risk

Rudiger Kiesel, William Perraudin and Alex Taylor

Abstract

This paper investigates the relation between credit and market risk over long investment horizons. We split credit risk into transition and spread risk so that results can be directly related to ratings-based credit risk models which adopt this decomposition. We find that spread risk for high credit quality exposures exhibits variable but generally negative correlation with interest rate changes. For low credit quality spreads, the correlation is markedly negative. Transition risk is also negatively correlated with interest rate changes in that VaRs are distinctly higher when calculated using a transition matrix based on years of data in which interest rates fall.

1 Introduction

1.1 VaRs for Market and Credit Risk

Since the mid-1990s, banks have made extensive use of VaR models for measuring and controlling the market risk of their trading portfolios. They have been encouraged in this by regulators who, since the 1997 Basel Accord Amendment on market risk (see Basel Committee on Banking Supervision (1996)) have permitted use of these models for the calculation of regulatory capital for trading books.

Recently, there has been much interest in a new generation of portfolio management models designed to measure the risks associated with portfolios of credit exposures. The frameworks proposed by JP Morgan (1997) (Creditmetrics), Credit Suisse Financial Products (1997) (CreditRisk+), and by the consulting firms KMV and McKinsey's have been widely discussed and analyzed.[1] Although they may be employed for different purposes (for example, portfolio allocation), the most common use of these models is to generate VaR estimates for credit-sensitive portfolios.

[1]See Basel Committee on Banking Supervision (1999) and Crouhy, Galai, and Mark (2000) for surveys, Gordy (2000) for simulations of simplified versions of Creditmetrics and CreditRisk+ on stylized portfolios, and Nickell, Perraudin, and Varotto (2001) for evaluation of the out-of-sample performance of Creditmetrics and an equity-based model resembling that of KMV.

Given these two developments, it is natural to ask whether market and credit risk models should be combined to obtain a complete picture of the risks faced by banks. So far, this has not been attempted by many firms. The main credit risk methodologies widely implemented, notably Creditmetrics and the model of KMV, ignore interest rate risk (the primary component of market risk for banks). In these models, it is supposed that future discount factors equal the currently-observed forward discount factors.

It is important to understand that bank risk management calculations in the two areas of credit and market risk are performed on a very different basis and for somewhat different purposes. Typically, firms implement trading book VaR models for short holding periods (one to ten days) and with small confidence levels (5% or 1%). For most banks, the objective of such high-frequency calculations for trading books is more directed at maintaining the quality of earnings than at avoiding catastrophic out-comes which could imperil the firm's solvency.[2]

By contrast, banks implement credit risk models for much longer time horizons (one year or more) and, typically, for smaller confidence levels (often 0.3% or lower). For senior management in banks, VaR calculations for credit risk are much more closely linked to the objective of avoiding serious deterioration in the balance sheet. Since the focus is so different, current market and credit risk calculations could never be directly integrated but the issue remains whether market and, in particular, interest rate risk should be included in the long horizon risk calculations that banks are beginning to perform for their credit books.

1.2 Interest Rate and Credit Risk Correlation

The main argument against incorporating interest risk is that the level of correlation between credit and interest rate risk is poorly understood. If two risks are perfectly correlated, calculating the VaRs and implied capital requirements separately and adding them together gives the same answer as if one performs an integrated calculation. If correlation is less than perfect, VaRs will generally be lower if the calculation is performed on an integrated basis.[3] Hence, a sensible, conservative approach, if the correlation between two risks is not known, is to conduct the VaR calculation on a piece-meal fashion.

Several papers have examined the degree of correlation between credit and interest rate risk. Longstaff and Schwartz (1995) employed bond price

[2]This is why firms have often used 5% confidence levels and a one day horizon. Regulators who focus solely on threats to solvency have insisted on 1% and a ten day horizon for regulatory capital calculations.

[3]Recent research on coherent risk measures (see Artzner, Delbaen, Eber, and Heath (1997) and Artzner *et al.*, this volume) emphasizes that this is not true if the positions involved imply enormous losses with small probabilities.

indices of which the constituent bonds possessed call features. The negative correlation between credit and interest rate risk that these authors find could reflect correlation between interest rates and the value of the call premia. A careful recent study by Duffee (1998) employs price data for straight bonds and finds that there is a significant negative correlation between interest rates and credit spreads for monthly data. These findings are in agreement with the bond pricing models of Merton (1974),Kim, Ramaswamy, and Sundaresan (1993), and Longstaff and Schwartz (1995) which predict negative correlation between interest rates and credit spreads.

A limitation of Duffee (1998)'s study if one's priority is to draw lessons for credit risk modeling is the monthly horizon over which he studies correlations. Interest rates and credit spreads may be negatively correlated for short horizons but exhibit positive correlation for the long horizons typically employed in credit risk models. Morris, Neal, and Rolph (1999) perform cointegration analysis on monthly returns on 10-year Treasury bonds and Moody's Aaa and Baa bond indices from 1960 to 1997, and argue that this is indeed the case.

1.3 Decomposing Credit Risk

Credit risk models decompose credit risk into various components. For example, Creditmetrics assume that credit risk comprises rating transition and recovery rate risk. Kiesel, Perraudin, and Taylor (2000) generalize Creditmetrics to include spread risk. Any one of the three types of risk, transition, recovery rate or spread risk could in principal be made correlated with interest rates. Wilson (1997) shows how one may include correlations between transition risk and random changes in macroeconomic variables including interest rates (see also Jarrow and Turnbull (2000)). In the pricing literature, Das and Tufano (1996) introduce recovery rates which are correlated with interest rates. Kiesel, Perraudin, and Taylor (2000) discuss how one might allow interest rates and spreads to be correlated within their framework.

In this paper, we examine correlations between constituent parts of credit risk, namely transition and spread risk in order to provide a guide as to how such correlation might be introduced within a credit risk model. In Section 2, we build on the analysis of Kiesel, Perraudin, and Taylor (2000) in estimating long-horizon correlations between spreads and interest rates. The spreads we employ are generic spreads for defaultable pure discount bonds issued by obligors with particular agency ratings (AAA, AA, A, BBB, BB and B). Our data comes from Bloomberg and from Perraudin and Taylor (1999) which applies cubic spline techniques to a large dataset of US-dollar-denominated international bonds.

2 Interest Rate-Credit Spread Correlations

2.1 Spread Data

We begin by examining correlation between interest rates and spreads, defined as the average spreads for pure discount bonds issued by obligors of given ratings. Typically, such spreads are obtained by fitting discount functions made up of splines or other flexible functional forms to price data on bonds with particular credit ratings. If agency ratings correctly captured credit standing in a timely fashion, abstracting from spread risk, one might expect that spreads for different ratings categories would vary little over time. However, ratings are sticky (see Delianedis and Geske (1998)) and spreads vary considerably over time (see Kiesel, Perraudin, and Taylor (2000)). A significant part of total credit risk is therefore made up of stochastic changes in average spreads.

To examine interest rate-spread correlations, we employ daily Bloomberg spread and US Treasury strip data covering the period April 1991 to November 1998. The spreads equal yields on notional zero coupon bonds with different credit ratings and maturities issued by US industrials minus the yields of US Treasury strips of the same maturity. The ratings for which data are available are: AAA, AA12, AA3, A1, A2, A3, BBB1, BBB2, BBB3, BB1, BB2, BB3, B1, B2, and B3. We focus on AAA, AA12, A2, BBB2, BB2, and B2, taking these to be representative of spreads for the coarser, non-numbered rating categories AAA, AA, A, BBB, BB, and B. For each rating category, we employ credit spreads for maturities of 2, 5 and 10 years. In total, the Bloomberg data series comprise 1,640 daily observations.

We also employ spread data estimated using cubic spline techniques from a large dataset of dollar-denominated international bond prices. Perraudin and Taylor (1999) describe the techniques employed and analyze different aspects of the spread data. This second set of spread data covers the period 21st April 1991 to 1st February 1998 and includes spreads for bonds in rating categories AAA, AA and A.

2.2 Long-Horizon Correlations

As mentioned in the Introduction, when studying correlation between different risks, it is important to distinguish between different time horizons. Duffee (1998) finds that interest rates and returns on risky bonds are negatively correlated over one-month horizons but Morris, Neal, and Rolph (1999) argue that this conclusion is highly sensitive to the horizon and that over long-horizons interest rates and credit spreads are positively correlated.

Morris, Neal, and Rolph (1999) use standard, linear cointegration methods which require that one assume a particular parametric form for the time series involved. In this paper, we employ a non-parametric approach initially developed by Cochrane (1988) and first applied for estimating long-horizon financial asset volatilities by Kiesel, Perraudin, and Taylor (2000). Essentially, this approach consists of estimating variances and covariances over long horizons by calculating the corresponding sample moments using *overlapping observations*.[4] For example the variance of the spread changes over k days, $\sigma^2(k)$, can be estimated from a time series of T daily spreads $S_0, S_1, ...S_T$:

$$\hat{\sigma}^2(k) \equiv \mathrm{Var}_n(S_{n+k} - S_n) = \frac{1}{T-k}\sum_{n=k}^{T}\left(S_n - S_{n-k} - \frac{k}{T}(S_T - S_0)\right)^2, \tag{2.1}$$

The use of over-lapping observations introduces small sample biases and so it is important to adjust for these. Cochrane (1988) devises suitable small sample adjustments for long-horizon variance estimators. Kiesel, Perraudin, and Taylor (2000) extend this to covariances and other moments.

2.3 Non-Parametric Estimates on International Bond Spreads

In Figure 1, we present non-parametric estimates of variance ratios and correlation coefficients between interest rates and credit spreads for different maturities (2, 5 and 10 years) and for ratings categories (AAA, AA and A). The range of horizons we consider is 1 to 250 working days. The credit spread data we employ in calculating the estimates reported in Figure 1 comes from cubic spline fits of an extensive dataset of dollar-denominated international bonds as described above (see also Perraudin and Taylor (1999)).

The top left panel in Figure 1 shows the ratio of small-sample-adjusted variance estimates for k days (for $1 \leq k \leq 250$) divided by an estimate of the 1-day variance multiplied by k. The ratios shown are for changes in 5-year maturity interest rates and credit spreads. For a pure unit root process, this ratio should equal unity for any k. For a first-difference stationary processes, the plot of the ratio should settle down to a constant as the horizon becomes long. As one may see, this is indeed the case both for interest rates and credit spreads. We also performed augmented Dickey–Fuller unit root tests for the interest rate and credit spread series and found that, at a 10% confidence level, we could not reject the presence of a unit root in any of our series.

The lower and right-hand panels in Figure 1 show interest-rate-credit-spread correlation coefficient estimates (adjusted for small sample bias) for 2,

---

[4]Since our data is of daily frequency and we are interested in annual investment horizons, the degree of overlap is very considerable. If a year consists of 250 working days, each pair of adjacent observations overlap by 2494 working days.

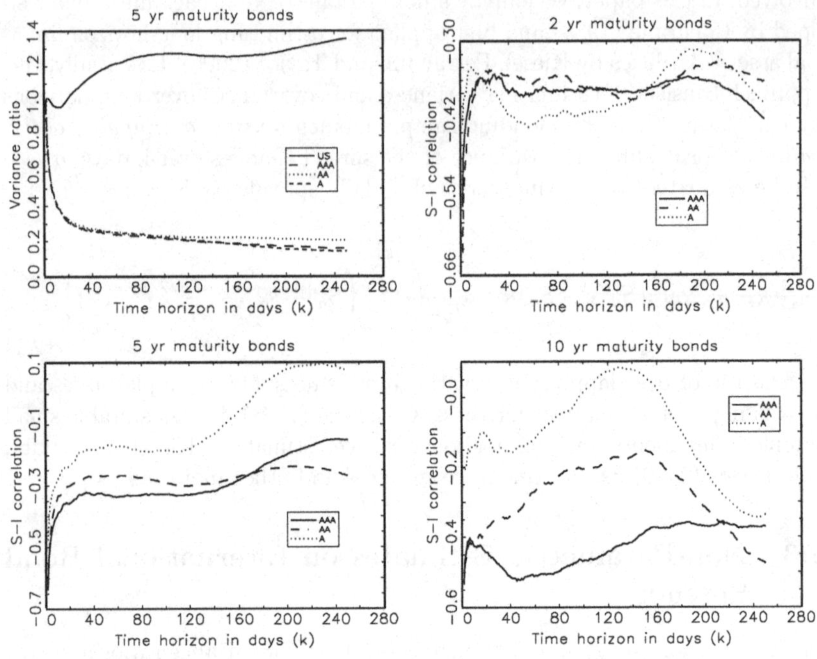

Figure 1: Spline fit spread data

5, and 10-year maturity interest rates and credit spreads.[5] A consistent pattern which emerges from the plots is that very short-horizon correlations are negative. For 2-year maturity spreads and interest rates, the degree of negative correlation is halved as one extends the horizon from 1 day to about a month. Beyond a month, it remains roughly constant as the horizon increases to a year. For 5-year maturity interest rates and spreads, again the correlation becomes much less negative in the first 1 to 2 months, but the decrease continues as the horizon grows larger. In the case of the single-A spread, the correlation actually becomes positive after about 8 months. The results for 10-year maturity spreads show wider variation in correlations across different rating categories with the single-A spreads again exhibiting positive correlation for a short range of horizons just over 6 months.

2.4 Non-Parametric Estimates on Bloomberg Spreads

Figure 2 shows estimates similar to those reported in Figure 1 but based on spreads for US industrials obtained from Bloomberg. Although estimated

[5]In each case, the correlation is between the spread on a pure discount bond spread and a Treasury strip yield of matching maturity.

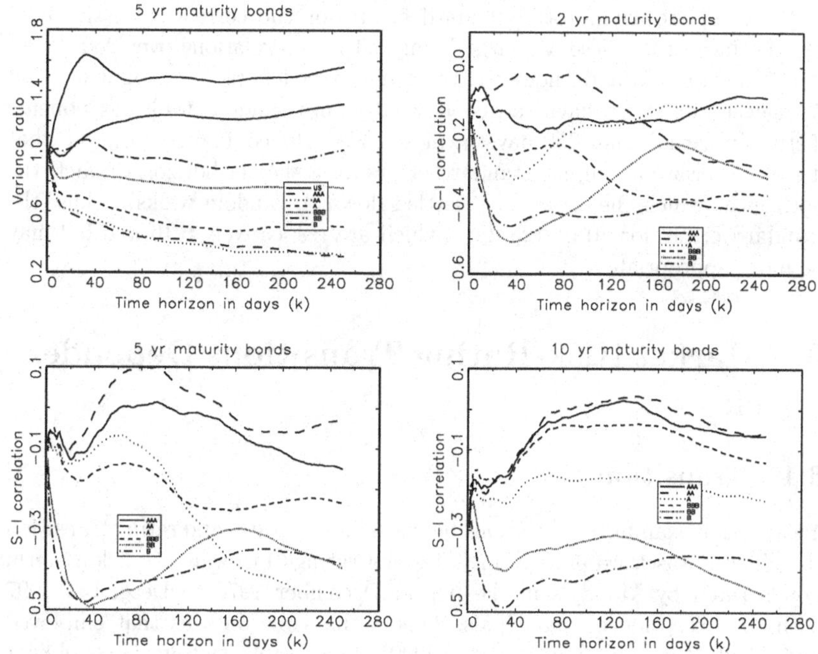

Figure 2: Bloomberg spread data

from a very large number of bonds, the Bloomberg spreads are in some ways less 'clean' than those based on our international bond price dataset since Bloomberg include option-adjusted prices of callable bonds in their spline fits. Such option-adjustments are only approximate.

The upper left hand plot in Figure 2 shows variance ratios. These again suggest that the interest rate and Bloomberg spread series are first difference stationary. The remaining plots in Figure 2 show correlation coefficient estimates for different maturities and ratings. Again, we find that correlations are primarily negative although high credit quality, ten-year maturity spreads do exhibit positive correlation for horizons greater than 6 months. This finding is interesting since the spreads for which Morris, Neal, and Rolph (1999) find positive long-horizon correlations are investment quality and of ten-year maturity.

However, a striking feature of the Bloomberg spread results is that the sub-investment-quality spreads (BB and B), which were not available in our international bond data set, are strongly negatively correlated with interest rates for all three maturities we examine. This is particularly interesting since neither Duffee (1998) nor Morris, Neal, and Rolph (1999) examine correlations for sub-investment-grade credit exposures.

We calculate asymptotic standard errors for the correlation estimates.[6] On the face of it, these are large being 0.3 for correlations over 250 days.[7] It is important to note, however, that the upper left plots in Figures 1 and 2 suggest our series have settled down to pure random walk components distinctly earlier than 250 days in most cases. Hence, one could argue that it is appropriate to employ standard errors for a shorter horizon (in fact, the horizon at which the series have settled down to random walks). Thus, the standard errors for 80 or 125 days which are respectively 0.16 and 0.2 may be more appropriate.

3 Market Risk-Rating Transitions Dependencies

3.1 Transition Data

This section examines correlations between interest rates and rating transition risk. The dataset we employ includes the ratings histories of all long-term bonds rated by Moody's in the period December 1970 to December 1997 with the exception of municipals. This is the same data as that employed by Nickell, Perraudin, and Varotto (2000). The sample contains 6,534 obligor ratings histories and the total number of obligor-years excluding with-drawn ratings (and hence observations in our sample) was 50,831. The ratings used are notional senior, unsecured ratings created by Moody's for all obligors who have issued Moody's rated long bonds at a given moment in time.[8]

An issue that arises in estimating ratings transition matrices is the appropriate treatment of withdrawn ratings. Ratings are withdrawn for a variety of reasons, for example because the bond is called or because the obligor ceases to continue paying Moody's the required annual fee. Typically, Aaa borrowers have an annual risk of ratings withdrawal of 4% while for B-rated issuers the risk is just over 10%. Carty (1997) argues that few ratings withdrawals (around 13%) are possibly correlated with changes in credit standing and hence that one should calculate ratings transition probabilities simply leaving the withdrawn ratings aside. This is the approach we follow.

We employ the coarser rating categories Aaa, Aa, A, Baa, Ba, B, and used by Moody's prior to 1982.[9] After that date, Moody's split the upper six

[6]See Kiesel, Perraudin, and Taylor (2000) for more details.

[7]The standard errors depend only on sample size and the degree of overlap k

[8]Lucas and Lonski (1992) mention that in their dataset, which resembles ours, 56% of ratings are based on directly observed senior, unsecured ratings. The remainder are derived from ratings on subordinated or secured bonds rated by Moody's. The approach taken by Moody's to inferring ratings is described in Carty (1997).

[9]We combine the Caa and C/Ca categories.

categories into numbered sub-categories.[10] We employ the coarser categories in this study because we wished to include data from 1970 onwards and wanted to have full data-comparability throughout our sample period. Also, one may doubt whether it is useful to employ the finer categorization in credit risk modeling. Credit spread data are not that reliable for finer ratings and the added complexity of having three times as many categories is probably not worthwhile.

3.2 Transition Matrices

Tables 1 and 2 contain estimates of transition matrices based on sub-samples of our 27 years of data. Each sub-sample consists of 9 years. Each year is measured from December 31st to December 31st of the following year. The matrices in Table 1 are estimated using sub-samples made up of years in which the average *level* of daily short-term interest rates was low, medium or high.[11] To estimate the transition matrices reported in Table 2, we construct sub-samples based on whether the year-on-year change in annual interest rates was small, large or in between.[12]

In principle, the sub-samples based on interest rate *changes* are more interesting. Our concern is with correlations between changes in asset values. Rating transitions in a particular year are associated with value changes as are *changes* in interest rates. So, it is natural to see whether large interest rate changes are associated with different average transitions than small interest rate changes.

The results in Tables 1 and 2 suggest that the changes in rating transition behaviour for years in which interest rates behave differently are not substantial except perhaps for some of the transition probabilities for lowly-rated bonds. However, in this latter case, there is relatively little data so the results are statistically less reliable.[13]

To facilitate comparisons between the transition matrices reported in Tables 1 and 2, in Table 3 we provide summary measures of the differences between pairs of matrices. For example, the mean absolute difference between entries in transition matrix estimates based on high and low interest

[10]Thus, for example, Aaa was split into Aaa1, Aaa2 and Aaa3, with Aaa1 being the highest credit quality.

[11]In other words, we ordered years according to the average level of interest rates and then split them into the first, second and third group of nine years.

[12]In this case, we ordered years according to the average year-on-year *change* in interest rates and took the first, second and third sets of 9 years as our sub-samples.

[13]The standard errors reported in the tables are calculated as $\sqrt{\hat{\pi}_{ij}(1-\hat{\pi}_{ij})/n_i}$ where n_i is the number of issuer years for the initial rating category and $\hat{\pi}_{ij}$ is the point estimate of the transition probability. These are lower than the true standard errors since they ignore dependence across different transitions which is likely to be significant cross-sectionally.

Table 1: Transition Matrices For Different Interest Rate Levels

Initial rating	Terminal rating Aaa	Aa	A	Baa	Ba	B	C	Def	Number issuer yrs
	Medium Interest Rate Years								
Aaa	93.9	6.0	0.1	– –	– –	– –	– –	– –	856
	(0.8)	(0.8)	(0.1)	– –	– –	– –	– –	– –	
Aa	1.6	87.3	10.4	0.5	0.2	0.0	– –	– –	2284
	(0.3)	(0.7)	(0.6)	(0.1)	(0.1)	(0.0)	– –	– –	
A	0.0	2.1	91.6	5.4	0.6	0.3	– –	– –	4498
	(0.0)	(0.2)	(0.4)	(0.3)	(0.1)	(0.1)	– –	– –	
Baa	0.0	0.3	5.0	88.5	4.9	0.9	0.1	0.2	3213
	(0.0)	(0.1)	(0.4)	(0.6)	(0.4)	(0.2)	(0.0)	(0.1)	
Ba	0.1	0.0	0.4	4.5	86.6	6.4	0.7	1.2	2816
	(0.1)	(0.0)	(0.1)	(0.4)	(0.6)	(0.5)	(0.2)	(0.2)	
B	– –	– –	0.2	0.6	5.7	84.1	4.9	4.4	1621
	– –	– –	(0.1)	(0.2)	(0.6)	(0.9)	(0.5)	(0.5)	
C	– –	– –	– –	1.9	– –	5.6	86.4	6.2	162
	– –	– –	– –	(1.1)	– –	(1.8)	(2.7)	(1.9)	
	High Interest Rate Years								
Aaa	88.1	9.7	2.0	– –	0.1	– –	– –	– –	884
	(1.1)	(1.0)	(0.5)	– –	(0.1)	– –	– –	– –	
Aa	1.2	90.4	8.0	0.3	0.2	– –	– –	– –	1732
	(0.3)	(0.7)	(0.7)	(0.1)	(0.1)	– –	– –	– –	
A	0.1	3.6	89.1	6.1	0.9	0.1	0.0	0.0	3756
	(0.0)	(0.3)	(0.5)	(0.4)	(0.2)	(0.0)	(0.0)	(0.0)	
Baa	0.0	0.3	7.4	84.5	6.7	0.8	0.2	0.1	2945
	(0.0)	(0.1)	(0.5)	(0.7)	(0.5)	(0.2)	(0.1)	(0.1)	
Ba	– –	0.1	0.6	5.3	82.3	9.7	0.5	1.4	2372
	– –	(0.1)	(0.2)	(0.5)	(0.8)	(0.6)	(0.1)	(0.2)	
B	– –	0.2	0.1	0.4	5.5	82.2	3.2	8.4	1029
	– –	(0.1)	(0.1)	(0.2)	(0.7)	(1.2)	(0.5)	(0.9)	
C	– –	– –	– –	1.3	5.3	5.3	61.3	26.7	75
	– –	– –	– –	(1.3)	(2.6)	(2.6)	(5.6)	(5.1)	
	Low Interest Rate Years								
Aaa	89.1	10.6	0.3	– –	– –	– –	– –	– –	774
	(1.1)	(1.1)	(0.2)	– –	– –	– –	– –	– –	
Aa	0.5	90.9	8.3	0.3	– –	– –	– –	– –	2386
	(0.1)	(0.6)	(0.6)	(0.1)	– –	– –	– –	– –	
A	0.1	1.5	94.5	3.8	0.1	– –	– –	– –	5351
	(0.0)	(0.2)	(0.3)	(0.3)	(0.0)	– –	– –	– –	
Baa	0.0	0.1	4.3	93.0	2.2	0.3	0.1	0.0	4069
	(0.0)	(0.0)	(0.3)	(0.4)	(0.2)	(0.1)	(0.0)	(0.0)	
Ba	– –	– –	0.6	6.3	87.9	4.9	0.2	0.1	2849
	– –	– –	(0.1)	(0.5)	(0.6)	(0.4)	(0.1)	(0.1)	
B	0.1	0.1	0.3	0.8	8.4	84.7	5.4	0.3	1905
	(0.1)	(0.1)	(0.1)	(0.2)	(0.6)	(0.8)	(0.5)	(0.1)	
C	– –	– –	– –	– –	1.4	8.3	88.7	1.7	423
	– –	– –	– –	– –	(0.6)	(1.3)	(1.5)	(0.6)	

Estimates are based on all Moody's notional, senior unsecured bond ratings between 31/12/70 and 31/12/97 measured on 31st December each year. We split the sample into three sets of 9 years in which interest rate *levels* are highest, lowest or in the middle and then estimate transition matrices for each sub-sample.

rate years is 2.74%.[14] The results suggest that years in which the change in interest rates is especially small (i.e., typically negative) exhibit the most unusual transition behaviour.

[14] We include entries in these mean calculations if one of the two corresponding entries from the two matrices is non-zero.

Table 2: Transition Matrices For Different Interest Rate Changes

Initial rating	Aaa	Aa	A	Baa	Ba	B	C	Def	Number issuer yrs
				Terminal rating					
				Medium Change Interest Rate Years					
Aaa	90.9	9.1	--	--	--	--	--	--	898
	(1.0)	(1.0)	--	--	--	--	--	--	
Aa	0.9	90.1	8.6	0.3	0.0	0.0	--	--	2491
	(0.2)	(0.6)	(0.6)	(0.1)	(0.0)	(0.0)	--	--	
A	0.1	2.1	92.4	4.9	0.4	0.2	--	--	5236
	(0.0)	(0.2)	(0.4)	(0.3)	(0.1)	(0.1)	--	--	
Baa	0.1	0.3	4.4	91.4	3.2	0.5	0.1	0.0	3817
	(0.0)	(0.1)	(0.3)	(0.5)	(0.3)	(0.1)	(0.1)	(0.0)	
Ba	0.0	0.0	0.5	5.6	85.7	6.8	0.5	0.8	3214
	(0.0)	(0.0)	(0.1)	(0.4)	(0.6)	(0.4)	(0.1)	(0.2)	
B	0.0	0.1	0.4	1.0	6.9	84.1	4.1	3.3	2189
	(0.0)	(0.1)	(0.1)	(0.2)	(0.5)	(0.8)	(0.4)	(0.4)	
C	--	--	--	0.6	0.9	7.9	85.5	5.0	317
	--	--	--	(0.4)	(0.5)	(1.5)	(2.0)	(1.2)	
				Large Change Interest Rate Years					
Aaa	94.7	4.6	0.6	--	0.1	--	--	--	809
	(0.8)	(0.7)	(0.3)	--	(0.1)	--	--	--	
Aa	1.2	92.2	6.3	0.2	0.2	--	--	--	1902
	(0.2)	(0.6)	(0.6)	(0.1)	(0.1)	--	--	--	
A	0.0	2.0	93.7	3.8	0.5	0.0	0.0	--	4415
	(0.0)	(0.2)	(0.4)	(0.3)	(0.1)	(0.0)	(0.0)	--	
Baa	0.0	--	5.0	90.4	3.7	0.5	0.1	0.1	3400
	(0.0)	--	(0.4)	(0.5)	(0.3)	(0.1)	(0.1)	(0.1)	
Ba	--	0.1	0.6	3.8	88.5	5.8	0.3	0.8	2497
	--	(0.1)	(0.2)	(0.4)	(0.6)	(0.5)	(0.1)	(0.2)	
B	--	--	0.1	0.2	6.0	86.1	4.1	3.5	1327
	--	--	(0.1)	(0.1)	(0.6)	(0.9)	(0.5)	(0.5)	
C	--	--	--	0.5	1.1	5.8	87.4	5.3	190
	--	--	--	(0.5)	(0.7)	(1.7)	(2.4)	(1.6)	
				Small Change Interest Rate Years					
Aaa	85.6	12.4	2.0	--	--	--	--	--	807
	(1.2)	(1.2)	(0.5)	--	--	--	--	--	
Aa	1.2	86.2	11.8	0.6	0.1	--	--	--	2009
	(0.2)	(0.8)	(0.7)	(0.2)	(0.1)	--	--	--	
A	0.0	2.9	89.8	6.4	0.6	0.2	--	0.0	3954
	(0.0)	(0.3)	(0.5)	(0.4)	(0.1)	(0.1)	--	(0.0)	
Baa	--	0.4	7.1	84.8	6.6	0.9	0.0	0.2	3010
	--	(0.1)	(0.5)	(0.7)	(0.5)	(0.2)	(0.0)	(0.1)	
Ba	0.0	0.0	0.4	6.7	82.9	8.0	0.7	1.1	2326
	(0.0)	(0.0)	(0.1)	(0.5)	(0.8)	(0.6)	(0.2)	(0.2)	
B	--	--	0.1	0.5	7.6	80.6	7.0	4.2	1039
	--	--	(0.1)	(0.2)	(0.8)	(1.2)	(0.8)	(0.6)	
C	--	--	--	0.7	3.3	7.8	81.0	7.2	153
	--	--	--	(0.7)	(1.4)	(2.2)	(3.2)	(2.1)	

Estimates are based on all Moody's notional, senior unsecured bond ratings between 31/12/70 and 31/12/97 measured on 31st December each year. We split the sample into three sets of 9 years in which interest rate *changes* are highest, lowest or in the middle and then estimate transition matrices for each sub-sample.

Table 3 also contains the mean difference in non-zero entries below the diagonal for pairs of transition matrices. These results suggest that small (i.e., negative) interest rate changes are associated with relatively many up-grades. This finding implies negative correlation between interest rate and spread risk. Also, years with high and low interest rate levels are associated with more up-grades than are medium interest rate years.

Table 3: Analysis of Transition Matrix Differences

Comparison	Num	Mn Ab Diff	Mn Diff LM	Med Ab Diff	Med Diff LM
High r - Low r	45	2.74	0.25	0.77	0.03
High r - Medium r	46	2.18	0.52	0.28	0.00
Low r - Medium r	44	1.34	0.25	0.63	0.01
Large Δr - Small Δr	44	1.93	-0.77	0.59	-0.19
Large Δr - Medium Δr	46	0.74	-0.31	0.19	-0.07
Small Δr - Medium Δr	45	1.36	0.38	0.43	0.01

Notes: Comparisons are between pairs of transition matrices calculated for different samples of years. The samples (each of 9 years are for high, low and medium interest rate levels and for large, small and medium changes in interest rates.

The column headings in the table are:

Num $\equiv$ number of cases in which a transition matrix entry is non-zero in one or other of the two matrices being compared.

Mn Ab Diff $\equiv$ mean absolute difference of entries Num non-zero entries.

Mn Diff LM $\equiv$ mean difference of lower off-diagonal non-zero entries.

Med Ab Diff $\equiv$ median absolute difference of entries Num non-zero entries.

Med Diff LM $\equiv$ median difference of lower off-diagonal non-zero entries.

3.3 VaR Calculations

To further elucidate the differences between the transition matrices reported in Table 2, we perform VaR calculations using a Creditmetrics framework. For a fuller account of such calculations, see Kiesel, Perraudin, and Taylor (2000). The VaRs are performed for portfolios of 500 credit exposures of equal size. We suppose that each exposure consisted of a 5-year pure discount bond issued by an obligor of a particular initial rating.

As in Kiesel, Perraudin, and Taylor (2000), the rating composition of the portfolios we examine is chosen to match the composition of real life bank

Table 4: VaRs for Portfolios Using Different Transition Matrices

Portfolio	VaR Confidence Level	Transition matrices for small Δr	medium Δr	large Δr
Investment	1%	1.77	1.00	1.34
	1/3%	0.82	0.47	0.60
High	1%	2.72	1.98	2.09
	1/3%	1.33	0.91	0.97
Average	1%	4.50	3.50	3.64
	1/3%	2.16	1.68	1.70

Note: The VaRs are calculated for portfolios of 500 credit exposures of equal size using the Creditmetrics approach. See JP Morgan (1997) or Kiesel, Perraudin, and Taylor (2000). The calculations are performed using three different rating transition matrices based on years in which interest rates increased by large, small or medium amounts. The VaRs are divided by the expected portfolio value and so are in %.

portfolios. Thus, our 'average' portfolio has the same rating composition as the aggregate portfolio of a sample of large US banks surveyed by the Federal Reserve Board (see Gordy (2000)). Our 'investment quality' portfolio has the a rating composition equal to that of the investment quality part of the 'average' portfolio. Finally, our 'high quality' portfolio has a composition equivalent to that of a sub-sample of high credit quality banks identified by the Federal Reserve in its survey.

We provide one-year VaR measures for the three portfolios at two confidence levels, 1% and 1/3%. The latter confidence level is closer to the levels commonly used by banks. The VaRs are effectively in percent of the portfolio value since we divide by the expected future value of the portfolio and multiply by 100.

The main finding that emerges from the VaRs in Table 4 is that portfolio VaRs for years of medium and large (generally positive) interest rate changes are similar whereas years in which interest rates fall sharply are associated with distinctly larger VaRs. This is consistent with negative correlation between interest rate and credit risk.

4 Conclusion

The results in this paper suggest that interest rate changes *are* correlated with changes in credit quality. For 2 and 5-year maturities, spreads changes are generally negatively associated with interest rate changes, although they are mostly less negatively correlated as the time horizon increases. For low, credit quality spreads, the correlation is substantially negative for a one-year horizon. For 10-year maturities (like those examined by Morris, Neal, and Rolph (1999)), correlations are negative for short horizons, but become positive for investment grade ratings as the horizon exceeds about six months. For sub-investment quality spreads, correlations are significantly negative for all horizons.

Turning to the relation between interest rates and rating transitions, we find that negative interest rate changes are associated with fewer up-grades. VaRs calculated with a transition matrix based on years of data in which interest rates have fallen are distinctly higher than VaRs based on transition matrices estimated from other years. Again, these findings suggest a negative correlation between interest rate and credit risk.

In general, our findings tend to confirm the results of Duffee (1998) who argues that the correlation between interest rates and risky bond returns are negative. His conclusions, however, are based on monthly data and he calculates correlations over monthly investment horizons. Morris, Neal, and Rolph (1999) argue that the negative correlation found by Duffee disappears if one considers longer time horizons. Our results are in line Duffee' s even though we are looking at a longer (one-year) horizon.

References

Artzner, P., F. Delbaen, J. Eber, and D. Heath (1997): 'Thinking Coherently,' *Risk* **10** (11).

Basel Committee on Banking Supervision (1996): 'Overview of the Amendment to the Capital Accord to Incorporate Market Risk,' unpublished mimeo, Bank for International Settlements, Basel.

——— (1999): 'Credit Risk Modelling: Current Practices and Applications,' Report 49, Bank for International Settlements, Basel, Switzerland.

Carty, L.V. (1997): 'Moody's Rating Migration and Credit Quality Correlation, 1920–1996,' Special comment, Moody's Investors Service, New York.

Cochrane, J.H. (1988): 'How Big Is the Random Walk in GNP?,' *Journal of Political Economy* **96** (5), 893–920.

Credit Suisse Financial Products (1997): 'Credit Risk+ : Technical Manual,' Discussion paper, CSFP.

Crouhy, M., D. Galai, and R. Mark (2000): 'A Comparative Analysis of Current Credit Risk Models,' *Journal of Banking and Finance* **4** (1–2), 59–117.

Das, S., and P. Tufano (1996): 'Pricing Credit Sensitive Debt When Interest Rates, Credit Ratings and Credit Spreads Are Stochastic,' *Journal of Financial Engineering* **5** 161–198.

Delianedis, G., and R. Geske (1998): 'Credit Risk and Risk Neutral Default Probabilities: Information about Ratings Migrations and Defaults,' Working paper, Anderson School, UCLA, Los Angeles.

Duffee, G.R. (1998): 'The Relationship Between Treasury Yields and Corporate Bond Yield Spreads,' *Journal of Finance* **53**, 2225–2242.

Gordy, M. (2000): 'A Comparative Anatomy of Credit Risk Models,' *Journal of Banking and Finance* **24** (1–2), 119–149.

Jarrow, R.A., and S.M. Turnbull (2000): 'The Intersection of Market and Credit Risk,' *Journal of Banking and Finance* **24** (1–2), 271–299.

JP Morgan (1997): *Creditmetrics-Technical Document.* JP Morgan, New York.

Kiesel, R., W.R. Perraudin, and A.P. Taylor (2000): 'The Structure of Credit Risk: Spread Volatility and Rating Transitions,' Working paper, Birkbeck College, London.

Kim, J., K. Ramaswamy, and S. Sundaresan (1993): 'Does Default Risk in Coupons Affect the Valuation of Corporate Bonds?: A Contingent Claims Model,' *Financial Management* **22** 117–131.

Longstaff, F.A., and E.S. Schwartz (1995): 'A Simple Approach to Valuing Risky and Floating Rate Debt,' *Journal of Finance* **50** 789–819.

Lucas, D.J., and J.G. Lonski (1992): 'Changes In Corporate Credit Quality 1970–1990,' *Journal of Fixed Income* 7–14.

Merton, R.C. (1974): 'On the Pricing of Corporate Debt: The Risk Structure of Interest Rates,' *Journal of Finance* **29** 449–70.

Morris, C., R. Neal, and D. Rolph (1999): 'Credit Spreads and Interest Rates,' mimeo, Indiana University, Indianapolis.

Nickell, P., W.R. Perraudin, and S. Varotto (2000): 'The Stability of Ratings Transitions,' *Journal of Banking and Finance* **24** (1–2), 1–14.

Nickell, P., W.R. Perraudin, and S. Varotto (2001): 'Ratings- Versus Equity-Based Credit Risk Models: An Empirical Study,' Working paper No. 132, Bank of England, London.

Perraudin, W.R., and A.P. Taylor (1999): 'On the Consistency of Ratings and Bond Market Spreads,' Working paper, Birkbeck College, London.

Wilson, T. (1997): 'Credit Risk Modelling: a New Approach,' unpublished mimeo, McKinsey Inc., New York.

Coherent Measures of Risk[1]

Philippe Artzner, Freddy Delbaen, Jean-Marc Eber and David Heath

Abstract

In this paper we study both market risks and non-market risks, without complete markets assumption, and discuss methods of measurement of these risks. We present and justify a set of four desirable properties for measures of risk, and call the measures satisfying these properties 'coherent'. We examine the measures of risk provided and the related actions required by SPAN, by the SEC/NASD rules and by quantile based methods. We demonstrate the universality of scenario-based methods for providing coherent measures. We offer suggestions concerning the SEC method. We also suggest a method to repair the failure of subadditivity of quantile-based methods.

1 Introduction

We provide in this paper a *definition* of risks (market risks as well as non-market risks) and present and justify a unified framework for the analysis, construction and implementation of *measures* of risk. We do not assume completeness of markets. These measures of risk can be used as (extra) capital requirements, to regulate the risk assumed by market participants, traders, insurance underwriters, as well as to allocate existing capital.

For these purposes, we:

(1) Define 'acceptable' future random net worths (see Section 2.1) and provide a set of axioms about the set of acceptable future net worths (Section 2.2);

(2) Define the measure of risk of an unacceptable position *once a reference, 'prudent', investment instrument has been specified*, as the minimum extra capital (see Section 2.3) which, invested in the reference instrument, makes the future value of the modified position become acceptable;

(3) State axioms on measures of risk and relate them to the axioms on acceptance sets. We argue that these axioms should hold for any risk measure which is to be used to effectively regulate or manage risks. We call risk measures which satisfy the four axioms *coherent*;

[1]First published in *Mathematical Finance* **9** 203–228 (1999). Reproduced with permission of Blackwell Publishers.

(4) Present, in Section 3, a (simplified) description of three existing methods for measuring market risk: the 'variance-quantile' method of value-at-risk (VaR), the margin system SPAN (Standard Portfolio Analysis of Risk) developed by the Chicago Mercantile Exchange, and the margin rules of the Securities and Exchanges Commission (SEC), which are used by the National Association of Securities Dealers (NASD);

(5) Analyze the existing methods in terms of the axioms and show that the last two methods are essentially the same (i.e., that when slightly modified they are mathematical duals of each other);

(6) Make a specific recommendation for the improvement of the NASD-SEC margin system (Section 3.2);

(7) Examine in particular the consequences of using value at risk for risk management (Section 3.3);

(8) Provide a general representation for all coherent risk measures in terms of 'generalized scenarios' (see Section 4.1), by applying a consequence of the separation theorem for convex sets already in the mathematics literature;

(9) Give conditions for extending into a coherent risk measure a measurement already agreed upon for a restricted class of risks (see Section 4.2);

(10) Use the representation results to suggest a specific coherent measure (see Section 5.1) called tail conditional expectation, as well as to give an example of construction of a coherent measure out of measures on separate classes of risks, for example credit risk and market risk (see Section 5.2).

(11) Our axioms are not restrictive enough to specify a unique risk measure. They instead characterize a large class of risk measures. The choice of precisely which measure to use (from this class) should presumably be made on the basis of additional economic considerations. Tail conditional expectation is, under some assumptions, the least expensive among these which are coherent and accepted by regulators since being more conservative than the value at risk measurement.

A non-technical presentation of part of this work is given in Artzner *et al.* (1997).

2 Definition of risk and of coherent risk measures

This section accomplishes the program set in (1), (2) and (3) above, in the presence of different regulations and different currencies.

2.1 Risk as the random variable: future net worth

Although several papers (including an earlier version of this one) define risk in terms of *changes* in values between two dates, we argue that because risk is related to the variability of the *future value* of a position, due to market changes or more generally to uncertain events, it is better to consider future values only. Notice indeed that there is no need for the initial costs of the components of the position to be determined from universally defined market prices (think of over-the-counter transactions). The principle of 'bygones are bygones' leads to this 'future wealth' approach.

The basic objects of our study shall therefore be the random variables on the set of states of nature at a future date, interpreted as possible future values of positions or portfolios currently held. A first, crude but crucial, measurement of the risk of a position will be whether its future value belongs or does not belong to the subset of *acceptable risks*, as decided by a supervisor like:

(a) a *regulator* who takes into account the unfavorable states when allowing a risky position which may draw on the resources of the government, for example as a guarantor of last resort;

(b) an *exchange's clearing firm* which has to make good on the promises to all parties, of transactions being securely completed;

(c) an investment *manager* who knows that his firm has basically given to its traders an exit option where the strike 'price' consists in being fired in the event of big trading losses on one's position.

In each case above, there is a trade-off between severity of the risk measurement, and level of activities in the supervised domain. The axioms and characterizations we shall provide do not single out a specific risk measure, and additional economic considerations have to play a role in the final choice of a measure.

For an *unacceptable risk* (i.e. a position with an unacceptable future net worth) one remedy may be to alter the position. Another remedy is to look for some commonly accepted instruments which, added to the current position, make its future value become acceptable to the regulator/supervisor.

The current cost of getting enough of this or these instrument(s) is a good candidate for a measure of risk of the initially unacceptable position.

For simplicity, we consider only one period of uncertainty $(0,T)$ between two dates 0 and T. The various currencies are numbered by i, $1 \leq i \leq I$ and, for each of them, one 'reference' instrument is given, which carries one unit of date 0 currency i into r_i units of date T currency i. Default free zero coupon bonds with maturity at date T may be chosen as particularly simple reference instruments in their own currency. Other possible reference instruments are mentioned in Section 2.3, right before the statement of Axiom T.

The period $(0,T)$ can be the period between hedging and rehedging, a fixed interval like two weeks, the period required to liquidate a position, or the length of coverage provided by an insurance contract.

We take the point of view of an investor subject to regulations and/or supervision in country 1. He considers a portfolio of securities in various currencies.

Date 0 exchange rates are supposed to be one, while e_i denotes the random number of units of currency 1 which one unit of currency i buys at date T.

An investor's initial portfolio consists of positions A_i, $1 \leq i \leq I$, (possibly within some institutional constraints, like the absence of short sales and a 'congruence' for each currency between assets and liabilities). The position A_i provides $A_i(T)$ units of currency i at date T. We call *risk* the investor's *future net worth* $\sum_{1 \leq i \leq I} e_i \cdot A_i(T)$.

Remark 2.1 The assumption of the position being held during the whole period can be relaxed substantially. In particular, positions may vary due to the agent's actions or those of counterparties. In general, we can consider the risk of following a strategy (which specifies the portfolio held at each date as a function of the market events and counterparties' actions) over an arbitrary period of time. Our current results in the simplified setting represent a first step.

2.2 Axioms on Acceptance Sets (Sets of Acceptable Future Net Worths)

We suppose that the set of all possible states of the world at the end of the period is known, but the probabilities of the various states occurring may be unknown or not subject to common agreement. When we deal with market risk, the state of the world might be described by a list of the prices of all securities and all exchange rates, and we assume that the set of all possible such lists is known. Of course, this assumes that markets at date T are liquid; if they are not, more complicated models are required, in which we can distinguish the risks of a position and of a future net worth, since, with illiquid markets, the mapping from the former to the latter may not be linear.

Notation

(a) We shall call Ω the set of states of nature, and assume it is finite. Considering Ω as the set of outcomes of an experiment, we compute the final net worth of a position for each element of Ω. It is a random variable denoted by X. Its negative part, $\max(-X, 0)$, is denoted by X^- and the supremum of X^- is denoted by $\|X^-\|$. The random variable identically equal to 1 is denoted by $\mathbf{1}$. The indicator function of state ω is denoted by $\mathbf{1}_{\{\omega\}}$.

(b) Let $\mathcal{G}$ be the set of all risks, that is the set of all real valued functions on Ω. Since Ω is assumed to be finite, $\mathcal{G}$ can be identified with R^n, where $n = card(\Omega)$. The cone of non-negative elements in $\mathcal{G}$ shall be denoted by L_+, its negative by L_-.

(c) We call $\mathcal{A}_{i,j}, j \in J_i$, a set of final net worths, expressed in currency i, which, in country i, are accepted by regulator/supervisor j.

(d) We shall denote $\mathcal{A}_i$ the intersection $\bigcap_{j \in J_i} \mathcal{A}_{i,j}$ and use the generic notation $\mathcal{A}$ in the listing of axioms below.

We shall now state axioms for acceptance sets. Some have an immediate interpretation while the interpretation of the third one will be more easy in terms of risk measure (see Axiom S in Section 2.3.) The rationale for Axioms 2.1 and 2.2 is that a final net worth which is always nonnegative does not require extra capital, while a net worth which is always (strictly) negative certainly does.

Axiom 2.1 *The acceptance set $\mathcal{A}$ contains L_+.*

Axiom 2.2 *The acceptance set $\mathcal{A}$ does not intersect the set L_{--} where*

$$L_{--} = \{X \mid \textit{ for each } \omega \in \Omega \;,\; X(\omega) < 0\}.$$

It will also be interesting to consider a stronger axiom:

Axiom 2.2′ *The acceptance set $\mathcal{A}$ satisfies $\mathcal{A} \cap L_- = \{0\}$.*

The next axiom reflects risk aversion on the part of the regulator, exchange director or trading room supervisor.

Axiom 2.3 *The acceptance set $\mathcal{A}$ is convex.*

A less natural requirement on the set of acceptable final net worths is stated in the next axiom.

Axiom 2.4 *The acceptance set $\mathcal{A}$ is a positively homogeneous cone.*

2.3 Correspondence between Acceptance Sets and Measures of Risk

Sets of acceptable future net worths are the primitive objects to be considered in order to describe acceptance or rejection of a risk. We present here how, *given* some 'reference instrument', there is a natural way to define a measure of risk by describing how close or how far from acceptance a position is.

Definition 2.1 A measure of risk is a mapping from $\mathcal{G}$ into R.

In Section 3 we shall speak of a *model-dependent* measure of risk when an explicit probability on Ω is used to construct it (see e.g. Sections 3.1 and 3.3), and of a *model-free* measure otherwise (see e.g. Section 3.2). Model-free measures can still be used in the case where only risks of positions are considered.

When positive, the number $\rho(X)$ assigned by the measure ρ to the risk X will be interpreted (see Definition 2.2 below) as the minimum extra cash the agent has to *add* to the risky position X, *and* to invest 'prudently', that is in the reference instrument, to be allowed to proceed with his plans. If it is negative, the cash amount $-\rho(X)$ can be withdrawn from the position, or received as restitution as in the case of organized markets for financial futures.

Remark 2.2 The reader may be surprised that we define a measure of risk on the *whole* of $\mathcal{G}$. Why, in particular, should we consider a risk, a final net worth, like the constant $-\mathbf{1}$? No one would or could willingly enter into a deal which for sure entails a negative of final net worth equal to 1! Let us provide three answers:

(a) We want to extend the accounting procedures dealing with future certain bad events (like loss in inventories, degradation [wear and tear] of physical plant), into measurement procedures for future uncertain bad events;

(b) Actual measurements used in practice seem to be indeed defined only for risks where *both* states with positive and states with negative final net worth exist. Section 4.2 shows that, under well-defined conditions, they can be extended without ambiguity to measurements for *all* functions in $\mathcal{G}$;

(c) Multiperiod models may naturally introduce at some intermediate date the prospect of such final net worths.

Remark 2.3 It has been pointed out to us that describing risk 'by a single number' involves a great loss of information. However, the actual decision about taking a risk or allowing one to take it is fundamentally binary, of the 'yes or no' type, and we claimed at the beginning of Section 2.1 that this is the actual origin of risk measurement.

Remark 2.4 The expression 'cash' deserves some discussion in the case of a publicly traded company. It refers to an increase in equity. The amount $\rho(X)$ may, for example, be used to lower the amount of debt in the balance sheet of the company.

We define a correspondence between acceptance sets and measures of risk.

Definition 2.2 *Risk measure associated to an acceptance set.* Given the total rate of return r on a reference instrument, the risk measure associated to the acceptance set $\mathcal{A}$ is the mapping from $\mathcal{G}$ to $\mathbb{R}$ denoted by $\rho_{\mathcal{A},r}$ and defined by

$$\rho_{\mathcal{A},r}(X) = \inf\{m \mid m \cdot r + X \in \mathcal{A}\}.$$

Remark 2.5 Acceptance sets allow us to address a question of importance to an international regulator and to the risk manager of a multinational firm, namely the invariance of acceptability of a position with respect to a change of currencies. If, indeed, we have for each currency $i, 1 \leq i \leq I$, $e_i \cdot \mathcal{A}_i = \mathcal{A}_1$ then, for each position providing an acceptable future net worth X in currency i, the same position provides a future net worth $e_i/e_j \cdot X$ in currency j, which is also acceptable. The situation is more complex for unacceptable positions. If a position requires an extra initial cash of $\rho_{\mathcal{A}_i,r_i}(X)$ units to be invested in the ith reference instrument, it is not necessarily true that this amount is equal to the number $\rho_{\mathcal{A}_j,r_j}(X)$ of initial units deemed sufficient by the regulation(s) in country j, if invested in the jth reference instrument, even though we supposed the initial exchange rate to be 1.

Definition 2.3 *Acceptance set associated to a risk measure.* The acceptance set associated to a risk measure ρ is the set denoted by $\mathcal{A}_\rho$ and defined by

$$\mathcal{A}_\rho = \{X \in \mathcal{G} \mid \rho(X) \leq 0\}.$$

We consider now several possible properties for a risk measure ρ defined on $\mathcal{G}$. They will be related, in Section 2.4, to the axioms stated above concerning acceptance sets. For clarity we label the new axioms with letters.

The first requirement ensures that the risk measure is stated in the same units as the final net worth, except for the use of the reference instrument. This particular asset is modeled as having the initial price 1 and a strictly positive price r (or *total return*) in any state of nature at date T. It is the regulator's (supervisor's) responsibility to accept for r possible random values as well as values smaller than 1.

Axiom T means that adding (resp. subtracting) the sure initial amount α to the initial position and investing it in the reference instrument, simply decreases (resp. increases) the risk measure by α.

Axiom T Translation invariance. *For all $X \in \mathcal{G}$ and all real numbers α, we have $\rho(X + \alpha \cdot r) = \rho(X) - \alpha$.*

Remark 2.6 Axiom T ensures that, for each X, $\rho(X + \rho(X) \cdot r) = 0$. This equality has a natural interpretation in terms of the acceptance set associated to ρ (see Definition 2.3 above).

Remark 2.7 By insisting on references to cash and to time, Axiom T clearly indicates that our approach goes much farther than the interpretation given by Wang of an earlier version of this paper: Wang (1996), page 3, indeed claims that 'the main function of a risk measure is to properly rank risks.'

Axiom S Subadditivity: for all X_1 and $X_2 \in \mathcal{G}$, $\rho(X_1+X_2) \leq \rho(X_1)+\rho(X_2)$.

We contend that this property, which could be stated in the following brisk form 'a merger does not create extra risk,' is a natural requirement:

(a) If an exchange's risk measure were to fail to satisfy this property, then, for example, an individual wishing to take the risk $X_1 + X_2$ may open two accounts, one for the risk X_1 and the other for the risk X_2, incurring the smaller margin requirement of $\rho(X_1) + \rho(X_2)$, a matter of concern for the *exchange*.

(b) If a firm were forced to meet a requirement of extra capital which did not satisfy this property, the firm might be motivated to break up into two separately incorporated affiliates, a matter of concern for the *regulator*.

(c) Bankruptcy risk inclines *society* to require less capital from a group without 'firewalls' between various business units than it does require when one 'unit' is protected from liability attached to failure of another 'unit'.

(d) Suppose that two desks in a firm compute, in a decentralized way, the measures $\rho(X_1)$ and $\rho(X_2)$ of the risks they have taken. If the function ρ is subadditive, the *supervisor* of the two desks can count on the fact that $\rho(X_1)+\rho(X_2)$ is a feasible guarantee relative to the global risk X_1+X_2. If indeed he has an amount m of cash available for their joint business, he knows that imposing limits m_1 and m_2 with $m = m_1 + m_2$, allows him to decentralise his cash constraint into two cash constraints, one per desk. Similarly, the firm can allocate its capital among managers.

Axiom PH Positive homogeneity. *For all $\lambda \geq 0$ and all $X \in \mathcal{G}$, $\rho(\lambda X) = \lambda\rho(X)$.*

Remark 2.8 If position size directly influences risk (for example, if positions are large enough that the time required to liquidate them depend on their sizes) then we should consider the consequences of lack of liquidity when computing the future net worth of a *position*. With this in mind, Axioms S and PH about mappings from *random variables* into the reals, remain reasonable.

Remark 2.9 Axiom S implies that $\rho(nX) \leq n\rho(X)$ for $n = 1, 2, \ldots$. In Axiom PH we have imposed the reverse inequality (and require equality for all positive λ) to model what a government or an exchange might impose in a situation where no netting or diversification occurs, in particular because the government does not prevent many firms to all take the same position.

Remark 2.10 Axioms T and PH imply that for each α, $\rho(\alpha \cdot (-r)) = \alpha$.

Axiom M Monotonicity. *For all X and $Y \in \mathcal{G}$ with $X \leq Y$, we have $\rho(Y) \leq \rho(X)$.*

Remark 2.11 Axiom M rules out the risk measure defined by $\rho(X) = -\mathbf{E}_{\mathbb{P}}[X] + \alpha \cdot \sigma_{\mathbb{P}}(X)$, where $\alpha > 0$ and where $\sigma_{\mathbb{P}}$ denotes the standard deviation operator, computed under the probability $\mathbb{P}$. Axiom S rules out the 'semi-variance' type risk measure defined by $\rho(X) = -\mathbf{E}_{\mathbb{P}}[X] + \sigma_{\mathbb{P}}((X - \mathbf{E}_{\mathbb{P}}[X])^-)$.

Axiom R Relevance. *For all $X \in \mathcal{G}$ with $X \leq 0$ and $X \neq 0$, we have $\rho(X) > 0$.*

Remark 2.12 This axiom is clearly necessary, but not sufficient, to prevent concentration of risks to remain undetected (see Section 4.3.)

We notice that for $\lambda > 0$, Axioms S, PH, M and R remain satisfied by the measure $\lambda \cdot \rho$, if satisfied by the measure ρ. It is not the case for Axiom T.

The following choice of required properties will define coherent risk measures.

Definition 2.4 *Coherence.* A risk measure satisfying the four axioms of translation invariance, subadditivity, positive homogeneity, and monotonicity, is called coherent.

2.4 Correspondence between the Axioms on Acceptance Sets and the Axioms on Measures of Risks

The reader has certainly noticed that we claimed the acceptance set to be the fundamental object, and discussed the axioms mostly in terms of the associated risk measure. The following propositions show that this was reasonable.

Proposition 2.1 *If the set $\mathcal{B}$ satisfies Axioms 2.1, 2.2, 2.3 and 2.4, the risk measure $\rho_{\mathcal{B},r}$ is coherent. Moreover $\mathcal{A}_{\rho_{\mathcal{B},r}} = \bar{\mathcal{B}}$, the closure of $\mathcal{B}$.*

Proof

(1) Axioms 2.2 and 2.3 ensure that for each X, $\rho_{\mathcal{B},r}(X)$ is a finite number.

(2) The equality $\inf\{p \mid X + (\alpha + p) \cdot r \in \mathcal{B}\} = \inf\{q \mid X + q \cdot r \in \mathcal{B}\} - \alpha$ proves that $\rho_{\mathcal{B},r}(X + r \cdot \alpha) = \rho(X) - \alpha$, and Axiom T is satisfied.

(3) The subadditivity of $\rho_{\mathcal{B}}$ follows from the fact that if $X + m \cdot r$ and $Y + n \cdot r$ both belong to $\mathcal{B}$, so does $X + Y + (m + n) \cdot r$ as Axioms 2.3 and 2.4 show.

(4) If $m > \rho_{\mathcal{B},r}(X)$ then for each $\lambda > 0$ we have $\lambda \cdot X + \lambda \cdot m \cdot r \in \mathcal{B}$, by Definition 2.3 and Axiom 2.4, and this proves that $\rho_{\mathcal{B},r}(\lambda \cdot X) \leq \lambda \cdot m$. If $m < \rho_{\mathcal{B},r}(X)$, then for each $\lambda > 0$ we have $\lambda \cdot X + \lambda \cdot m \cdot r \notin \mathcal{B}$, and this proves that $\rho_{\mathcal{B},r}(\lambda \cdot X) \geq \lambda \cdot m$. We conclude that $\rho_{\mathcal{B},r}(\lambda \cdot X) = \lambda \cdot \rho_{\mathcal{B},r}(X)$.

(5) Monotonicity of $\rho_{\mathcal{B},r}$ follows from the fact that if $X \leq Y$ and $X + m \cdot r \in \mathcal{B}$ then $Y + m \cdot r \in \mathcal{B}$ by use of Axioms 2.3 and 2.1, and of Definition 2.3.

(6) For each $X \in \mathcal{B}$, $\rho_{\mathcal{B},r}(X) \leq 0$ hence $X \in \mathcal{A}_{\rho_{\mathcal{B},r}}$. Proposition 2.2 and points (1) through (5) above ensure that $\mathcal{A}_{\rho_{\mathcal{B},r}}$ is closed, which proves that $\mathcal{A}_{\rho_{\mathcal{B},r}} = \bar{\mathcal{B}}$.

Proposition 2.2 *If a risk measure ρ is coherent, then the acceptance set $\mathcal{A}_\rho$ is closed and satisfies Axioms 2.1, 2.2, 2.3 and 2.4. Moreover $\rho = \rho_{\mathcal{A}_\rho,r}$.*

Proof

(1) Subadditivity and positive homogeneity ensure that ρ is a convex function on $\mathcal{G}$, hence continuous, and that the set $\mathcal{A}_\rho = \{X \mid \rho(X) \leq 0\}$ is a closed, convex and homogeneous cone.

(2) Positive homogeneity implies that $\rho(0) = 0$. Together with monotonicity this ensures that the set $\mathcal{A}_\rho$ contains the positive orthant L_+.

(3) Let X be in L_{--} with $\rho(X) < 0$. Axiom M ensures that $\rho(0) < 0$, a contradiction. If $\rho(X) = 0$, then we find $\alpha > 0$ such that $X + \alpha \cdot r \in L_{--}$, which provides, by use of Axiom T, the relation $-\alpha \geq 0$, a contradiction. Hence $\rho(X) > 0$, that is $X \notin \mathcal{A}_\rho$, which establishes Axiom 2.2.

(4) For each X, let δ be any number with $\rho_{\mathcal{A}_\rho,r}(X) < \delta$. Then $X + \delta \cdot r \in \mathcal{A}_\rho$, hence $\rho(X + \delta \cdot r) \leq 0$, hence $\rho(X) \leq \delta$, which proves that $\rho(X) \leq \rho_{\mathcal{A}_\rho,r}(X)$, that is $\rho \leq \rho_{\mathcal{A}_\rho,r}$.

(5) For each X, let δ be any number with $\delta > \rho(X)$. Then $\rho(X + \delta \cdot r) < 0$ and $X + \delta \cdot r \in \mathcal{A}_\rho$, hence $\rho_{\mathcal{A}_\rho,r}(X + \delta \cdot r) \leq 0$. This proves that $\rho_{\mathcal{A}_\rho,r}(X) \leq \delta$ and that $\rho_{\mathcal{A}_\rho,r}(X) \leq \rho(X)$, hence $\rho_{\mathcal{A}_\rho,r} \leq \rho$.

Proposition 2.3 *If a set $\mathcal{B}$ satisfies Axioms 2.1, 2.2′, 2.3 and 2.4, then the coherent risk measure $\rho_{\mathcal{B},r}$ satisfies the relevance axiom. If a coherent risk measure ρ satisfies the relevance axiom, then the acceptance set $\mathcal{A}_{\rho_{\mathcal{B}},r}$ satisfies Axiom 2.2′.*

Proof

(1) For an X like in the statement of Axiom R we know that $X \in L_-$ and $X \neq 0$, hence, by Axiom 2.2′, $X \notin \mathcal{B}$, which means $\rho_{\mathcal{B},r}(X) > 0$.

(2) For $X \in L_-$ and $X \neq 0$ Axiom R provides $\rho(X) > 0$ and $X \notin \mathcal{B}$.

3 Three Currently Used Methods of Measuring Market Risk

In this section, we give a (simplified) description of three currently used methods of measuring market risk:

1. SPAN (1995) developed by the Chicago Mercantile Exchange;
2. the Securities Exchange Commission (SEC) rules used by the National Association of Securities Dealers (see NASD (1996) and Federal Reserve System (1994)), similar to rules used by the Pacific Exchange and the Chicago Board of Options Exchange;
3. the quantile-based Value at Risk (or VaR) method BCBS (1996), Dowd (1998), Duffie and Pan (1997), Derivatives Policy Group (1995), Risk Magazine (1996), RiskMetrics (1995).

We examine the relationship of these three methods with the abstract approach provided in Section 2. We also suggest slightly more general forms for some of the methods. It will be shown that the distinction made above between model-free and model-dependent measures of risk *actually* shows up.

3.1 An Organized Exchange's Rules: The SPAN Computations

To illustrate the SPAN margin system SPAN (1995) (see also MATIF (1993), pages 7–8), we consider how the initial margin is calculated for a simple portfolio consisting of units of a futures contract and of several puts and calls with a common expiration date on this futures contract. The SPAN margin for such a portfolio is computed as follows: First, fourteen 'scenarios' are considered. Each scenario is specified by an up or down move of volatility combined with no move, or an up move, or a down move of the futures price by 1/3, 2/3 or 3/3 of a specified 'range.' Next, two additional scenarios relate to 'extreme' up or down moves of the futures price. The measure of risk is the *maximum* loss incurred, using the full loss for the first fourteen scenarios and only 35% of the loss for the last two 'extreme' scenarios. A specified model, typically the Black model, is used to generate the corresponding prices for the options under each scenario.

The calculation can be viewed as producing the maximum of the expected loss under each of sixteen probability measures. For the first fourteen scenarios the probability measures are point masses at each of the fourteen points in the space Ω of securities prices. The cases of extreme moves correspond to taking the convex combination $(0.35, 0.65)$ of the losses at the 'extreme move' point under study and at the 'no move at all' point (i.e., prices remain the same). We shall call these probability measures 'generalized scenarios'.

The account of the investor holding a portfolio is required to have sufficient current net worth to support the maximum expected loss. If it does not, then extra cash is required as margin call, in an amount equal to the 'measure of risk' involved. This is completely in line with our interpretation of Definition 2.3.

The following definition generalizes the SPAN computation and presents it in our framework:

Definition 3.1 The risk measure defined by a non-empty set $\mathcal{P}$ of probability measures or 'generalized scenarios' on the space Ω and the total return r on a reference instrument, is the function $\rho_{\mathcal{P}}$ on $\mathcal{G}$ defined by

$$\rho_{\mathcal{P}}(X) = \sup\{\mathbf{E}_{\mathbb{P}}[-X/r] \mid \mathbb{P} \in \mathcal{P}\}.$$

The scenario-based measures from Definition 3.1 are coherent risk measures:

Proposition 3.1 *Given the total return r on a reference instrument and the non-empty set $\mathcal{P}$ of probability measures, or 'generalized scenarios', on the set Ω of states of the world, the risk measure $\rho_{\mathcal{P}}$ of Definition 3.1 is a coherent risk measure. It satisfies the relevance axiom if and only if the union of the supports of the probabilities $\mathbb{P} \in \mathcal{P}$ is equal to the set Ω.*

Proof Axioms PH and M ensure that a coherent risk measure satisfies Axiom R if and only if the negative of each indicator function $\mathbf{1}_{\{\omega\}}$ has a (strictly) positive risk measure. This is equivalent to the fact that any state belongs to at least one of the supports of the probabilities found in the set $\mathcal{P}$.

Section 4.1 shows that *each* coherent risk measure is obtained by way of scenarios.

3.2 Some Model-Free Measures of Risks: the SEC rules on Final Net Worth

The second example of a risk measure used in practice is found in the rules of the Securities and Exchange Commission and the National Association of Securities Dealers. Their common approach is to consider portfolios as formal *lists* of securities and impose 'margin' requirements on *them*, in contrast to the SPAN approach which takes the random *variables* — gains and losses of the portfolios of securities — as basic objects to measure. In the terminology of BCBS (1996) we have here something similar to a 'standardized measurement method'.

Certain spread positions like a long call and a short call of higher exercise price on the same underlying stock, both calls having same maturity date, are described in NASD (1996), page 8133, SEC rule 15c3–1a,(11), as requiring no margin (no 'deduction'). No justification is given for this specification. We shall use the paper Rudd and Schroeder (1982) as the basis for explaining, for a simple example, the computation of margin according to these common rules.

Let A be a portfolio consisting of two long calls with strike 10, two short calls with strike 20, three short calls with strike 30, four long calls with strike 40 and one short call with strike 50. For simplicity assume all calls European and exercise dates equal to the end of the holding period. A simple graph shows that the final value of this position is never below -10, which should entail a margin deposit of *at most* 10.

Under the SEC method, the position A is represented or 'decomposed' as a portfolio of long call spreads. No margin is required for a spread if the strike of the long side is less than the strike of the short side. A margin of $K - H$ is required for the spread consisting of a long call with strike K and a short call with strike H, when $H \leq K$. The margin resulting from a representation or 'decomposition' is the sum of the margins attached to each call spread. The investor is presumably able to choose the best possible representation. A simple linear programming computation will show that 30 is the resulting minimum, that is much more than the negative of the worst possible future value of the position!

Remark 3.1 This 30 seems to be the result of an attempt to bound the largest payout which the investor might have to make at the end of the period. In this method, the current value of his account must be at least as large as the current value of the calls plus 30.

Remark 3.2 A careful reading of the SEC rules reveals that one must:

(i) first mark the account (reference instruments *plus* calls) to market;

(ii) deduct the market value of the calls (long or short),

(iii) then deduct the various 'margins' required for the spreads in the chosen decomposition (we shall call the total as the 'margin');

(iv) and then check that this is at least 0.

In the framework of Definition 2.3, this bears some analogy to

(i) marking to market *both* the positions in the 'risky' instruments as well as in the reference one,

(ii) subtract the market value of the risky part,

(iii) make sure that the difference is positive.

We now formalize the special role played by the call spreads, which we call 'standard risks,' and the natural margin requirements on them in the SEC rules approach to risk measurement, following the lines of Rudd and Schroeder (1982), Section 4 (see also Cox and Rubinstein (1985), pages 107–109). Given some underlying security, we denote by C_K the European call with exercise price K and exercise date equal to the end of the holding period, and by $S_{H,K}$ the spread portfolio consisting of 'one long C_H, one short C_K', which we denote by $C_H - C_K$. These spreads shall be 'standard risks' for which a simple rule of margin requirement is given. They are then used to 'support' general portfolios of calls and provide conservative capital requirements.

We describe the extra capital requirement for a portfolio A consisting of a_H calls $C_H, H \in \mathcal{H}$, $\mathcal{H}$ a finite set of strikes. For simplicity we assume that $\sum_H a_H = 0$, i.e., we have no net long or short position. The exchange allows one to compute the margin for such a portfolio A by solving the linear programming problem:

$$\inf_{n_{H,K}} \sum_{H,K,H \neq K} n_{H,K}(H-K)^+ \tag{3.1}$$

under the conditions that

$$\text{for all } H, K, H \neq K, \text{ we have } n_{H,K} \geq 0 \text{ and } A = \sum_{H,K,H \neq K} n_{H,K} S_{H,K}.$$

This program provides the holder of portfolio A with the cheapest decomposition ensuring that *each* spread showing in it has a non-negative net worth at date T.

Going one step farther than Rudd and Schroeder (pages 1374–1376) we write explicitly the dual program:

$$\sup_{\nu_K} \sum_K \nu_K a_K, \tag{3.2}$$

where the sup is taken over all (ν_K) satisfying: $\nu_H - \nu_K \leq (H-K)^+$.

For the interpretation of this *dual* problem, we rewrite the preceding program with the negative π_K of the dual variables, getting

$$\inf_{\pi_K} \sum_K \pi_K a_K \tag{3.3}$$

under the conditions that

$$\pi_H - \pi_K \geq -(H - K)^+$$

or

$$\pi_H - \pi_K \geq 0 \text{ if } H < K$$

and

$$\pi_H - \pi_K \geq K - H \text{ if } H > K,$$

the last inequalities being rewritten as

$$\pi_K - \pi_H \leq H - K \text{ if } H > K. \tag{3.4}$$

Notice that if we interpret π_H as the cash flows associated with the call C_H at expiration date T, the objective function in (3.3) is the cash flow of the portfolio A at expiration. The duality theorem of linear programming ensures that the worst payout to the holder of portfolio A, under all scenarios satisfying the constraints specified in problem (3.3), cannot be larger than the lowest margin accepted by the exchange. The exchange is therefore sure than the investor commitments will be fulfilled.

It is remarkable that the primal problem (3.1) did not seem to refer to a model of distribution for future prices of the call. Yet the duality results in an implicit set of states of nature consisting of call prices, with a surprise! Our example of portfolio A in the beginning of this Section has shown indeed that the exchange is, in some way, *too* secure, as we now explain.

That the cash flows of the calls must satisfy the constraints (3.4) specified for problem (3.3) (and indeed many other constraints such as convexity as a function of strike, see Merton (1973), Theorem 8.4) is well known. For the specific portfolio A studied in Section 3.2, the set of strikes is $\mathcal{H} = \{10, 20, 30, 40, 50\}$, and an optimal primal solution is given by $n^*_{10,20} = 2$, $n^*_{40,50} = 1$, $n^*_{40,30} = 3$, all others $n^*_{H,K} = 0$, for a minimal margin of 30. The cash flows are given by $\pi^*_{10} = \pi^*_{20} = \pi^*_{30} = 10$ and $\pi^*_{40} = \pi^*_{50} = 0$, which provides the value -30 for the minimal cash flow of the portfolio at expiration. However, this minimal cash flow corresponds to cash flows for the individual options which cannot arise for any stock price. Indeed, if the stock price at expiration is S, the cash flow of C_H is $(S - H)^+$, which is obviously convex in H. Thus since $\pi^*_{20} + \pi^*_{40} < 2\pi^*_{30}$, these $\pi's$ cannot arise as cash flows for any terminal stock price. Briefly, there are *too many* scenarios considered, because some of them are *impossible* scenarios.

The convexity of the call price as function of the strike can be derived from the fact that a long 'butterfly' portfolio as $B_{20} = C_{10} - 2C_{20} + C_{30}$ must have a positive price. Therefore, we submit this butterfly to the decomposition method and write it as a sum of spreads $S_{10,20} + S_{30,20}$, which requires a margin of 10. If we instead take the approach of Section 2, looking at random variables, more precisely at the random net worth at the end of the holding period, we realize that the butterfly never has negative net worth, or, equivalently, that the net loss it can suffer is never larger than its initial net worth. The butterfly portfolio should therefore be margin free, which would imply a margin of only 10 for the original portfolio $A = 2B_{20} + 2B_{30} - B_{40}$. In our opinion it is not coherent, in this setting, to have only the spreads $S_{H,K}$ (for $H \leq K$) as margin free portfolios. The method uses *too few* standard risks.

In Section 4.2 we present a framework for extensions of risk measurements of 'standard risks' and give conditions under which our construction actually produces coherent measures of risk. The results of Section 4.1 on scenario representation of coherent measures will allow to interpret the extension in terms of scenarios attached to the original measurement.

3.3 Some Model-Dependent Rules based on Quantiles

The last example of measures of risk used in practice is the 'Value at Risk' (or VaR) measure. It is usually defined in terms of net wins or P/L and therefore ignores the difference between money at one date and money at a different date, which, for small time periods and a single currency, may be acceptable. It uses quantiles, which requires us to pay attention to discontinuities and *intervals* of quantile numbers.

Definition 3.2 *Quantiles.* Given $\alpha \in \,]0,1[$, the number q is an α-quantile of the random variable X under the probability distribution $\mathbb{P}$ if one of the three equivalent properties below is satisfied:

(i) $\mathbb{P}[X \leq q] \geq \alpha \geq \mathbb{P}[X < q]$;

(ii) $\mathbb{P}[X \leq q] \geq \alpha$ and $\mathbb{P}[X \geq q] \geq 1 - \alpha$;

(iii) $F_X(q) \geq \alpha$ and $F_X(q-) \leq \alpha$ with $F_X(q_-) = \lim_{x \to q, x<q} F(x)$, where F_X is the cumulative distribution function of X.

Remark The set of such α-quantiles is a closed interval. Since Ω is finite, there is a finite left-(resp. right-) end point q_α^- (resp. q_α^+) which satisfies $q_\alpha^- = \inf\{x \mid \mathbb{P}[X \leq x] \geq \alpha\}$ [equivalently $\sup\{x \mid \mathbb{P}[X \leq x] < \alpha\}$] (resp. $q_\alpha^+ = \inf\{x \mid \mathbb{P}[X \leq x] > \alpha\}$). With the exception of at most countably many α the equality $q_\alpha^- = q_\alpha^+$ holds. The quantile q_α^- is the number $F^{\leftarrow}(\alpha) = \inf\{x \mid \mathbb{P}\{X \leq x\} \geq \alpha\}$ defined in Embrechts *et al.* (1997) Definition 3.3.5 (see also Duffie and Pan (1997)).

We formally define VaR in the following way:

Definition 3.3 *Value at risk measurement.* Given $\alpha \in]0,1[$, and a reference instrument r, the value-at-risk VaR_α at level α of the final net worth X with distribution $\mathbb{P}$, is the negative of the quantile q_α^+ of X/r, that is

$$\mathrm{VaR}_\alpha(X) = -\inf\{x \mid \mathbb{P}[X \leq x \cdot r] > \alpha\}.$$

Remark 3.4 Notice that what we are using for defining VaR_α is really the amount of additional capital that a VaR_α type calculation entails.

Remark 3.5 We have here what is called an 'internal' model in BCBS (1996), and it is not clear whether the (estimated) physical probability or a 'well-chosen' subjective probability should be used.

We will now show that, while satisfying properties T, PH and M, VaR_α *fails* to satisfy the subadditivity property.

Consider as an example, the following two digital options on a stock, with the same exercise date T, the end of the holding period. The first option denoted by A (initial price u) pays 1000 if the value of the stock at time T is more than a given U, and nothing otherwise, while the second option denoted by B (initial price l) pays 1000 if the value of the stock at T is less than L (with $L < U$), and nothing otherwise.

Choosing L and U such that $\mathbb{P}\{S_T < L\} = \mathbb{P}\{S_T > U\} = 0.008$ we look for the 1% values at risk of the future net worths of positions taken by two traders writing respectively 2 options A and 2 options B. They are $-2 \cdot u$ and $-2 \cdot l$ respectively (r supposed to be one). By contrast, the positive number $1000 - l - u$ is the 1% value at risk of the future net worth of the position taken by a trader writing $A + B$. This implies that the set of acceptable net worths (in the sense of Definition 2.4 applied to the value at risk measure) is not convex. Notice that this is even a worse feature than the non-subadditivity of the measurement. We give below one more example of non-subadditivity.

Remark 3.6 We note that if quantiles are computed under a distribution for which all prices are jointly normally distributed, then the quantiles do satisfy subadditivity as long as probabilities of excedence are smaller than 0.5. Indeed, $\sigma_{X+Y} \leq \sigma_X + \sigma_Y$ for each pair (X, Y) of random variables. Since for a normal random variable X we have

$$\mathrm{VaR}_\alpha(X) = -(\mathbf{E}_{\mathbb{P}}[X] + \Phi^{-1}(\alpha) \cdot \sigma_{\mathbb{P}}(X)),$$

with Φ the cumulative standard normal distribution and since $\Phi^{-1}(0.5) = 0$, the proof of subadditivity follows.

Remark 3.7 Several works on quantile-based measures (see Dowd (1998), Risk Magazine (1996), RiskMetrics (1995)) consider mainly the *computational* and *statistical* problems they raise, without first considering the implications of this method of measuring risks.

Remark 3.8 Since the beginning of this century, casualty actuaries have been involved in computation and use of quantiles. The choice of initial capital controls indeed the probability of ruin at date T. Loosely speaking, 'ruin' is defined in (retrospective) terms by the negativity, at date T, of the *surplus*, defined to be:

$$\begin{aligned} Y &= \text{capital at date } 0 + \text{premium received} \\ &\qquad - \text{claims paid (from date 0 to date } T). \end{aligned}$$

Imposing an upper bound $1 - \alpha$ on the probability of Y being negative determines the initial capital via a quantile calculation (for precise information, see the survey article Bühlmann (1990)).

Under some circumstances, related to Remark 3.6 above, (see Daykin *et al.* (1994), pages 157, 168), this 'capital at risk' is a measure which possesses the subadditivity property. For some models the surplus represents the net worth of the insurance firm at date T. In general, the difficulty of assigning a market value to insurance liabilities forces us to distinguish surplus and net worth.

Remark 3.9 We do not know of organized exchanges using value at risk as the basis of risk measurement for margin requirements.

For a second example of non-subadditivity, briefly allow an infinite set Ω and consider two *independent* identically distributed random variables X_1 and X_2 having the same density 0.90 on the interval $[0, 1]$, the same density 0.05 on the interval $[-2, 0]$. Assume that each of them represents a future random net worth with positive expected value, that is a possibly interesting risk. Yet, in terms of quantiles, the 10% values at risk of X_1 and X_2 being equal to 0, whereas an easy calculation showing that the 10% value at risk of $X_1 + X_2$ is certainly larger than 0, we conclude that the individual controls of these risks do not allow directly a control of their sum, if we were to use the 10% value at risk.

Value at risk measurement also *fails* to recognise *concentration* of risks. A remarkably simple example concerning credit risk is due to Claudio Albanese (see Albanese (1997)). Assume that the base rate of interest is zero, and that the spreads on all corporate bonds is 2%, while these bonds default, independently from company to company, with a (physical) probability of 1%. If an amount of $1,000,000$ borrowed at the base rate is invested in the bonds of a single company, the 5% value at risk of the resulting position is negative, namely $-20,000$, and there is 'no risk'.

If, in order to diversify, the whole amount is invested equally into bonds of one hundred different companies, the following happens in terms of value at risk. Since the probability of at least two companies defaulting is greater than 0.18 it follows that the portfolio of bonds leads to a negative future net worth with a probability greater than 0.05: *diversification* of the original portfolion has *increased* the measure of risk, while the 'piling-up' of risky bonds issued by the same company had remained undetected. We should not rely on such 'measure'.

Value at risk also *fails* to encourage a reasonable *allocation* of risks among agents, as can be seen from the following simple example. Let Ω consists of three states $\omega_1, \omega_2, \omega_3$ with respective probabilities $0.94, 0.03, 0.03$. Let two agents have the same future net worth X with $X(\omega_1) \geq 0, X(\omega_2) = X(\omega_3) = -100$. If one uses the 5% value at risk measure, one would not find sufficient an extra capital (for each agent) of 80. But this same capital would be found more than sufficient, for each agent, if, by a risk exchange, the two agree on the modified respective future net worths Y and Z, where $Y(\omega_1) = Z(\omega_1) = X(\omega_1), Y(\omega_2) = Z(\omega_3) = -120, Y(\omega_3) = Z(\omega_2) = -80$. This is not reasonable since the allocation $(X+80, X+80)$ Pareto dominates the allocation $(Y+80, Z+80)$ if the agents are risk averse.

In conclusion, the basic reasons to reject the value at risk measure of risks are the following:

(a) value at risk does not behave nicely with respect to addition of risks, even independent ones, creating severe aggregation problems.

(b) the use of value at risk does not encourage and, indeed, sometimes prohibits diversification, because value at risk does not take into account the *economic consequences* of the events the probabilities of which it controls.

4 Representation Theorems for Coherent Risk Measures

This section provides two representations of coherent risk measures. The first corresponds exactly to the SPAN example of Section 3.1 and the second is the proper generalisation of the NASD/SEC examples of Section 3.2. These representation results are used in Section 5.2 to provide an example of algorithm to measure risks in trades involving two different sources of randomness, once coherent measures of risks for trades dealing with only one of these sources have been agreed upon.

4.1 Representation of Coherent Risk Measures by Scenarios

In this section we show that Definition 3.1 provides the most general coherent risk measure: *any* coherent risk measure arises as the supremum of the expected negative of final net worth for some collection of 'generalized scenarios' or probability measures on states of the world. We continue to suppose that Ω is a finite set, otherwise we would also get finitely additive measures as scenarios.

The σ-algebra, 2^{Ω}, is the class of all subsets of Ω. Initially there is no particular probability measure on Ω.

Proposition 4.1 *Given the total return r on a reference investment, a risk measure ρ is coherent if and only if there exists a family $\mathcal{P}$ of probability measures on the set of states of nature, such that*

$$\rho(X) = \sup\{\mathbf{E}_{\mathbb{P}}[-X/r] \mid \mathbb{P} \in \mathcal{P}\}.$$

Remark 4.1 We note that ρ can also be seen as an insurance premium principle. In that case, denoting by $\mathbb{R}$ the physical measure, we find that the condition $\mathbb{R} \in \mathcal{P}$ (or in the convex hull of this set), is of great importance. This condition is translated as follows: for all $X \leq 0$ we have $\mathbf{E}_{\mathbb{R}}[-X/r] \leq \rho(X)$.

Remark 4.2 The more scenarios considered, the more conservative (i.e. the larger) is the risk measure obtained.

Remark 4.3 We remind the reader about Proposition 3.1. It will prove that Axiom R is satisfied by ρ if and only if the union of the supports of the probabilities in $\mathcal{P}$ is the whole set Ω of states of nature.

Proof of Proposition 4.1

(1) We thank a referee for pointing out that the mathematical content of Proposition 4.1, which we had proved on our own, is already in the book Huber (1981). We therefore simply identify the terms in Proposition 2.1, Chapter 10 of Huber (1981) with these of our terminology of risks and risk measure.

(2) The sets Ω and $\mathcal{M}$ of Huber (1981) are our set Ω and the set of probabilities on Ω. Given a risk measure ρ we associate to it the functional E^* by $E^*(X) = \rho(-r \cdot X)$. Axiom M for ρ is equivalent to Property (2.7) of Huber (1981) for E^*, Axioms PH and T together are equivalent to Property (2.8) for E^*, and Axiom S is Property (2.9).

(3) The 'if' part of our Proposition 4.1 is obvious. The 'only if' part results from the 'representability' of E^*, since Proposition 2.1 of Huber (1981) states that

$$\rho(X) = E^*(-X/r) = \sup\{\mathbf{E}_{\mathbb{P}}[-X/r] \mid \mathbb{P} \in \mathcal{P}_\rho\},$$

where $\mathcal{P}_\rho$ is defined as the set

$$\{\mathbb{P} \in \mathcal{M} \mid \text{ for all } X \in \mathcal{G},\ \mathbf{E}_{\mathbb{P}}[X] \leq E^*(X) = \rho(-r \cdot X)\}$$
$$= \{\mathbb{P} \in \mathcal{M} \mid \text{ for all } Y \in \mathcal{G},\ \mathbf{E}_{\mathbb{P}}[-Y/r] \leq \rho(Y)\}.$$

Remark 4.4 *Model risk* can be taken into account by including into the set $\mathcal{P}$ a family of distributions for the future prices, possibly arising from other models.

Remark 4.5 Professor Bühlmann kindly provided us with references to works by Hattendorf (1868), Kanner (1867), and Wittstein (1867), which he had mentioned in his Göttingen presentation (Bühlmann (1995)). These authors consider, in the case of insurance risks, possible losses *only*, neglecting the case of gains. For example, risk for a company providing annuities is linked to the random *excess* number of survivors over the expected number given by the lifetable. Several of these references, for example Hattendorf (1868), §3, page 5, contain an example of a risk measure used in life insurance, namely the 'mittlere Risico' constructed out of one scenario, related to the life table used by a company. It is defined as the mathematical expectation of the *positive part* of the loss, as 'die Summe aller möglichen Verluste, jeden multipliciert in die Wahrscheinlichkeit seines Eintretens'. This procedure defines a risk measure satisfying Axioms S, PH, M.

Remark 4.6 It is important to distinguish between a point mass scenario and a simulation trial: the first is chosen by the investor or the supervisor, while the second is chosen randomly according to a *distribution* they have prescribed beforehand.

Conclusion The result in Proposition 4.1 completely explains the occurrence of the first type of actual risk measurement, the one based on scenarios, as described in Section 3.1. Any coherent risk measure appears therefore as given by a 'worst case method', in a framework of generalized scenarios. At this point we emphasize that scenarios should be announced to all traders within the firm (by the manager) or to all firms (by the regulator). In the first case, we notice that decentralization of risk management within the firm is available only *after* these announcements. Yet, in quantile-based methods, even after the announcements of individual limits, there remains a problem preventing decentralized risk management: two operators ignorant of each other's actions may well each comply with their individual quantile limits and yet no automatic procedure provides for an interesting upper bound for the measure of the joint risk due to their actions. As for the regulation case we allow ourselves to interpret a sentence from Stulz (1996): 'regulators like Value at Risk, because they can regulate it' as pointing to the formidable task of building and announcing a reasonable set of scenarios.

4.2 Construction of Coherent Risk Measures by Extension of Certain Risk Measurements

We now formalize the attempts described in Section 3.2 to measure risks. Their basis is to impose margin requirements on certain basic portfolios considered as 'standard risks', to use combinations of those risks to 'support' other risks and then bound from above required capital, using the margins required for standard risks.

Definition 4.1 *Supports of a risk.* Given a set $\mathcal{Y}$ of functions on Ω, we consider a family, indexed by $\mathcal{Y}$, of nonnegative numbers $\mu = (\mu_Y)_{Y \in \mathcal{Y}}$, all of them but a finite number being zero, and we say that the couple (μ, γ), where γ is a real number, 'supports' X, for $X \in \mathcal{G}$, provided

$$X \geq \sum_{Y \in \mathcal{Y}} \mu_Y Y + \gamma \cdot r.$$

The set of all such (μ, γ) which support X will be denoted by $S_{\mathcal{Y}}(X)$.

The idea is now to use these 'supports', made of 'standard risks', to bound above possible extensions of a function Ψ defined on a subset of $\mathcal{G}$. A consistency condition is required to avoid supports leading to infinitely negative values.

Condition 4.1 *Given a set $\mathcal{Y}$ of functions on Ω, and a function $\Psi \colon \mathcal{Y} \longrightarrow \mathbb{R}$, we say that Ψ fulfills Condition 4.1 if for each support (μ, γ) of 0, we have the inequality $\sum_{Y \in \mathcal{Y}} \mu_Y \Psi(Y) - \gamma \geq 0$.*

Proposition 4.2 *Given a set $\mathcal{Y}$ of functions on Ω and a function $\Psi \colon \mathcal{Y} \longrightarrow \mathbb{R}$, the equality*

$$\rho_\Psi(X) = \inf_{(\mu,\gamma) \in S_{\mathcal{Y}}(X)} \sum_{Y \in \mathcal{Y}} \mu_Y \Psi(Y) - \gamma$$

defines a coherent risk measure ρ_Ψ, if and only if Ψ fulfills Condition 4.1. If so, ρ_Ψ is the largest coherent measure ρ such that $\rho \leq \Psi$ on $\mathcal{Y}$.

Proof

(1) The necessity of Condition 4.1 is obvious.

(2) Since $(0, 0)$ is a support of the element $X = 0$ of $\mathcal{G}$ and since Condition 4.1 ensures that any support of 0 provides a nonnegative number, we find that $\rho_\Psi(0) = 0$. Notice that if Condition 4.1 is violated, then we would get $\rho_\Psi(0) = -\infty$.

(3) Axiom S required from a coherent risk measure follows here from the relation $S_{\mathcal{Y}}(X_1 + X_2) \supset S_{\mathcal{Y}}(X_1) + S_{\mathcal{Y}}(X_2)$, and Axiom PH is satisfied since, given $\lambda > 0$, (μ, γ) supports X if and only if $(\lambda \cdot \mu, \lambda \cdot \gamma)$ supports $\lambda \cdot X$.

(4) For a support (μ, γ) of a risk X let us call the number $\sum_{Y \in \mathcal{Y}} \mu_Y \Psi(Y) - \gamma$ the 'cost' of the support. By noticing for each risk X and each real α, that the support (μ, γ) for $X + \alpha \cdot r$ provides the support $(\mu, \gamma - \alpha)$ for X, at a cost lower by the amount α than the cost of the support of $X + \alpha \cdot r$, we find that $\rho_\Psi(X) = \rho_\Psi(X + \alpha \cdot r) + \alpha$. Axiom T is therefore satisfied by ρ_Ψ.

(5) Since for $X \leq Z$ we have $S_{\mathcal{Y}}(Z) \supset S_{\mathcal{Y}}(X)$, Axiom M is satisfied by ρ_Ψ.

(6) For any coherent measure ρ with $\rho \leq \Psi$ on $\mathcal{Y}$ we must have, for any support (μ, γ) of X, the inequality $\rho(X) \leq \sum_{Y \in \mathcal{Y}} \mu_Y \Psi(Y) - \gamma$ and therefore $\rho(X) \leq \rho_\Psi(X)$.

Remark 4.7 As opposed to the case of scenarios based measures of risks, the fewer initial standard risks are considered, the more conservative is the coherent risk measure obtained. This is similar to what happens with the SEC rules since Section 3.2 showed us that too many scenarios, and dually, too few standard risks, were considered.

Condition 4.1 allows one to consider the function ρ_Ψ in particular on the set $\mathcal{Y}$, the set of prespecified risks. There, it is clearly bounded above by the original function Ψ. An extra consistency condition will prove helpful to figure out whether ρ_Ψ is actually equal to Ψ on $\mathcal{Y}$.

Condition 4.2 *Given a set $\mathcal{Y}$ of functions on Ω and a function $\Psi: \mathcal{Y} \longrightarrow \mathbb{R}$, we say that Condition 4.2 is satisfied by Ψ if for each element $Z \in \mathcal{Y}$ and each support (μ, γ) of Z we have $\Psi(Z) \leq \sum_{Y \in \mathcal{Y}} \mu_Y \Psi(Y) - \gamma$.*

Remark 4.8 It is an easy exercise to prove that Condition 4.2 implies Condition 4.1.

Proposition 4.3 *Given a set $\mathcal{Y}$ of functions on Ω and a function $\Psi: \mathcal{Y} \longrightarrow \mathbb{R}_+$ satisfying Condition 4.2, the coherent risk measure ρ_Ψ is the largest possible extension of the function Ψ to a coherent risk measure.*

Proof

(1) Condition 4.2 just ensures that the value at $Z \in \mathcal{Y}$ of the original function Ψ is bounded above by the sum obtained with any support of Z, hence also by their infimum $\rho_\Psi(Z)$, which proves that $\rho_\Psi = \Psi$ on $\mathcal{Y}$.

(2) Let ρ be any coherent risk measure, which is also an extension of Ψ. Since $\rho \leq \Psi$ on $\mathcal{Y}$, Proposition 4.3 ensures that $\rho \leq \rho_\Psi$.

Propositions 4.2 and 4.3 above applied to $(\mathcal{Y}, \Psi) = (\mathcal{G}, \rho)$, provide a statement similar to Proposition 4.1 about representation of coherent risk measures.

Proposition 4.4 *A risk measure ρ is coherent if and only if it is of the form ρ_Ψ for some Ψ fulfilling Condition 4.1.*

Remark 4.9 It can be shown that for a coherent risk measure ρ built as a ρ_Ψ, the following set of probabilities

$$\mathcal{P}_\Psi = \{\mathbb{P} \mid \text{ for all } X \in \mathcal{G}, \mathbf{E}_{\mathbb{P}}[-X/r] \leq \Psi(X)\}$$

is non-empty and verifies the property

$$\rho(X) = \sup\{\mathbf{E}_{\mathbb{P}}[-X/r] \mid \mathbb{P} \in \mathcal{P}_\Psi\}.$$

4.3 Relation between Scenario Probabilities and Pricing Measures

The representation result in Proposition 4.1 allows us to approach the problem of risk concentration for *coherent* risk measures.

If the position consisting of the short Arrow–Debreu security corresponding to state of nature ω, has a non-positive measure of risk, that is bankruptcy in the state ω is 'allowed', the market price of this security should also be non-positive. To formalize this observation we suppose an arbitrage free market, and denote by $\mathcal{Q}_r$ the closed convex set of pricing probability measures on Ω, using the instrument r as *numeraire*. Given the coherent risk measure $\rho_{\mathcal{B},r}$ associated to r and to an acceptance set B, simply denoted by ρ_r (see Proposition 2.2), it will be natural to assume the following condition:

Condition 4.3 The closed convex set $\mathcal{P}_{\rho_r}$ of probability measures defining the coherent risk measure ρ_r has a non-empty intersection with the closed convex set $\mathcal{Q}_r$ of probability pricing measures.

When Condition 4.3 is satisfied, there is some $\mathbb{Q} \in \mathcal{Q}_r$ such that for any future net worth Y , $\mathbf{E}_{\mathbb{Q}}[-Y/r] \leq \rho_r(Y)$, hence if Y has a strictly negative price under $\mathbb{Q}$ it cannot be accepted. We interpret this fact in the following manner: if a firm can, by trading, add a position Y to its portfolio *and* receive cash at the same time, *without* having any extra capital requirement, then there is a bound to the quantity of Y which the firm can add this way without trigging a request for extra capital.

If Condition 4.3 is not satisfied, then there exists a future net worth Y such that

$$\sup\{\mathbf{E}_{\mathbb{Q}}[Y/r] \mid \mathbb{Q} \in \mathcal{Q}_r\} < \inf\{\mathbf{E}_{\mathbb{S}}[Y/r] \mid \mathbb{S} \in \mathcal{P}_{\rho_r}\}.$$

Hence for each pricing measure $\mathbb{Q}$ we have $\mathbf{E}_{\mathbb{Q}}[-Y/r] > \rho_r(Y)$ and therefore the future net worth $Z = Y + \rho_r(Y) \cdot r$ satisifies both conditions $\rho_r(Z) = 0$ and $\mathbf{E}_{\mathbb{Q}}[Z/r] < 0$. We have therefore an acceptable position with strictly negative price, a situation which may well lead to an undetected accumulation of risk.

5 Two Applications of Representations of Coherent Risk Measures

5.1 A Proposal: the 'Worst Conditional Expectation' Measure of Risk

Casualty actuaries have been working for long computing pure premium for policies with deductible, using the conditional average of claim size, *given* that the claim exceeds the deductible, see Hogg and Klugman (1984). In the same manner, reinsurance treaties have involved the conditional distribution of a claim for a policy (or of the total claim for a portfolio of policies), *given* that it is above the ceding insurer's retention level. In order to tackle the question of 'how bad is bad', which is *not* addressed by the value at risk measurement, some actuaries (see Albrecht (1993), Embrechts (1995)) have first identified the deductible (or retention level) with the quantile used in the field of financial risk measurement. We prove below that one of the suggested methods gets us close to *coherent* risk measures.

Considering the 'lower partial moment' or expectation of the 'shortfall', the presentation in Albrecht (1993) would translate, with our paper's notations, into measuring a risk X by the number $\mathbf{E}_{\mathbb{P}}\left[\min\left(0, -\mathrm{VaR}_{\alpha}(X) - X\right)\right]$.

The presentations in Bassi *et al.* (1998), Embrechts (1995), use instead the *conditional* expectation of the shortfall given that it is positive. The quoted texts (see also Embrechts *et al.* (1997), Definition 3.4.6 as well as the methods indicated there to *estimate* the whole conditional distribution) present the terminology 'mean excess function'. We suggest the term tail conditional expectation since we do not consider the excess but the whole of the variable X:

Definition 5.1 *Tail conditional expectation (or* 'TailVaR'*).* Given a base probability measure $\mathbb{P}$ on Ω, a total return r on a reference instrument and a level α, the tail conditional expectation is the measure of risk defined by

$$\mathrm{TCE}_{\alpha}(X) = -\mathbf{E}_{\mathbb{P}}\left[X/r \mid X/r \leq -\mathrm{VaR}_{\alpha}(X)\right].$$

Definition 5.2 *Worst conditional expectation.* Given a base probability measure $\mathbb{P}$ on Ω, a total return r on a reference instrument and a level α, the worst conditional expectation is the coherent measure of risk defined by

$$\mathrm{WCE}_{\alpha}(X) = -\inf\{\mathbf{E}_{\mathbb{P}}\left[X/r \mid A\right] \mid \mathbb{P}\left[A\right] > \alpha\}.$$

Remark 5.1 TCE_{α} has been suggested as a possible ingredient of reinsurance treaties (see Amsler (1991)).

Proposition 5.1 *We have the inequality* $\mathrm{TCE}_{\alpha} \leq \mathrm{WCE}_{\alpha}$.

Proof

(1) Let us denote X/r by Y. If $F_Y(q_{\alpha}^{+}(Y)) > \alpha$, the set $A = \{\omega \mid Y(\omega) \leq q_{\alpha(Y)}^{+}\}$ is one used in the definition of WCE_{α}, hence the claim is true.

(2) If $F_Y(q_{\alpha}^{+}(Y)) = \alpha$ it follows from the definition of q_{α}^{+} and the monotonicity of F_Y that for each $\varepsilon > 0$, $F_Y(\varepsilon + q_{\alpha}^{+}(Y)) > \alpha$. Hence, setting $A_{\varepsilon} = \{\omega \mid Y(\omega) \leq \varepsilon + q_{\alpha}^{+}(Y)\}$, we get

$$\mathrm{WCE}_{\alpha}(X) \geq -\mathbf{E}_{\mathbb{P}}\left[Y \mid A_{\varepsilon}\right] = -\frac{\mathbf{E}_{\mathbb{P}}\left[Y \cdot \mathbf{1}_{A_{\varepsilon}}\right]}{\mathbb{P}\left[A_{\varepsilon}\right]}.$$

Since F_Y is right-continuous, $\lim_{\varepsilon\to 0}\mathbb{P}[A_\varepsilon] = F_Y(q_\alpha^+(Y))$ and $A_\varepsilon \downarrow A_0$ so the right hand side has the limit $-\mathbf{E}_{\mathbb{P}}[Y \mid A_0] = TCE_\alpha(X)$.

The paper Albanese (1997) makes numerical studies of portfolios built out of collection of risky bonds. It looks for a coherent measure which dominates the Value at Risk measurement and yet gets close to it on a specific bond portfolio.

We interpret and generalize this search as the problem of a firm constrained by the supervisors along the lines of the quantile risk measurement. Nevertheless, the firm wishes at the same time to operate on a coherent basis, at the lowest possible cost. Proposition 5.4 will provide circumstances where the firm's problem has a clear-cut solution.

Proposition 5.2 *For each risk X one has the equality*

$$\mathrm{VaR}_\alpha(X) = \inf\{\rho(X) \mid \rho \textit{ coherent and } \rho \geq \mathrm{VaR}_\alpha\}.$$

The proof will use the following

Lemma 5.1 *If ρ is the coherent risk measure defined by a set $\mathcal{P}$ of probability measures, then $\rho \geq \mathrm{VaR}_\alpha$ if and only if for each B with $\mathbb{P}[B] > \alpha$ and each $\varepsilon > 0$ there is a $\mathbb{Q} \in \mathcal{P}$ with $\mathbb{Q}[B] > 1 - \varepsilon$.*

Proof

(1) Necessity: take $X = -r\cdot\mathbf{1}_B$ where $\mathbb{P}[B] > \alpha$. Clearly $\mathrm{VaR}_\alpha(-r\cdot\mathbf{1}_B) = 1$ and hence $\rho(-r\cdot\mathbf{1}_B) \geq 1$. This implies that for each $\varepsilon > 0$ there exists $\mathbb{Q} \in \mathcal{P}$ with $\mathbb{Q}[B] \geq 1 - \varepsilon$.

(2) Sufficiency: let $-k = \mathrm{VaR}_\alpha(X)$, then $\mathbb{P}[X \leq k\cdot r] \geq \alpha$ and for each $\delta > 0$ we have $\mathbb{P}[X \leq (k+\delta)\cdot r] > \alpha$.

Let $\mathbb{Q} \in \mathcal{P}$ be chosen such that $\mathbb{Q}[X \leq (k+\delta)\cdot r] \geq 1-\delta$. We obtain $\mathbf{E}_{\mathbb{Q}}[-X/r] \geq (-k-\delta)\cdot(1-\delta) - \delta\cdot\|X/r\|$. Since $\delta > 0$ was arbitrary we find that $\rho(X) \geq -k$.

Proof

(1) Given any risk X let again $-k = \mathrm{VaR}_\alpha(X)$. Then $\mathbb{P}[X \leq k\cdot r] \geq \alpha$ and for each $\delta > 0, \mathbb{P}[X \leq (k+\delta)\cdot r] > \alpha$. We will construct a coherent risk measure ρ such that $\rho \geq \mathrm{VaR}_\alpha$ and $\rho(X) \leq \mathrm{VaR}_\alpha(X) + \delta$.

(2) For any set B with $\mathbb{P}[B] > \alpha$, we must have $\mathbb{P}[B \cap \{X \geq k\cdot r\}] > 0$ and we can define h_B as $\mathbf{1}_{B\cap\{X\geq k\cdot r\}}/\mathbb{P}[B \cap \{X \geq k\cdot r\}]$ and $\mathbb{Q}_B = h_B\cdot\mathbb{P}$. Lemma 5.1 shows that the measure ρ built with all the $\mathbb{Q}_B$ dominates VaR_α, but for X we obtain $\rho(X) = \sup_{\mathbb{Q}_B}\mathbf{E}_{\mathbb{Q}_B}[-X/r] \leq -k = \mathrm{VaR}_\alpha(X)$.

Definition 3.1 and Proposition 3.1 allow one to address a question by Ch. Petitmengin, Société Générale, about the coherence of the TCE_α measure.

Proposition 5.3 *Assume that the base probability $\mathbb{P}$ on Ω is uniform. If X is a risk such that no two values of the discounted risk $Y = X/r$ in different states are ever equal, then* $\mathrm{TCE}_\alpha(X) = \mathrm{WCE}_\alpha(X)$.

Proof

(1) Given $\alpha \in]0,1[$ let us denote $-\mathrm{VaR}_\alpha(X)$ by q, the set $\{X \le q \cdot r\}$ by B, and the various values of $Y = X/r$ by $y_1 < y_2 < \cdots < y_n$.

(2) Let k be the integer with $0 \le k < n$ such that $\alpha \in [\frac{k}{n}, \frac{k+1}{n})$. We will prove that $-\mathrm{VaR}_\alpha(X) = q_\alpha^+(Y) = q = y_{k+1}$.

(3) For each $u > q$ we have

$$\frac{\#\{i \mid y_i \le u\}}{n} > \alpha,$$

hence the integer $\#\{i \mid y_i \le u\}$ being strictly greater than $\alpha \cdot n$ is at least $k+1$.

(4) By taking $u = y_{k+1}$ we actually minimize the integer $\#\{i \mid y_i \le u\}$ and therefore prove the point stated in (2).

(5) The set $Y(B)$ is the set $\{y_1, \ldots, y_{k+1}\}$ and

$$\mathrm{TCE}_\alpha(X) = -\mathbf{E}\,[X/r \mid X \le q \cdot r] = -\frac{y_1 + \cdots + y_{k+1}}{k+1}.$$

(6) Any set C containing at least $k+1$ states of nature and different from B will provide values for $-Y$ averaging to strictly less than $\mathrm{TCE}_\alpha(X)$, which therefore equals $\mathrm{WCE}_\alpha(X)$.

Proposition 5.4 *Assume that the base probability $\mathbb{P}$ on Ω is uniform. If a coherent risk measure ρ only depends on the distribution of the discounted risk and is greater than the risk measure VaR_α, then it is greater than the WCE_α (coherent) risk measure.*

Proof

(1) Given a risk X, we denote $-\mathrm{VaR}_\alpha(X)$ simply by q and X/r by Y. The set $A = \{\omega \mid Y(\omega) \le q\}$ has cardinality $p > n \cdot \alpha$ and A is written after possible renumbering as $A = \{\omega_1, \omega_2, \ldots, \omega_p\}$ with $Y(\omega_i) \le Y(\omega_{i+1})$ for $1 \le i \le p-1$.

(2) Define $\bar{Y}(\omega_i)$ for $i \le p$ as $y^* = (Y(\omega_1) + \cdots + Y(\omega_p))/p = \mathbf{E}\,[Y \mid Y \le q]$ and as $Y(\omega_i)$ otherwise.

(3) For a permutation σ of the first p integers, we define Y^σ by $Y^\sigma(\omega_i) = Y(\omega_{\sigma(i)})$ for $1 \le i \le p$, and $Y^\sigma(\omega_j) = Y(\omega_j)$ for $p+1 \le j \le n$. We then find that $\bar{Y}$ is also the average of the $p!$ random variables Y^σ.

(4) The assumption that for each risk $Z\,, \rho(Z)$ only depends on the distribution of Z/r implies that all the $\rho(r \cdot Y^\sigma)$ are equal to $\rho(X)$. The convexity of the function ρ then implies the inequality $\rho(X) \ge \rho(r \cdot \bar{Y})$.

(5) The last assumption made on ρ implies that $\rho(r \cdot \bar{Y}) \ge \mathrm{VaR}_\alpha(r \cdot \bar{Y})$.

(6) We have $\mathrm{VaR}_\alpha(r \cdot \bar{Y}) = -y^* = \mathbf{E}\,[-Y \mid Y \le q]$ since for $i \le p\,, \bar{Y}(\omega_i) \le Y(\omega_p)$. Hence $\rho(X) \ge \mathbf{E}\,[-X/r \mid X \le q \cdot r]$.

(7) For a dense set of random variables X on the finite state space Ω we know by Proposition 5.3 that $\mathbf{E}\,[-X/r \mid X \le q \cdot r] = \mathrm{WCE}_\alpha(X)$, hence the inequality $\rho(X) \ge \mathrm{WCE}_\alpha(X)$ holds for a dense set of elements X of $\mathcal{G}$.

(8) Both risk measures ρ and WCE_α are coherent, hence continuous functions on $\mathcal{G}$. The inequality $\rho \ge \mathrm{WCE}_\alpha$ is therefore true on the whole of $\mathcal{G}$.

5.2 Construction of a Measure Out of Measures on Separate Classes of Risks

It is important to realize that Proposition 4.3 can be applied to a set $\mathcal{Y}$ of risks having no structure. It can be the union of a family $(\mathcal{Y}_j)_{j \in J}$ of sets of risks, where, for each j a function (Ψ_j) is given on $\mathcal{Y}_j$, in such a way that $\Psi_j = \Psi_{j'}$ on $\mathcal{Y}_j \cap \mathcal{Y}_{j'}$. The function Ψ is then defined by its restrictions to each of the $\mathcal{Y}_j$.

The different sets $\mathcal{Y}_j$ may be exchange based risks on the one hand and over the counter risks on the other hand, or market risks and credit risks in a framework where a *joint* internal model would be looked for. Similarly, multi-line aggregated combined risk optimisation tools (see Shimpi (1998)) would call for a combined measure of risks. The functions Ψ_j may come from preliminary rules given by exchanges and/or by regulators (see BCBS (1996)). Assuming that Condition 4.2 is being satisfied, which will depend on inequalities satisfied by the Ψ_j, Proposition 4.3 allows one to *mechanically* compute a coherent risk measure extending the family of the Ψ_j and dominating *any* other possible coherent risk measure *chosen* by exchanges and/or by regulators to extend the family of the Ψ_j. It therefore provides a conservative coherent tool for risk management.

In the special case of $\Omega = \Omega_1 \times \Omega_2$ with given coherent risk measures ρ_i, $i = 1, 2$, on $\mathcal{G}_i$, we define $\mathcal{Y}_i$ as the set of all functions on Ω which are of the form $f_i \circ pr_i$ where f_i is any function on Ω_i, and where pr_i is the projection of

Ω on its ith factor. We also define Ψ_i on $\mathcal{Y}_i$ by the equality $\Psi_i(f_i \circ pr_i) = \rho_i(f_i)$. Since $\mathcal{Y}_1 \cap \mathcal{Y}_2$ consists of the constants, the functions Ψ_1 and Ψ_2 are equal on it and they define a function Ψ on $\mathcal{Y} = \mathcal{Y}_1 \cup \mathcal{Y}_2$ which satisfies Condition 4.2.

Let $\mathcal{P}_i$ be the set of scenarios defining ρ_i and let $\mathcal{P}$ be the set of probabilities on Ω with marginals in $\mathcal{P}_1$ and $\mathcal{P}_2$ respectively. We claim that the risk measure ρ_Ψ on the set $\mathcal{G}$ of functions on Ω, that is the largest coherent risk measure extending both Ψ_1 and Ψ_2, is equal to the risk measure $\rho_\mathcal{P}$, generated, as in Definition 3.1, by the scenarios in $\mathcal{P}$.

Proposition 5.5 *The two coherent risk measures $\rho_\mathcal{P}$ and ρ_Ψ are equal.*

Proof The restriction of $\rho_\mathcal{P}$ to $\mathcal{Y}_i$ equals Ψ_i since for each function f_i on Ω_i we have

$$\begin{aligned}
\rho_\mathcal{P}(f_i \circ pr_i) &= \sup\{\mathbf{E}_\mathbb{P}[-f_i \circ pr_i/r] \mid \mathbb{P} \circ pr_1^{-1} \in \mathcal{P}_1 , \mathbb{P} \circ pr_2^{-1} \in \mathcal{P}_2\} \\
&= \sup\{\mathbf{E}_{\mathbb{P} \circ pr_i^{-1}}[-f_i/r] \mid \mathbb{P} \circ pr_i^{-1} \in \mathcal{P}_i\} \\
&= \rho_i(f_i) = \Psi_i(f_i \circ pr_i),
\end{aligned}$$

which proves that $\rho_\mathcal{P} \leq \rho_\Psi$.

To prove the reverse inequality we use point (3) in the proof of Proposition 4.1 and show that if a probability $\mathbb{Q}$ on Ω is such that for each function X on Ω, $\mathbf{E}_\mathbb{Q}[-X/r] \leq \rho_\Psi(X)$, then $\mathbb{Q}$ has its marginals $\mathbb{Q}_1$ and $\mathbb{Q}_2$ in $\mathcal{P}_1$ and $\mathcal{P}_2$ respectively. Choose indeed $X = f_i \circ pr_i$. We find that $\mathbf{E}_\mathbb{Q}[-f_i \circ pr_i/r] = \mathbf{E}_{\mathbb{Q}_i}[-f_i/r]$ which proves that for each $f_i \in \mathcal{G}_i$ one has $\mathbf{E}_{\mathbb{Q}_i}[-f_i/r] \leq \rho_\Psi(f_i \circ pr_i)$, and therefore $\mathbb{Q}_i \in \mathcal{P}_i$.

Acknowledgments

The authors acknowledge financial support from Société Générale for this work. The views expressed are those of the authors. They thank for useful discussions on a related paper, participants of the following meetings: Boston University Mathematical Finance Day, March 31, 1996, University of Waterloo Insurance and Finance Conference, May 24, 1996, StudienZentrum Gerzensee Symposium in Financial Markets, July 15–26, 1996, Latsis Symposium, ETH Zürich, September 24–25, 1996, Universität Karlsruhe Geld, Finanzwirtschaft, Banken, Versicherungen Tagung, December 11–13, 1996, Aarhus University Workshop, February 25–March 1, 1997. They also thank D. Madan, discussant at the French Finance Association International Meeting, Geneva, June 1996, F. Diebold, discussant at the Federal Reserve Bank of Atlanta 1997 Financial Markets Conference, R. Bliss, Ph. Boyle, V. Brousseau, P. Embrechts, A. Hoffman, W. Neuefeind, Ch. Petitmengin, P. Poncet, J. Renegar, E. Shiu as well as a referee of an earlier version of this paper.

References

Albanese, C. (1997) 'Credit exposure, diversification risk and coherent VaR'. Working Paper, Department of Mathematics, University of Toronto.

Albrecht, P. (1993) 'Normal and lognormal shortfall risk'. In *Proceedings 3rd AFIR International Colloquium*, Roma, **2**, 417–430.

Amsler, M.-H. (1991) 'Réassurance du risque de ruine', *Mitteilungen der Schweiz. Vereinigung der Versicherungsmathematiker* **1**, 33–49.

Artzner, Ph., F. Delbaen, J.-M. Eber, and D. Heath (1997) 'Thinking coherently', *RISK* **10**, November, 68–71.

BCBS (1996) *Amendment to the Capital Accord to Incorporate Market Risks*, Basle Committee on Banking Supervision.

Bassi, F., P. Embrechts, and M. Kafetzaki (1998) 'Risk management and quantile estimation'. In *Practical Guide to Heavy Tails*, edited by R. Adler, R.T. Feldman and M. Taqqu, Birkhäuser.

Bühlmann, H. (1990) 'Tendencies of development in risk theory'. In *Proceedings, 1989 Centennial Celebration of the Actuarial Profession in North America*, Society of Actuaries, Schaumburg, IL, 499–522.

Bühlmann, H. (1995) 'Der Satz von Hattendorf und Hattendorf's Originalarbeit', *Vortrag auf der Jubiläumstagung, Göttingen: 'Hundert Jahre Versicherungsmathematik an Universitäten'*.

Cox, J., and M. Rubinstein (1985) *Options Markets*, Prentice-Hall.

Daykin, C., T. Pentikainen, and T. Pesonen (1994) *Practical Risk Theory for Actuaries*, Chapman & Hall.

Derivatives Policy Group (1995) *Framework for Voluntary Oversight of the OTC Derivatives Activities of Securities Firm Affiliates to Promote Confidence and Stability in Financial Markets*, New York.

Dowd, K. (1998) *Beyond Value at Risk: The New Science of Risk Management*, Wiley.

Duffie, D., and J. Pan (1997) 'An overview of value at risk', *Journal of Derivatives* **4**, 7–49.

Embrechts, P., (1995) *A Survival Kit to Quantile Estimation*, UBS Quant Workshop, Zürich.

Embrechts, P., C. Klüppelberg, and T. Mikosch (1997) *Modelling Extremal Events*, Springer.

Federal Reserve System (1994) *Securities Credit Transactions Regulation T, Margin Credit extended by Brokers and Dealers*, as amended effective November 25, 1994. Board of Governors of the Federal Reserve System.

Hattendorff, K. (1868) 'Über die Berechnung der Reserven des Risico bei der Lebensversicherung', *Rundschau der Versicherungen* **18**, 1–18.

Hogg, R., and S. Klugman (1984) *Loss Distributions*, Wiley.

Huber, P. (1981) *Robust Statistics*, Wiley.

Kanner, M. (1867) 'Bestimmung des mittleren Risico bei Lebensversicherungen', Deutsche Versicherungs-Zeitung.

MATIF (1993) *Initial Margin Calculation Method*, MATIF, SA, Paris.

Merton, R. (1973) 'Theory of rational option pricing', *Bell J. Econ. Mgmt. Sci.* **4**, 141–183.

NASD (1996) Reprint of the *Manual of the National Association of Securities Dealers*, July 1996, CCR, Chicago.

Risk Magazine (1996) 'Value at Risk', *Risk Magazine Special Supplement*, June, 68–71.

RiskMetrics (1995) Technical Document, Morgan Guarantee Trust Company, Global Research, New York.

Rudd, A., and M. Schroeder (1982) 'The calculation of minimum margin', *Mgmt. Sci.* **28**, 1369–1379.

Shimpi, P. (1998) 'The convergence of insurance and financial markets, a practitioner's view'. Presentation at the Conference on the Interplay between Finance, Insurance and Statistics, Cortona, June 8–12.

SPAN (1995) *Standard Portfolio Analysis of Risk*, Chicago Mercantile Exchange, Chicago.

Stulz, R. (1996) 'Rethinking risk management'. Presentation at the Guest Speaker Conference French Finance Association Meeting, Geneva, June 24.

Wang, T. (1996) 'A characterisation of dynamic risk measures'. Working Paper, Faculty of Commerce, University of British Columbia.

Wittstein, T. (1867) *Mathematische Statistik und deren Anwendung auf National-Ökonomie und Versicherungswissenschaft*, Hahn'sche Buchhandlung, Hannover.

Correlation and Dependence in Risk Management: Properties and Pitfalls

Paul Embrechts, Alexander J. McNeil and Daniel Straumann

Abstract

Modern risk management calls for an understanding of stochastic dependence going beyond simple linear correlation. This article deals with the static (non-time-dependent) case and emphasizes the copula representation of dependence for a random vector. Linear correlation is a natural dependence measure for multivariate normally, and more generally, elliptically distributed risks but other dependence concepts like comonotonicity and rank correlation should also be understood by the risk management practitioner. Using counterexamples the falsity of some commonly held views on correlation is demonstrated; in general, these fallacies arise from the naive assumption that dependence properties of the elliptical world also hold in the non-elliptical world. In particular, the problem of finding multivariate models which are consistent with prespecified marginal distributions and correlations is addressed. Pitfalls are highlighted and simulation algorithms avoiding these problems are constructed.

1 Introduction

1.1 Correlation in finance and insurance

In financial theory the notion of correlation is central. The Capital Asset Pricing Model (CAPM) and the Arbitrage Pricing Theory (APT) (Campbell, Lo & MacKinlay 1997) use correlation as a measure of dependence between different financial instruments and employ an elegant theory, which is essentially founded on an assumption of multivariate normally distributed returns, in order to arrive at an optimal portfolio selection. Although insurance has traditionally been built on the assumption of independence and the law of large numbers has governed the determination of premiums, the increasing complexity of insurance and reinsurance products has led recently to increased actuarial interest in the modelling of dependent risks (Wang 1997); an example is the emergence of more intricate multi-line products. The current quest for a sound methodological basis for integrated risk management also raises the issue of correlation and dependence. Although contemporary financial risk management revolves around the use of correlation to describe dependence between risks, the inclusion of non-linear derivative products invalidates many of the distributional assumptions underlying the use of correlation. In insurance these assumptions are even more problematic because of the typical skewness and heavy-tailedness of insurance claims data.

Recently, within the actuarial world, dynamic financial analysis (DFA) and dynamic solvency testing (DST) have been heralded as a way forward for integrated risk management of the investment and underwriting risks to which an insurer (or bank) is exposed. DFA, for instance, is essentially a Monte Carlo or simulation-based approach to the joint modelling of risks (see e.g. CAS (1997) or Lowe & Stanard (1997)). This necessitates model assumptions that combine information on marginal distributions together with ideas on interdependencies. The correct implementation of a DFA-based risk management system certainly requires a proper understanding of the concepts of dependence and correlation.

1.2 Correlation as a source of confusion

But correlation, as well as being one of the most ubiquitous concepts in modern finance and insurance, is also one of the most misunderstood concepts. Some of the confusion may arise from the literary use of the word to cover any notion of dependence. To a mathematician correlation is only one particular measure of stochastic dependence among many. It is the canonical measure in the world of multivariate normal distributions, and more generally for spherical and elliptical distributions. However, empirical research in finance and insurance shows that the distributions of the real world are seldom in this class.

As motivation for the ideas of this article we include Figure 1. This shows 1000 bivariate realisations from two *different* probability models for (X, Y). In both models X and Y have identical gamma marginal distributions and the linear correlation between them is 0.7. However, it is clear that the dependence between X and Y in the two models is qualitatively quite different and, if we consider the random variables to represent insurance losses, the second model is the more dangerous model from the point of view of an insurer, since extreme losses have a tendency to occur together. We will return to this example later in the article (see Section 5); for the time-being we note that the dependence in the two models cannot be distinguished on the grounds of correlation alone.

The main aim of the article is to collect and clarify the essential ideas of dependence, linear correlation and rank correlation that anyone wishing to model dependent phenomena should know. In particular, we highlight a number of important fallacies concerning correlation which arise when we work with models other than the multivariate normal. Some of the pitfalls which await the end-user are quite subtle and perhaps counter-intuitive.

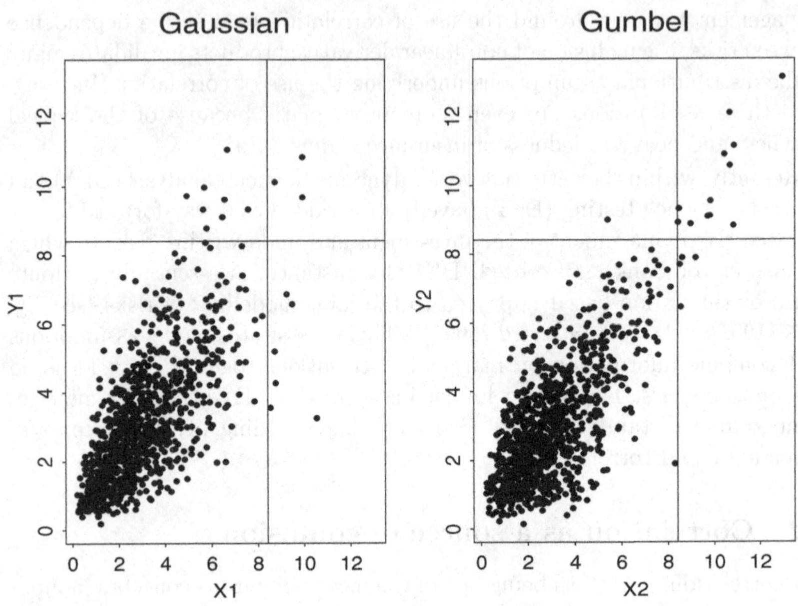

Figure 1: 1000 random variates from two distributions with identical Gamma(3,1) marginal distributions and identical correlation $\rho = 0.7$, but *different* dependence structures.

We are particularly interested in the problem of constructing multivariate distributions which are consistent with given marginal distributions and correlations, since this is a question that anyone wanting to simulate dependent random vectors, perhaps with a view to DFA, is likely to encounter. We look at the existence and construction of solutions and the implementation of algorithms to generate random variates. Various other ideas recur throughout the article. At several points we look at the effect of dependence structure on the Value-at-Risk or VaR under a particular probability model, i.e. we measure and compare risks by looking at quantiles. We also relate these considerations to the idea of a coherent measure of risk as introduced by Artzner, Delbaen, Eber & Heath (1999).

We concentrate on the *static* problem of describing dependence between a pair or within a group of random variables. There are various other problems concerning the modelling and interpretation of serial correlation in stochastic processes and cross-correlation between processes; see Boyer, Gibson & Loretan (1999) for problems related to this. We do not consider the statistical problem of estimating correlations and rank correlation, where a great deal could also be said about the available estimators, their properties and their robustness, or the lack of it.

Another statistical aspect which we do not cover in this article is the issue of fitting copulas to data. For this important practical question there are a number of references. Frees & Valdez (1998), and Klugman & Parsa (1999) take an actuarial point of view, whereas Genest & Rivest (1993), Genest, Ghoudi & Rivest (1995), and Capéraà, Fougères & Genest (1997) develop the general statistical theory of fitting copulas.

1.3 Organization of article

In Section 2 we begin by discussing joint distributions and the use of copulas as descriptions of dependence between random variables. Although copulas are a much more recent and less well known approach to describing dependence than correlation, we introduce them first for two reasons. First, they are the principal tool we will use to illustrate the pitfalls of correlation and second, they are the approach which in our opinion affords the best understanding of the general concept of dependence.

In Section 3 we examine linear correlation and define spherical and elliptical distributions, which constitute, in a sense, the natural environment of the linear correlation. We mention both some advantages and shortcomings of correlation. Section 4 is devoted to a brief discussion of some alternative dependence concepts and measures including comonotonicity and rank correlation. Three of the most common fallacies concerning linear correlation and dependence are presented in Section 5. In Section 6 we explain how vectors of dependent random variables may be simulated using correct methods.

2 Copulas

Probability-integral and quantile transforms play a fundamental role when working with copulas. In the following proposition we collect together some essential facts that we use repeatedly in this article. The notation $X \sim F$ means that the random variable X has distribution function F.

Proposition 1. *Let X be a random variable with distribution function F. Let F^{-1} be the quantile function of F, i.e.*

$$F^{-1}(\alpha) = \inf\{x | F(x) \geq \alpha\},$$

$\alpha \in (0,1)$. *Then*

1. *For any standard-uniformly distributed $U \sim U(0,1)$ we have $F^{-1}(U) \sim F$. This gives a simple method for simulating random variates with distribution function F.*

2. *If F is continuous then the random variable $F(X)$ is standard-uniformly distributed, i.e. $F(X) \sim U(0,1)$.*

Proof. In most elementary texts on probability. □

2.1 What is a copula?

The dependence between the real-valued random variables $X_1, \ldots, X_n$ is completely described by their joint distribution function

$$F(x_1, \ldots, x_n) = \mathbb{P}[X_1 \leq x_1, \ldots, X_n \leq x_n].$$

The idea of separating F into a part which describes the dependence structure and parts which describe the marginal behaviour only, has led to the concept of a copula.

Suppose we transform the random vector $\mathbf{X} = (X_1, \ldots, X_n)^t$ componentwise to have standard-uniform marginal distributions, $\mathrm{U}(0,1)$[1]. For simplicity we assume to begin with that $X_1, \ldots, X_n$ have continuous marginal distributions $F_1, \ldots, F_n$, so that from Proposition 1 this can be achieved by using the probability-integral transformation $T : \mathbb{R}^n \to \mathbb{R}^n$, $(x_1, \ldots, x_n)^t \mapsto (F_1(x_1), \ldots, F_n(x_n))^t$. The joint distribution function C of $(F_1(X_1), \ldots, F_n(X_n))^t$ is then called the copula of the random vector $(X_1, \ldots, X_n)^t$ or the multivariate distribution F. It follows that

$$\begin{aligned} F(x_1, \ldots, x_n) &= \mathbb{P}[F_1(X_1) \leq F_1(x_1), \ldots, F_n(X_n) \leq F_n(x_n)] \\ &= C(F_1(x_1), \ldots, F_n(x_n)). \end{aligned} \tag{2.1}$$

Definition 1. A copula is the distribution function of a random vector $\mathbf{X} = (x_1, \ldots, x_n)^t$ in $\mathbb{R}^n$ with uniform-$(0,1)$ marginals. Alternatively a copula is any function $C : [0,1]^n \to [0,1]$ which has the three properties:

1. $C(x_1, \ldots, x_n)$ is increasing in each component x_i.

2. $C(1, \ldots, 1, x_i, 1, \ldots, 1) = x_i$ for all $i \in \{1, \ldots, n\}$, $x_i \in [0,1]$.

3. For all $(a_1, \ldots, a_n), (b_1, \ldots, b_n) \in [0,1]^n$ with $a_i \leq b_i$ we have:

$$\sum_{i_1=1}^{2} \cdots \sum_{i_n=1}^{2} (-1)^{i_1+\cdots+i_n} C(x_{1i_1}, \ldots, x_{ni_n}) \geq 0, \tag{2.2}$$

where $x_{j1} = a_j$ and $x_{j2} = b_j$ for all $j \in \{1, \ldots, n\}$.

These two alternative definitions can be shown to be equivalent. It is a particularly easy matter to verify that the first definition in terms of a multivariate distribution function with standard uniform marginals implies the three properties above: property 1 is clear; property 2 follows from the fact that the marginals are uniform-$(0,1)$; property 3 is true because the sum (2.2) can be interpreted as $\mathbb{P}[a_1 \leq X_1 \leq b_1, \ldots, a_n \leq X_n \leq b_n]$, which is non-negative.

[1] Alternatively one could transform to any other distribution, but $\mathrm{U}(0,1)$ is particularly easy.

For any continuous multivariate distribution the representation (2.2) holds for a *unique* copula C. If $F_1, \ldots, F_n$ are not all continuous it can still be shown (see Schweizer & Sklar (1983), Chapter 6) that the joint distribution function can always be expressed as in (2.2), although in this case C is *no longer* unique and we refer to it as a *possible* copula of F.

The representation (2.2), and some invariance properties which we will show shortly, suggest that we interpret a copula associated with $(X_1, \ldots X_n)^t$ as being *the dependence structure*. This makes particular sense when all the F_i are continuous and the copula is unique; in the discrete case there will be more than one way of writing the dependence structure. Pitfalls related to non-continuity of marginal distributions are presented in Marshall (1996). A recent, very readable introduction to copulas is Nelsen (1999).

2.2 Examples of copulas

For independent random variables the copula trivially takes the form

$$C_{ind}(x_1, \ldots, x_n) = x_1 \cdot \ldots \cdot x_n. \tag{2.3}$$

We now consider some particular copulas for non-independent pairs of random variables (X, Y) having continuous distributions. The Gaussian or normal copula is

$$C_\rho^{Ga}(x, y) = \int_{-\infty}^{\Phi^{-1}(x)} \int_{-\infty}^{\Phi^{-1}(y)} \frac{1}{2\pi(1-\rho^2)^{1/2}} \exp\left\{\frac{-(s^2 - 2\rho st + t^2)}{2(1-\rho^2)}\right\} dsdt, \tag{2.4}$$

where $-1 < \rho < 1$ and Φ is the univariate standard normal distribution function. Variables with standard normal marginal distributions and this dependence structure, i.e. variables with d.f. $C_\rho^{Ga}(\Phi(x), \Phi(y))$, are standard bivariate normal variables with correlation coefficient ρ. Another well-known copula is the Gumbel or logistic copula

$$C_\beta^{Gu}(x, y) = \exp\left[-\left\{(-\log x)^{1/\beta} + (-\log y)^{1/\beta}\right\}^\beta\right], \tag{2.5}$$

where $0 < \beta \leq 1$ is a parameter which controls the amount of dependence between X and Y; $\beta = 1$ gives independence and the limit of C_β^{Gu} for $\beta \to 0+$ leads to perfect dependence, as will be discussed in Section 4. This copula, unlike the Gaussian, is a copula which is consistent with bivariate extreme value theory and could be used to model the limiting dependence structure of component-wise maxima of bivariate random samples (Joe (1997), Galambos (1987)).

The following is a simple method for generating a variety of copulas which will be used later in the article. Let $f, g : [0,1] \to \mathbb{R}$ with $\int_0^1 f(x)dx =$

$\int_0^1 g(y)dy = 0$ and $f(x)g(y) \geq -1$ for all $x, y \in [0,1]$. Then $h(x,y) = 1 + f(x)g(y)$ is a bivariate density function on $[0,1]^2$. Consequently,

$$C(x,y) = \int_0^x \int_0^y h(u,v)dudv = xy + \left(\int_0^x f(u)du\right)\left(\int_0^y g(v)dv\right) \tag{2.6}$$

is a copula. If we choose $f(x) = \alpha(1-2x)$, $g(y) = (1-2y)$, $|\alpha| \leq 1$, we obtain, for example, the Farlie–Gumbel–Morgenstern copula $C(x,y) = xy[1 + \alpha(1-x)(1-y))]$. Many copulas and methods to construct them can be found in the literature; see for example Hutchinson & Lai (1990) or Joe (1997).

2.3 Invariance

The following proposition shows one attractive feature of the copula representation of dependence, namely that the dependence structure as summarized by a copula is *invariant* under increasing and continuous transformations of the marginals.

Proposition 2. *If $(X_1, \dots, X_n)^t$ has copula C and $T_1, \dots, T_n$ are increasing continuous functions, then $(T_1(X_1), \dots, T_n(X_n))^t$ also has copula C.*

Proof. Let $(U_1, \dots, U_n)^t$ have distribution function C (in the case of continuous marginals F_{X_i} take $U_i = F_{X_i}(X_i)$). We may write

$$\begin{aligned}
&C(F_{T_1(X_1)}(x_1), \dots, F_{T_n(X_n)}(x_n)) \\
&= \mathbb{P}[U_1 \leq F_{T_1(X_1)}(x_1), \dots, U_n \leq F_{T_n(X_n)}(x_n)] \\
&= \mathbb{P}[F^{-1}_{T_1(X_1)}(U_1) \leq x_1, \dots, F^{-1}_{T_n(X_n)}(U_n) \leq x_n] \\
&= \mathbb{P}[T_1 \circ F^{-1}_{X_1}(U_1) \leq x_1, \dots, T_n \circ F^{-1}_{X_n}(U_n) \leq x_n] \\
&= \mathbb{P}[T_1(X_1) \leq x_1, \dots, T_n(X_n) \leq x_n].
\end{aligned}$$

□

Remark 1. The continuity of the transformations T_i is necessary for general random variables $(X_1, \dots, X_n)^t$ since, in that case, $F^{-1}_{T_i(X_i)} = T_i \circ F^{-1}_{X_i}$. In the case where all marginal distributions of $\mathbf{X}$ are continuous it suffices that the transformations are increasing (see also Chapter 6 of Schweizer & Sklar (1983)).

As a simple illustration of the relevance of this result, suppose we have a probability model (multivariate distribution) for dependent insurance losses of various kinds. If we decide that our interest now lies in modelling the logarithm of these losses, the copula will not change. Similarly if we change from a model of returns on several financial assets to a model of logarithmic returns, the copula will not change, only the marginal distributions.

3 Linear Correlation

3.1 What is correlation?

We begin by considering pairs of real-valued, non-degenerate random variables X, Y with finite variances.

Definition 2. The linear correlation coefficient between X and Y is

$$\rho(X,Y) = \frac{\mathrm{Cov}[X,Y]}{\sqrt{\sigma^2[X]\sigma^2[Y]}},$$

where $\mathrm{Cov}[X,Y]$ is the covariance between X and Y, $\mathrm{Cov}[X,Y] = \mathbb{E}[XY] - \mathbb{E}[X]\mathbb{E}[Y]$ and $\sigma^2[X], \sigma^2[Y]$ denote the variances of X and Y respectively.

The linear correlation is a measure of *linear* dependence. In the case of independent random variables, $\rho(X,Y) = 0$ since $\mathrm{Cov}[X,Y] = 0$. In the case of perfect linear dependence, i.e. $Y = aX + b$ a.s. or $\mathbb{P}[Y = aX + b] = 1$ for $a \in \mathbb{R} \setminus \{0\}$, $b \in \mathbb{R}$, we have $\rho(X,Y) = \pm 1$. This is shown by considering the representation

$$\rho(X,Y)^2 = \frac{\sigma^2[Y] - \min_{a,b} \mathbb{E}[(Y - (aX+b))^2]}{\sigma^2[Y]}. \tag{3.1}$$

In the case of imperfect linear dependence, $-1 < \rho(X,Y) < 1$, and this is the case when misinterpretations of correlation are possible, as will later be seen in Section 5. Equation (3.1) shows the connection between correlation and simple linear regression. The right hand side can be interpreted as the relative reduction in the variance of Y by linear regression on X. The regression coefficients a_R, b_R, which minimise the squared distance $\mathbb{E}[(Y - (aX+b))^2]$ are given by

$$\begin{aligned} a_R &= \frac{\mathrm{Cov}[X,Y]}{\sigma^2[X]}, \\ b_R &= \mathbb{E}[Y] - a_R\mathbb{E}[X]. \end{aligned}$$

Correlation satisfies the linearity property

$$\rho(\alpha X + \beta, \gamma Y + \delta) = \mathrm{sgn}(\alpha \cdot \gamma)\rho(X,Y),$$

when $\alpha, \gamma \in \mathbb{R} \setminus \{0\}$, $\beta, \delta \in \mathbb{R}$. Correlation is thus invariant under positive, i.e. strictly increasing, affine transformations.

The generalisation of correlation to more than two random variables is straightforward. Consider vectors of random variables $\mathbf{X} = (X_1, \ldots, X_n)^t$ and $\mathbf{Y} = (Y_1, \ldots, Y_n)^t$ in $\mathbb{R}^n$. We can summarise all pairwise covariances and correlations in $n \times n$ matrices $\mathrm{Cov}[\mathbf{X}, \mathbf{Y}]$ and $\rho(\mathbf{X}, \mathbf{Y})$. As long as the corresponding variances are finite we define

$$\begin{aligned} \mathrm{Cov}[\mathbf{X}, \mathbf{Y}]_{ij} &:= \mathrm{Cov}[X_i, Y_j], \\ \rho(\mathbf{X}, \mathbf{Y})_{ij} &:= \rho(X_i, Y_j)\ 1 \leq i, j \leq n. \end{aligned}$$

It is well known that these matrices are symmetric and positive semi-definite. Often one considers only pairwise correlations between components of a single random vector; for this purpose when $\mathbf{Y} = \mathbf{X}$ we consider $\rho(\mathbf{X}) := \rho(\mathbf{X}, \mathbf{X})$ or $\mathrm{Cov}[\mathbf{X}] := \mathrm{Cov}[\mathbf{X}, \mathbf{X}]$.

The popularity of linear correlation can be explained in several ways. Correlation is often straightforward to calculate. For many bivariate distributions it is a simple matter to calculate second moments (variances and covariances) and hence to derive the correlation coefficient. Alternative measures of dependence, which we will encounter in Section 4 may be more difficult to calculate.

Moreover, correlation and covariance are easy to manipulate under linear operations. Under linear transformations $\mathbb{R}^n \to \mathbb{R}^m$ defined by $x \mapsto Ax + a$ and $x \mapsto Bx + b$ for $A, B \in \mathbb{R}^{m \times n}$, $a, b \in \mathbb{R}^m$ we have

$$\mathrm{Cov}[A\mathbf{X} + a, B\mathbf{Y} + b] = A\mathrm{Cov}[\mathbf{X}, \mathbf{Y}]B^t.$$

A special case is the following elegant relationship between variance and covariance for a random vector. For every linear combination of the components $\alpha^t\mathbf{X}$ with $\alpha \in \mathbb{R}^n$,

$$\sigma^2[\alpha^t\mathbf{X}] = \alpha^t\mathrm{Cov}[\mathbf{X}]\alpha.$$

Thus, the variance of any linear combination is fully determined by the pairwise covariances between the components. This fact is commonly exploited in portfolio theory.

A third reason for the popularity of correlation is its naturalness as a measure of dependence in multivariate normal distributions and, more generally, in multivariate spherical and elliptical distributions, as will shortly be discussed. First, we mention a few disadvantages of correlation.

3.2 Shortcomings of correlation

We consider again the case of two jointly distributed real-valued random variables X and Y.

- The variances of X and Y must be finite or the linear correlation is not defined. This is not ideal for a dependence measure and causes problems when we work with heavy-tailed distributions. For example, the covariance and the correlation between the two components of a bivariate t_ν-distributed random vector are not defined for $\nu \leq 2$. Non-life actuaries who model losses in different business lines with infinite variance distributions must be aware of this.

- Independence of two random variables implies they are uncorrelated (linear correlation equal to zero) but zero correlation does not in general imply independence. A simple example where the covariance disappears despite strong dependence between random variables is obtained by taking $X \sim \mathcal{N}(0,1)$, $Y = X^2$, since the third moment of the standard normal distribution is zero. Only in the case of the *multivariate* normal is it permissible to interpret uncorrelatedness as implying independence. This implication is no longer valid when only the marginal distributions are normal and the *joint* distribution is non-normal, which will also be demonstrated below in Example 1 in Section 5. The class of spherical distributions model uncorrelated random variables but are not, except in the case of the multivariate normal, the distributions of independent random variables.
- Linear correlation has the serious deficiency that it is *not* invariant under non-linear strictly increasing transformations $T : \mathbb{R} \to \mathbb{R}$. For two real-valued random variables we have in general
$$\rho(T(X), T(Y)) \neq \rho(X, Y).$$
If we take the bivariate standard normal distribution with correlation ρ and the transformation $T(x) = \Phi(x)$ (the standard normal distribution function) we have
$$\rho(T(X), T(Y)) = \frac{6}{\pi} \arcsin\left(\frac{\rho}{2}\right), \tag{3.2}$$
see Joag-dev (1984). In general one can also show (see Kendall & Stuart (1979), page 600) for bivariate normally-distributed vectors and arbitrary transformations $T, \widetilde{T} : \mathbb{R} \to \mathbb{R}$ that
$$|\rho(T(X), \widetilde{T}(Y))| \leq |\rho(X, Y)|,$$
which is implied by (3.2) in particular.

3.3 Spherical and elliptical distributions

The spherical distributions extend the standard multivariate normal distribution $\mathcal{N}_n(0, I)$, i.e. the distribution of independent standard normal variables. They provide a family of symmetric distributions for uncorrelated random vectors with mean zero.

Definition 3. A random vector $\mathbf{X} = (X_1, \ldots, X_n)^t$ has a spherical distribution if for every orthogonal linear matrix $U \in \mathbb{R}^{n\times n}$ (i.e. matrices satisfying $UU^t = U^tU = I_{n\times n}$)

$$U\mathbf{X} =_{\mathrm{d}} \mathbf{X}.$$ [2]

[1]We use $=_{\mathrm{d}}$ to denote equality in distribution, as is standard.

The characteristic function $\psi(\mathbf{t}) = \mathbb{E}[\exp(i\mathbf{t}^t\mathbf{X})]$ of such distributions takes a particularly simple form. There exists a function $\phi : \mathbb{R}_+ \to \mathbb{R}$ such that $\psi(\mathbf{t}) = \phi(\mathbf{t}^t\mathbf{t}) = \phi(t_1^2 + \cdots + t_n^2)$. For example, for the multivariate Gaussian distribution with uncorrelated standard normal components,

$$\phi(t) = \exp[-(t_1^2 + \cdots + t_n^2)/2]. \tag{3.3}$$

The function ϕ is the *characteristic generator* of the spherical distribution and we write

$$\mathbf{X} \sim S_n(\phi).$$

If $\mathbf{X}$ has a density $f(\mathbf{x}) = f(x_1, \ldots, x_n)$ then this is equivalent to $f(\mathbf{x}) = g(\mathbf{x}^t\mathbf{x}) = g(x_1^2 + \cdots + x_n^2)$ for some function $g : \mathbb{R}_+ \to \mathbb{R}_+$, so that the spherical distributions are best interpreted as those distributions whose density is constant on spheres. Some other examples of densities in the spherical class are those of the multivariate t-distribution with ν degrees of freedom $f(\mathbf{x}) = c(1 + \mathbf{x}^t\mathbf{x}/\nu)^{-(n+\nu)/2}$ and the logistic distribution $f(\mathbf{x}) = c\exp(-\mathbf{x}^t\mathbf{x})/[1 + \exp(-\mathbf{x}^t\mathbf{x})]^2$, where c is a generic normalizing constant. Note that these are the distributions of uncorrelated random variables but, contrary to the normal case, *not* the distributions of independent random variables. In the class of spherical distributions the multivariate normal is the only distribution of independent random variables; see Fang, Kotz & Ng (1987), page 106.

The spherical distributions admit an alternative stochastic representation. $\mathbf{X} \sim S_n(\phi)$ if and only if

$$\mathbf{X} =_{\mathrm{d}} R \cdot \mathbf{U}, \tag{3.4}$$

where the random vector $\mathbf{U}$ is uniformly distributed on the unit hypersphere $S_{n-1} = \{\mathbf{x} \in \mathbb{R}^n | \mathbf{x}^t\mathbf{x} = 1\}$ in $\mathbb{R}^n$ and $R \geq 0$ is a positive random variable, independent of $\mathbf{U}$ (Fang *et al.* (1987), page 30). Spherical distributions can thus be interpreted as mixtures of uniform distributions on spheres of differing radius in $\mathbb{R}^n$. For example, in the case of the standard multivariate normal distribution the *generating variate* satisfies $R \sim \sqrt{\chi_n^2}$, and in the case of the multivariate t-distribution with ν degrees of freedom $R^2/n \sim F(n, \nu)$ holds, where $F(n, \nu)$ denotes an F-distribution with n and ν degrees of freedom.

Elliptical distributions extend the multivariate normal $\mathcal{N}_n(\mu, \Sigma)$, i.e. the distribution with mean μ and covariance matrix Σ. Mathematically they are the affine maps of spherical distributions in $\mathbb{R}^n$.

Definition 4. Let $T : \mathbb{R}^n \to \mathbb{R}^n, \mathbf{x} \mapsto A\mathbf{x} + \mu$, $A \in \mathbb{R}^{n \times n}$, $\mu \in \mathbb{R}^n$ be an affine map. $\mathbf{X}$ has an elliptical distribution if $\mathbf{X} = T(\mathbf{Y})$ and $\mathbf{Y} \sim S_n(\phi)$.

Since the characteristic function can be written as

$$\begin{aligned}\psi(\mathbf{t}) &= \mathbb{E}[\exp(i\mathbf{t}^t\mathbf{X})] = \mathbb{E}[\exp(i\mathbf{t}^t(A\mathbf{Y} + \mu))] \\ &= \exp(i\mathbf{t}^t\mu)\,\mathbb{E}\big[\exp(i(A^t\mathbf{t})^t\mathbf{Y})\big] = \exp(i\mathbf{t}^t\mu)\phi(\mathbf{t}^t\Sigma\mathbf{t}),\end{aligned}$$

where $\Sigma := AA^t$, we denote the elliptical distributions

$$\mathbf{X} \sim \mathrm{E}_n(\mu, \Sigma, \phi).$$

For example, $\mathcal{N}_n(\mu, \Sigma) = \mathrm{E}_n(\mu, \Sigma, \phi)$ with ϕ given by (3.4). If $\mathbf{Y}$ has a density $f(\mathbf{y}) = g(\mathbf{y}^t\mathbf{y})$ and if A is regular (i.e. $\det(A) \neq 0$ so that Σ is strictly positive-definite), then $\mathbf{X} = A\mathbf{Y} + \mu$ has density

$$h(\mathbf{x}) = \frac{1}{\sqrt{\det(\Sigma)}} g((\mathbf{x} - \mu)^t \Sigma^{-1} (\mathbf{x} - \mu)),$$

and the contours of equal density are now ellipsoids. For example for the multivariate Gaussian distribution $g(t) = \exp(-t^2/2)/\sqrt{2\pi}$.

Knowledge of the distribution of $\mathbf{X}$ does not completely determine the elliptical representation $\mathrm{E}_n(\mu, \Sigma, \phi)$; it uniquely determines μ but Σ and ϕ are only determined up to a positive constant[3]. In particular Σ can be chosen so that it is directly interpretable as the covariance matrix of $\mathbf{X}$, although this is not always standard. Let $\mathbf{X} \sim \mathrm{E}_n(\mu, \Sigma, \phi)$, so that $\mathbf{X} =_{\mathrm{d}} \mu + A\mathbf{Y}$ where $\Sigma = AA^t$ and $\mathbf{Y}$ is a random vector satisfying $\mathbf{Y} \sim S_n(\phi)$. Equivalently $\mathbf{Y} =_{\mathrm{d}} R \cdot \mathbf{U}$, where $\mathbf{U}$ is uniformly distributed on S^{n-1} and R is a positive random variable independent of $\mathbf{U}$. If $\mathbb{E}[R^2] < \infty$ it follows that $\mathbb{E}[\mathbf{X}] = \mu$ and $\mathrm{Cov}[\mathbf{X}] = AA^t\mathbb{E}[R^2]/n = \Sigma\mathbb{E}[R^2]/n$ since $\mathrm{Cov}[\mathbf{U}] = I_{n \times n}/n$. By starting with the characteristic generator $\widetilde{\phi}(u) := \phi(u/c)$ with $c = n/\mathbb{E}[R^2]$ we ensure that $\mathrm{Cov}[\mathbf{X}] = \Sigma$. An elliptical distribution is thus fully described by its mean, its covariance matrix and its characteristic generator.

We now consider some of the reasons why correlation and covariance are natural measures of dependence in the world of elliptical distributions. First, many of the properties of the multivariate normal distribution are shared by the elliptical distributions. Linear combinations, marginal distributions and conditional distributions of elliptical random variables can largely be determined by linear algebra using knowledge of covariance matrix, mean and generator. This is summarized in the following properties.

- Any affine transformation of an elliptically distributed random vector is also elliptical with the same characteristic generator ϕ. If $\mathbf{X} \sim \mathrm{E}_n(\mu, \Sigma, \phi)$ and $B \in \mathbb{R}^{m \times n}$, $b \in \mathbb{R}^m$, then

$$B\mathbf{X} + b \sim \mathrm{E}_m(B\mu + b, B\Sigma B^t, \phi).$$

  It is immediately clear that the components $X_1, \ldots, X_n$ are all symmetrically distributed random variables of the same type[4].

[3] If X is elliptical and non-degenerate there exists μ, A and $\mathbf{Y} \sim S_n(\phi)$ so that $\mathbf{X} =_{\mathrm{d}} A\mathbf{Y} + \mu$, but for any $\lambda \in \mathbb{R} \setminus \{0\}$ we also have $\mathbf{X} =_{\mathrm{d}} (A/\lambda)\lambda\mathbf{Y} + \mu$ where $\lambda\mathbf{Y} \sim S_n(\widetilde{\phi})$ and $\widetilde{\phi}(u) := \phi(\lambda^2 u)$. In general, if $X \sim \mathrm{E}_n(\mu, \Sigma, \phi) = \mathrm{E}_n(\widetilde{\mu}, \widetilde{\Sigma}, \widetilde{\phi})$ then $\mu = \widetilde{\mu}$ and there exists $c > 0$ so that $\widetilde{\Sigma} = c\Sigma$ and $\widetilde{\phi}(u) = \phi(u/c)$ (see Fang *et al.* (1987), page 43).

[4] Two random variables X und Y are of the same type if we can find $a > 0$ and $b \in \mathbb{R}$ so that $Y =_{\mathrm{d}} aX + b$.

- The marginal distributions of elliptical distributions are also elliptical with the same generator. Let $\mathbf{X} = \begin{pmatrix}\mathbf{X}_1\\ \mathbf{X}_2\end{pmatrix} \sim \mathrm{E}_n(\Sigma, \mu, \phi)$ with $\mathbf{X}_1 \in \mathbb{R}^p$, $\mathbf{X}_2 \in \mathbb{R}^q$, $p + q = n$. Let $\mathbb{E}[\mathbf{X}] = \mu = \begin{pmatrix}\mu_1\\ \mu_2\end{pmatrix}$, $\mu_1 \in \mathbb{R}^p$, $\mu_2 \in \mathbb{R}^q$ and $\Sigma = \begin{pmatrix}\Sigma_{11} & \Sigma_{12}\\ \Sigma_{21} & \Sigma_{22}\end{pmatrix}$, accordingly. Then
$$\mathbf{X}_1 \sim \mathrm{E}_p(\mu_1, \Sigma_{11}, \phi), \ \mathbf{X}_2 \sim \mathrm{E}_q(\mu_2, \Sigma_{22}, \phi).$$
- We assume that Σ is strictly positive-definite. The conditional distribution of $\mathbf{X}_1$ given $\mathbf{X}_2$ is also elliptical, although in general with a different generator ϕ:
$$\mathbf{X}_1|\mathbf{X}_2 \sim \mathrm{E}_p(\mu_{1.2}, \Sigma_{11.2}, \widetilde{\phi}), \tag{3.5}$$
where $\mu_{1.2} = \mu_1 + \Sigma_{12}\Sigma_{22}^{-1}(\mathbf{X}_2 - \mu_2)$, $\Sigma_{11.2} = \Sigma_{11} - \Sigma_{12}\Sigma_{22}^{-1}\Sigma_{21}$. The distribution of the generating variable $\widetilde{R}$ in (3.4) corresponding to $\widetilde{\phi}$ is the conditional distribution is
$$\sqrt{(\mathbf{X} - \mu)^t\Sigma^{-1}(\mathbf{X} - \mu) - (\mathbf{X}_2 - \mu_2)^t\Sigma_{22}^{-1}(\mathbf{X}_2 - \mu_2)},$$
given $\mathbf{X}_2$. Since in the case of multivariate normality uncorrelatedness is equivalent to independence we have $\widetilde{R} =_{\mathrm{d}} \sqrt{\chi_p^2}$ and $\widetilde{\phi} = \phi$, so that the conditional distribution is of the same type as the unconditional; for general elliptical distributions this is not true. From (3.5) we see that
$$\mathbb{E}[\mathbf{X}_1|\mathbf{X}_2] = \mu_{1.2} = \mu_1 + \Sigma_{12}\Sigma_{22}^{-1}(\mathbf{X}_2 - \mu_2),$$
so that the best prediction of $\mathbf{X}_1$ given $\mathbf{X}_2$ is linear in $\mathbf{X}_2$ and is simply the linear regression of $\mathbf{X}_1$ on $\mathbf{X}_2$. In the case of multivariate normality we have additionally
$$\mathrm{Cov}[\mathbf{X}_1|\mathbf{X}_2] = \Sigma_{11.2} = \Sigma_{11} - \Sigma_{12}\Sigma_{22}^{-1}\Sigma_{21},$$
which is independent of $\mathbf{X}_2$. The independence of the conditional covariance from $\mathbf{X}_2$ is also a characterisation of the multivariate normal distribution in the class of elliptical distributions (Kelker 1970).

Since the type of all marginal distributions is the same, we see that an elliptical distribution is *uniquely* determined by its mean, its covariance matrix and knowledge of this type. Alternatively the dependence structure (copula) of a continuous elliptical distribution is *uniquely* determined by the correlation matrix and knowledge of this type. For example, the copula of the bivariate t-distribution with ν degrees of freedom and correlation ρ is
$$C_{\nu,\rho}^t(x,y) = \int_{-\infty}^{t_\nu^{-1}(x)}\int_{-\infty}^{t_\nu^{-1}(y)} \frac{1}{2\pi(1-\rho^2)^{1/2}}\left\{1 + \frac{(s^2 - 2\rho st + t^2)}{\nu(1-\rho^2)}\right\}^{-(\nu+2)/2} ds\, dt, \tag{3.6}$$

where $t_\nu^{-1}(x)$ denotes the inverse of the distribution function of the standard univariate t-distribution with ν degrees of freedom. This copula is seen to depend only on ρ and ν.

An important question is: which univariate types are possible for the marginal distribution of an elliptical distribution in $\mathbb{R}^n$ for any $n \in \mathbb{N}$? Without loss of generality, it is sufficient to consider the spherical case (Fang *et al.* (1987), pages 48–51). F is the marginal distribution of a spherical distribution in $\mathbb{R}^n$ for any $n \in \mathbb{N}$ if and only if F is a mixture of centred normal distributions. In other words, if F has a density f, the latter is of the form,

$$f(x) = \frac{1}{\sqrt{2\pi}} \int_0^\infty \frac{1}{\sigma} \exp\Big(-\frac{x^2}{2\sigma^2}\Big) G(d\sigma),$$

where G is a distribution function on $[0, \infty)$ with $G(0) = 0$. The corresponding spherical distribution has the alternative stochastic representation

$$X =_{\mathrm{d}} S \cdot Z,$$

where $S \sim G$, $Z \sim \mathcal{N}_n(0, I_{n\times n})$ and S and Z are independent. For example, the multivariate t-distribution with ν degrees of freedom can be constructed by taking $S \sim \sqrt{\nu}/\sqrt{\chi^2_\nu}$.

3.4 Covariance and elliptical distributions in risk management

A further important feature of the elliptical distributions is that these distributions are amenable to the *standard* approaches of risk management. They support both the use of *Value-at-Risk* as a measure of risk and the mean-variance (Markowitz) approach (see e.g. Campbell *et al.* (1997)) to risk management and portfolio optimization.

Suppose that $\mathbf{X} = (X_1, \dots, X_n)^t$ represents n risks with an elliptical distribution with mean 0 and that we consider *linear* portfolios of such risks

$$\Big\{ Z = \sum_{i=1}^n \lambda_i X_i \mid \lambda_i \in \mathbb{R} \Big\}$$

with distribution F_Z. The Value-at-Risk (VaR) of portfolio Z at probability level α is given by

$$\mathrm{VaR}_\alpha(Z) = F_Z^{-1}(\alpha) = \inf\{z \in \mathbb{R} : F_Z(z) \geq \alpha\};$$

i.e. it is simply an alternative notation for the quantile function of F_Z evaluated at α and we will thus often use $\mathrm{VaR}_\alpha(Z)$ and $F_Z^{-1}(\alpha)$ interchangeably.[5]

[5]We interpret large positive values of the random variable Z as *losses* and concentrate on quantiles in the right-hand tail of F_Z. This is in contrast to Artzner *et al.* (1999), who interpret *negative* values as losses.

In the elliptical world the use of VaR as a measure of the risk of a portfolio Z makes sense because VaR is a coherent risk measure in this world. A coherent risk measure in the sense of Artzner *et al.* (1999) is a real-valued function ϱ on the space of real-valued random variables only depending on the probability distribution which fulfills the following (sensible) properties:

A1. (Monotonicity). For any two random variables $X \geq Y$: $\varrho(X) \geq \varrho(Y)$.
A2. (Subadditivity). For any two random variables X and Y we have
$$\varrho(X+Y) \leq \varrho(X) + \varrho(Y).$$
A3. (Positive homogeneity). For $\lambda \geq 0$ we have that $\varrho(\lambda X) = \lambda \varrho(X)$.
A4. (Translation invariance).
For any $a \in \mathbb{R}$ we have that $\varrho(X + a) = \varrho(X) + a$.

In the elliptical world the use of any positive homogeneous, translation-invariant measure of risk to rank risks or to determine optimal risk-minimizing portfolio weights under the condition that a certain *return* is attained, is equivalent to the Markowitz approach where the variance is used as risk measure. Alternative risk measures such as VaR_α or expected shortfall, $\mathbb{E}[Z|Z > \mathrm{VaR}_\alpha(Z)]$, give different numerical values, but have no effect on the management of risk. We make these assertions more precise in Theorem 1.

Throughout this article for notational and pedagogical reasons we use VaR in its most simplistic form, i.e. disregarding questions of appropriate horizon, estimation of the underlying profit-and-loss distribution, etc. However, the key messages stemming from this oversimplified view carry over to more concrete VaR calculations in practice.

Theorem 1. *Suppose* $\mathbf{X} \sim E_n(\mu, \Sigma, \phi)$ *with* $\sigma^2[X_i] < \infty$ *for all* i. *Let*

$$\mathcal{P} = \left\{ Z = \sum_{i=1}^{n} \lambda_i X_i \mid \lambda_i \in \mathbb{R} \right\}$$

be the set of all linear portfolios. Then the following are true.

1. *(Subadditivity of VaR.) For any two portfolios* $Z_1, Z_2 \in \mathcal{P}$ *and* $0.5 \leq \alpha < 1$,

$$\mathit{VaR}_\alpha(Z_1 + Z_2) \leq \mathit{VaR}_\alpha(Z_1) + \mathit{VaR}_\alpha(Z_2).$$

2. *(Equivalence of variance and positive homogeneous risk measurement.) Let* ϱ *be a real-valued risk measure on the space of real-valued random variables which depends only on the distribution of each random variable* X. *Suppose this measure satisfies A3. Then for* $Z_1, Z_2 \in \mathcal{P}$

$$\varrho(Z_1 - \mathbb{E}[Z_1]) \leq \varrho(Z_2 - \mathbb{E}[Z_2]) \iff \sigma^2[Z_1] \leq \sigma^2[Z_2].$$

3. *(Markowitz risk-minimizing portfolio.) Let ϱ be as in 2 and assume that A4 is also satisfied. Let*

$$\mathcal{E} = \left\{ Z = \sum_{i=1}^{n} \lambda_i X_i \mid \lambda_i \in \mathbb{R}, \sum_{i=1}^{n} \lambda_i = 1, \mathbb{E}[Z] = r \right\}$$

be the subset of portfolios giving expected return r. Then

$$\text{argmin}_{Z\in\mathcal{E}}\ \varrho(Z) = \text{argmin}_{Z\in\mathcal{E}}\ \sigma^2[Z].$$

Proof. The main observation is that $(Z_1, Z_2)^t$ has an elliptical distribution so Z_1, Z_2 and $Z_1 + Z_2$ all have distributions of the same type.

1. Let q_α be the α-quantile of the standardised distribution of this type. Then

$$\begin{aligned}
\text{VaR}_\alpha(Z_1) &= \mathbb{E}[Z_1] + \sigma[Z_1]q_\alpha, \\
\text{VaR}_\alpha(Z_2) &= \mathbb{E}[Z_2] + \sigma[Z_2]q_\alpha, \\
\text{VaR}_\alpha(Z_1 + Z_2) &= \mathbb{E}[Z_1 + Z_2] + \sigma[Z_1 + Z_2]q_\alpha.
\end{aligned}$$

   Since $\sigma[Z_1 + Z_2] \leq \sigma[Z_1] + \sigma[Z_2]$ and $q_\alpha \geq 0$ the result follows.

2. Since Z_1 and Z_2 are random variables of the same type, there exists an $a > 0$ such that $Z_1 - \mathbb{E}[Z_1] =_{\mathrm{d}} a(Z_2 - \mathbb{E}[Z_2])$. It follows that

$$\varrho(Z_1 - \mathbb{E}[Z_1]) \leq \varrho(Z_2 - \mathbb{E}[Z_2]) \iff a \leq 1 \iff \sigma^2[Z_1] \leq \sigma^2[Z_2].$$

3. Follows from 2 and the fact that we optimize over portfolios with identical expectation.

□

While this theorem shows that in the *elliptical world* the Markowitz variance-minimizing portfolio minimizes popular risk measures like VaR and expected shortfall (both of which are coherent in this world), it can also be shown that the Markowitz portfolio minimizes some other risk measures which do not satisfy A3 and A4. The partial moment measures of downside risk provide an example. The kth (upper) partial moment of a random variable X with respect to a threshold τ is defined to be

$$\text{UPM}_{k,\tau}(X) = \mathbb{E}\left[\{(X-\tau)_+\}^k\right],\ k \geq 0, \tau \in \mathbb{R}.$$

Suppose we have portfolios $Z_1, Z_2 \in \mathcal{E}$ and assume additionally that $r \leq \tau$, so that the threshold is set above the expected return r. Using a similar approach to the preceding theorem it can be shown that

$$\sigma^2[Z_1] \leq \sigma^2[Z_2] \iff (Z_1 - \tau) =_{\mathrm{d}} a(Z_2 - \tau) - (1-a)(\tau - r),$$

with $0 < a \leq 1$. It follows that

$$\mathrm{UPM}_{k,\tau}(Z_1) \leq \mathrm{UPM}_{k,\tau}(Z_2) \iff \sigma^2[Z_1] \leq \sigma^2[Z_2],$$

from which the equivalence to Markowitz is clear. See Harlow (1991) for an empirical case study of the change in the optimal asset allocation when $\mathrm{UPM}_{1,\tau}$ (target shortfall) and $\mathrm{UPM}_{2,\tau}$ (target semi-variance) are used.

4 Alternative dependence concepts

We begin by clarifying what we mean by the notion of perfect dependence. We go on to discuss other measures of dependence, in particular rank correlation. We concentrate on pairs of random variables.

4.1 Comonotonicity

For every copula the well-known Fréchet bounds apply (Fréchet (1957))

$$\underbrace{\max\{x_1 + \cdots + x_n + 1 - n, 0\}}_{C_\ell(x_1,\dots,x_n)} \leq C(x_1, \dots, x_n) \leq \underbrace{\min\{x_1, \dots, x_n\}}_{C_u(x_1,\dots,x_n)}; \quad (4.1)$$

these follow from the fact that every copula is the distribution function of a random vector $(U_1, \dots, U_n)^t$ with $U_i \sim \mathrm{U}(0,1)$. In the case $n = 2$ the bounds C_ℓ and C_u are themselves copulas since, if $U \sim \mathrm{U}(0,1)$, then

$$\begin{aligned} C_\ell(x_1, x_2) &= \mathbb{P}[U \leq x_1, 1 - U \leq x_2] \\ C_u(x_1, x_2) &= \mathbb{P}[U \leq x_1, U \leq x_2], \end{aligned}$$

so that C_ℓ and C_u are the bivariate distribution functions of the vectors $(U, 1-U)^t$ and $(U, U)^t$ respectively.

The distribution of $(U, 1-U)^t$ has all its mass on the diagonal between $(0,1)$ and $(1,0)$, whereas that of $(U,U)^t$ has its mass on the diagonal between $(0,0)$ and $(1,1)$. In these cases we say that C_ℓ and C_u describe perfect negative and perfect positive dependence respectively. This is formalized in the following theorem.

Theorem 2. *Let $(X,Y)^t$ have one of the copulas C_ℓ or C_u.*[6] *(In the former case this means $F(x_1, x_2) = \max\{F_1(x_1) + F_2(x_2) - 1, 0\}$; in the latter $F(x_1, x_2) = \min\{F_1(x_1), F_2(x_2)\}$.) Then there exist two monotonic functions $u, v : \mathbb{R} \to \mathbb{R}$ and a real-valued random variable Z so that*

$$(X,Y)^t =_d (u(Z), v(Z))^t,$$

with u increasing and v decreasing in the former case and with both increasing in the latter. The converse of this result is also true.

[6] If there are discontinuities in F_1 or F_2 so that the copula is not unique, then we interpret C_ℓ and C_u as being *possible copulas*.

Proof. The proof for the second case is given essentially in Wang & Dhaene (1998). A geometrical interpretation of Fréchet copulas is given in Mikusinski, Sherwood & Taylor (1992). We consider only the first case $C = C_\ell$, the proofs being similar. Let U be a U(0,1)-distributed random variable. We have

$$(X,Y)^t =_{\rm d} (F_1^{-1}(U), F_2^{-1}(1-U))^t = (F_1^{-1}(U), F_2^{-1} \circ g\,(U))^t,$$

where $F_i^{-1}(q) = \inf_{x \in \mathbb{R}}\{F_i(x) \geq q\}$, $q \in (0,1)$ is the quantile function of F_i, $i = 1, 2$, and $g(x) = 1 - x$. It follows that $u := F_1^{-1}$ is increasing and $v := F_2^{-1} \circ g$ is decreasing. For the converse assume

$$(X,Y)^t =_{\rm d} (u(Z), v(Z))^t,$$

with u and v increasing and decreasing respectively. We define $A := \{Z \in u^{-1}((-\infty, x])\}$, $B := \{Z \in v^{-1}((-\infty, y])\}$. If $A \cap B \neq \emptyset$ then the monotonicity of u and v imply that with respect to the distribution of Z

$$\mathbb{P}[A \cup B] = \mathbb{P}[\Omega] = 1 = \mathbb{P}[A] + \mathbb{P}[B] - \mathbb{P}[A \cap B]$$

and hence $\mathbb{P}[A \cap B] = \mathbb{P}[u(Z) \leq x, v(Z) \leq y] = F_1(x) + F_2(y) - 1$. If $A \cap B = \emptyset$, then $F_1(x) + F_2(y) - 1 \leq 0$. In all cases we have

$$\mathbb{P}[u(Z) \leq x, v(Z) \leq y] = \max\{F_1(x) + F_2(y) - 1, 0\}.$$

□

We introduce the following terminology.

Definition 5. [Yaari (1987)] If $(X,Y)^t$ has the copula C_u (see again footnote 6) then X and Y are said to be *comonotonic*; if it has copula C_ℓ they are said to be *countermonotonic.*

In the case of continuous distributions F_1 and F_2 a stronger version of the result can be stated:

$$C = C_\ell \iff Y = T(X) \text{ a.s., } T = F_2^{-1} \circ (1 - F_1) \text{ decreasing,} \quad (4.2)$$
$$C = C_u \iff Y = T(X) \text{ a.s., } T = F_2^{-1} \circ F_1 \text{ increasing.} \quad (4.3)$$

4.2 Desired properties of dependence measures

A measure of dependence, like linear correlation, summarises the dependence structure of two random variables in a single number. We consider the properties that we would like to have from this measure. Let $\delta(\cdot,\cdot)$ be a dependence measure which assigns a real number to any pair of real-valued random variables X and Y. Ideally, we desire the following properties:

P1. $\delta(X,Y) = \delta(Y,X)$ (symmetry).
P2. $-1 \le \delta(X,Y) \le 1$ (normalisation).
P3. $\delta(X,Y) = 1 \iff X,Y$ comonotonic;
$\delta(X,Y) = -1 \iff X,Y$ countermonotonic.
P4. For $T : \mathbb{R} \to \mathbb{R}$ strictly monotonic on the range of X:

$$\delta(T(X),Y) = \begin{cases} \delta(X,Y) & \text{T increasing,} \\ -\delta(X,Y) & \text{T decreasing.} \end{cases}$$

Linear correlation fulfills properties P1 and P2 only. In the next Section we see that rank correlation also fulfills P3 and P4 if X and Y are continuous. These properties obviously represent a selection and the list could be altered or extended in various ways (see Hutchinson & Lai (1990), Chapter 11). For example, we might like to have the property

P5. $\delta(X,Y) = 0 \iff X,Y$ are independent.

Unfortunately, this contradicts property P4 as the following shows.

Proposition 3. *There is no dependence measure satisfying P4 and P5.*

Proof. Let $(X,Y)^t$ be uniformly distributed on the unit circle S^1 in $\mathbb{R}^2$, so that $(X,Y)^t = (\cos\phi, \sin\phi)^t$ with $\phi \sim \mathrm{U}(0,2\pi)$. Since $(-X,Y)^t =_{\mathrm{d}} (X,Y)^t$, we have

$$\delta(-X,Y) = \delta(X,Y) = -\delta(X,Y),$$

which implies $\delta(X,Y) = 0$ although X and Y are dependent. With the same argumentation it can be shown that the measure is zero for any *spherical* distribution in $\mathbb{R}^2$. □

If we require P5, then we can consider dependence measures which only assign positive values to pairs of random variables. For example, we can consider the amended properties,

P2b. $0 \le \delta(X,Y) \le 1$.
P3b. $\delta(X,Y) = 1 \iff X,Y$ comonotonic or countermonotonic.
P4b. For $T : \mathbb{R} \to \mathbb{R}$ strictly monotonic $\delta(T(X),Y) = \delta(X,Y)$.

If we restrict ourselves to the case of continuous random variables there are dependence measures which fulfill all of P1, P2b, P3b, P4b and P5, although they are in general measures of theoretical rather than practical interest. We introduce them briefly in the next section. A further measure which satisfies all of P1, P2b, P3b, P4b and P5 (with the exception of the implication $\delta(X,Y) = 1 \Longrightarrow X,Y$ comonotonic or countermonotonic) is *monotone correlation*,

$$\delta(X,Y) = \sup_{f,g} \rho(f(X), g(Y)),$$

where ρ represents linear correlation and the supremum is taken over all monotonic f and g such that $0 < \sigma^2(f(X)), \sigma^2(g(Y)) < \infty$ (Kimeldorf & Sampson 1978). The disadvantage of all of these measures is that they are constrained to give non-negative values and as such cannot differentiate between positive and negative dependence and that it is often not clear how to estimate them. An overview of dependence measures and their statistical estimation is given by Tjøstheim (1996).

4.3 Rank correlation

Definition 6. Let X and Y be random variables with distribution functions F_1 and F_2 and joint distribution function F. Spearman's rank correlation is defined by

$$\rho_S(X,Y) = \rho(F_1(X), F_2(Y)), \tag{4.4}$$

where ρ is the usual linear correlation. Let (X_1, Y_1) and (X_2, Y_2) be two independent pairs of random variables drawn from F; then Kendall's rank correlation is defined by

$$\rho_\tau(X,Y) = \mathbb{P}[(X_1 - X_2)(Y_1 - Y_2) > 0] - \mathbb{P}[(X_1 - X_2)(Y_1 - Y_2) < 0]. \tag{4.5}$$

For the remainder of this Section we assume that F_1 and F_2 are continuous distributions, although some of the properties of rank correlation that we derive could be partially formulated for discrete distributions. Spearman's rank correlation is then seen to be the correlation of the copula C associated with $(X,Y)^t$. Both ρ_S and ρ_τ can be considered to be measures of the degree of monotonic dependence between X and Y, whereas linear correlation measures the degree of linear dependence only. The generalisation of ρ_S and ρ_τ to $n > 2$ dimensions can be done analogously to that of linear correlation: we write pairwise correlations in a $n \times n$-matrix.

We collect together the important facts about ρ_S and ρ_τ in the following theorem.

Theorem 3. *Let X and Y be random variables with continuous distributions F_1 and F_2, joint distribution F and copula C. The following are true:*

1. $\rho_S(X,Y) = \rho_S(Y,X)$, $\rho_\tau(X,Y) = \rho_\tau(Y,X)$.
2. *If X and Y are independent then* $\rho_S(X,Y) = \rho_\tau(X,Y) = 0$.
3. $-1 \le \rho_S(X,Y), \rho_\tau(X,Y) \le +1$.
4. $\rho_S(X,Y) = 12 \int_0^1 \int_0^1 \{C(x,y) - xy\} dx dy$.
5. $\rho_\tau(X,Y) = 4 \int_0^1 \int_0^1 C(u,v) dC(u,v) - 1$.

6. *For $T : \mathbb{R} \to \mathbb{R}$ strictly monotonic on the range of X, both ρ_S and ρ_τ satisfy P4.*

7. $\rho_S(X,Y) = \rho_\tau(X,Y) = 1 \iff C = C_u \iff Y = T(X)$ *a.s. with T increasing.*

8. $\rho_S(X,Y) = \rho_\tau(X,Y) = -1 \iff C = C_\ell \iff Y = T(X)$ *a.s. with T decreasing.*

Proof. 1., 2. and 3. are easily verified.

4. Use the identity, due to Höffding (1940),
$$\mathrm{Cov}[X,Y] = \int_{-\infty}^{\infty}\int_{-\infty}^{\infty} \{F(x,y) - F_1(x)F_2(y)\}\,dxdy \tag{4.6}$$
which is found, for example, in Dhaene & Goovaerts (1996). Recall that $(F_1(X), F_2(Y))^t$ have joint distribution C.

5. Calculate
$$\begin{aligned}\rho_\tau(X,Y) &= 2\mathbb{P}[(X_1 - X_2)(Y_1 - Y_2) > 0] - 1\\ &= 2\cdot 2\iiiint_{\mathbb{R}^4} \mathbf{1}_{\{x_1>x_2\}}\mathbf{1}_{\{y_1>y_2\}}\,dF(x_2,y_2)\,dF(x_1,y_1) - 1\\ &= 4\iint_{\mathbb{R}^2} F(x_1,y_1)\,dF(x_1,y_1) - 1\\ &= 4\iint C(u,v)\,dC(u,v) - 1.\end{aligned}$$

6. Follows since both ρ_τ and ρ_S can be expressed in terms of the copula which is invariant under strictly increasing transformations of the marginals.

7. From 4. it follows immediately that $\rho_S(X,Y) = +1$ iff $C(x,y)$ is maximized iff $C = C_u$ iff $Y = T(X)$ a.s. Suppose $Y = T(X)$ a.s. with T increasing, then the continuity of F_2 ensures $\mathbb{P}[Y_1 = Y_2] = \mathbb{P}[T(X_1) = T(X_2)] = 0$, which implies $\rho_\tau(X,Y) = \mathbb{P}[(X_1 - X_2)(Y_1 - Y_2) > 0] = 1$. Conversely $\rho_\tau(X,Y) = 1$ means $\mathbb{P}\otimes\mathbb{P}[(\omega_1,\omega_2) \in \Omega\times\Omega|(X(\omega_1) - X(\omega_2))(Y(\omega_1) - Y(\omega_2)) > 0\}] = 1$. Let us define sets $A = \{\omega \in \Omega|X(w) \le x\}$ and $B = \{\omega \in \Omega|Y(w) \le y\}$. Assume $\mathbb{P}[A] \le \mathbb{P}[B]$. We have to show $\mathbb{P}[A \cap B] = \mathbb{P}[A]$. If $\mathbb{P}[A \setminus B] > 0$ then also $\mathbb{P}[B \setminus A] > 0$ and $(X(\omega_1) - X(\omega_2))(Y(\omega_1) - Y(\omega_2)) < 0$ on the set $(A\setminus B)\times(B\setminus A)$, which has measure $\mathbb{P}[A\setminus B]\cdot\mathbb{P}[B\setminus A] > 0$, and this is a contradiction. Hence $\mathbb{P}[A\setminus B] = 0$, from which one concludes $\mathbb{P}[A\cap B] = \mathbb{P}[A]$.

8. We use a similar argument to 7.

□

In this result we have verified that rank correlation does have the properties P1, P2, P3 and P4. As far as P5 is concerned, the spherical distributions again provide examples where pairwise rank correlations are zero, despite the presence of dependence.

Theorem 3 (part 4) shows that ρ_S is a scaled version of the signed volume enclosed by the surfaces $S_1 : \; z = C(x, y)$ and $S_2 : \; z = xy$. The idea of measuring dependence by defining suitable distance measures between the surfaces S_1 and S_2 is further developed in Schweizer & Wolff (1981), where the three measures

$$\begin{aligned}
\delta_1(X,Y) &= 12 \int_0^1 \int_0^1 |C(u,v) - uv| du dv \\
\delta_2(X,Y) &= \left(90 \int_0^1 \int_0^1 |C(u,v) - uv|^2 du dv \right)^{1/2} \\
\delta_3(X,Y) &= 4 \sup_{u,v \in [0,1]} |C(u,v) - uv|
\end{aligned}$$

are proposed. These are the measures that satisfy our amended set of properties including P5 but are constrained to give non-negative measurements and as such cannot differentiate between positive and negative dependence. A further disadvantage of these measures is statistical. Whereas statistical estimation of ρ_S and ρ_τ from data is straightforward (see Gibbons (1988) for the estimators and Tjøstheim (1996) for asymptotic estimation theory) it is much less clear how we estimate measures like $\delta_1, \delta_2, \delta_3$.

The main advantages of rank correlation over ordinary correlation are the invariance under monotonic transformations and the sensible handling of perfect dependence. The main disadvantage is that rank correlations do not lend themselves to the same elegant variance-covariance manipulations that were discussed for linear correlation; they are not moment-based correlations. As far as calculation is concerned, there are cases where rank correlations are easier to calculate and cases where linear correlations are easier to calculate. If we are working, for example, with multivariate normal or t-distributions then calculation of linear correlation is easier, since first and second moments are easily determined. If we are working with a multivariate distribution which possesses a simple closed-form copula, like the Gumbel or Farlie–Gumbel–Morgenstern, then moments may be difficult to determine and calculation of rank correlation using Theorem 3 (parts 4 and 5) may be easier.

4.4 Tail Dependence

If we are particularly concerned with extreme values an asymptotic measure of tail dependence can be defined for pairs of random variables X and Y. If the marginal distributions of these random variables are continuous then this dependence measure is also a function of their copula, and is thus invariant under strictly increasing transformations.

Definition 7. Let X and Y be random variables with distribution functions F_1 and F_2. The *coefficient of (upper) tail dependence* of X and Y is

$$\lim_{\alpha \to 1-} \mathbb{P}[Y > F_2^{-1}(\alpha) \mid X > F_1^{-1}(\alpha)] = \lambda,$$

provided a limit $\lambda \in [0,1]$ exists. If $\lambda \in (0,1]$ X and Y are said to be *asymptotically dependent* (in the upper tail); if $\lambda = 0$ they are *asymptotically independent.*

As for rank correlation, this definition makes most sense in the case that F_1 and F_2 are continuous distributions. In this case it can be verified, under the assumption that the limit exists, that

$$\begin{aligned} &\lim_{\alpha \to 1-} \mathbb{P}[Y > F_2^{-1}(\alpha) \mid X > F_1^{-1}(\alpha)] \\ &= \lim_{\alpha \to 1-} \mathbb{P}[Y > \mathrm{VaR}_\alpha(Y) \mid X > \mathrm{VaR}_\alpha(X)] = \lim_{\alpha \to 1-} \frac{\overline{C}(\alpha,\alpha)}{1-\alpha}, \end{aligned}$$

where $\overline{C}(u,u) = 1 - 2u + C(u,u)$ denotes the survivor function of the unique copula C associated with $(X,Y)^t$. Tail dependence is best understood as an asymptotic property of the copula.

Calculation of λ for particular copulas is straightforward if the copula has a simple closed form. For example, for the Gumbel copula introduced in (2.5) it is easily verified that $\lambda = 2 - 2^\beta$, so that random variables with this copula are asymptotically dependent provided $\beta < 1$.

For copulas without a simple closed form, such as the Gaussian copula or the copula of the bivariate t-distribution, an alternative formula for λ is more useful. Consider a pair of uniform random variables $(U_1, U_2)^t$ with distribution $C(x,y)$. Applying l'Hospital's rule we obtain

$$\lambda = -\lim_{x \to 1-} \frac{d\overline{C}(x,x)}{dx} = \lim_{x \to 1-} \Pr[U_2 > x \mid U_1 = x] + \lim_{x \to 1-} \Pr[U_1 > x \mid U_2 = x].$$

Furthermore, if C is an *exchangeable* copula, i.e. $(U_1,U_2)^t =_{\mathrm{d}} (U_2,U_1)^t$, then

$$\lambda = 2 \lim_{x \to 1-} \Pr[U_2 > x \mid U_1 = x].$$

It is often possible to evaluate this limit by applying the same quantile transform F_1^{-1} to both marginals to obtain a bivariate distribution for which the

$\nu \setminus \rho$	-0.5	0	0.5	0.9	1
2	0.06	0.18	0.39	0.72	1
4	0.01	0.08	0.25	0.63	1
10	0.0	0.01	0.08	0.46	1
∞	0	0	0	0	1

Table 1: Values of λ for the copula of the bivariate t-distribution for various values of ν, the degrees of freedom, and ρ, the correlation. Last row represents the Gaussian copula.

conditional probability is known. If F_1 is a distribution function with infinite right endpoint then

$$\begin{aligned}\lambda = 2 \lim_{x \to 1-} \Pr[U_2 > x \mid U_1 = x] &= 2 \lim_{x \to \infty} \Pr[F_1^{-1}(U_2) > x \mid F_1^{-1}(U_1) = x] \\ &= 2 \lim_{x \to \infty} \Pr[Y > x \mid X = x],\end{aligned}$$

where $(X,Y)^t \sim C(F_1(x), F_1(y))$.

For example, for the Gaussian copula C_ρ^{Ga} we would take $F_1 = \Phi$ so that $(X,Y)^t$ has a standard bivariate normal distribution with correlation ρ. Using the fact that $Y \mid X = x \sim N(\rho x, 1-\rho^2)$, it can be calculated that

$$\lambda = 2 \lim_{x \to \infty} \overline{\Phi}(x\sqrt{1-\rho}/\sqrt{1+\rho}).$$

Thus the Gaussian copula gives asymptotic independence, provided that $\rho < 1$. Regardless of how high a correlation we choose, if we go far enough into the tail, extreme events appear to occur independently in each margin. See Sibuya (1961) or Resnick (1987), Chapter 5, for alternative demonstrations of this fact.

The bivariate t-distribution provides an interesting contrast to the bivariate normal distribution. If $(X,Y)^t$ has a standard bivariate t-distribution with ν degrees of freedom and correlation ρ then, conditional on $X = x$,

$$\left(\frac{\nu+1}{\nu+x^2}\right)^{1/2} \frac{Y - \rho x}{\sqrt{1-\rho^2}} \sim t_{\nu+1}.$$

This can be used to show that

$$\lambda = 2\overline{t}_{\nu+1}\left(\sqrt{\nu+1}\sqrt{1-\rho}/\sqrt{1+\rho}\right),$$

where $\overline{t}_{\nu+1}$ denotes the tail of a univariate t-distribution. Provided $\rho > -1$ the copula of the bivariate t-distribution is asymptotically dependent. In Table 1 we tabulate the coefficient of tail dependence for various values of ν

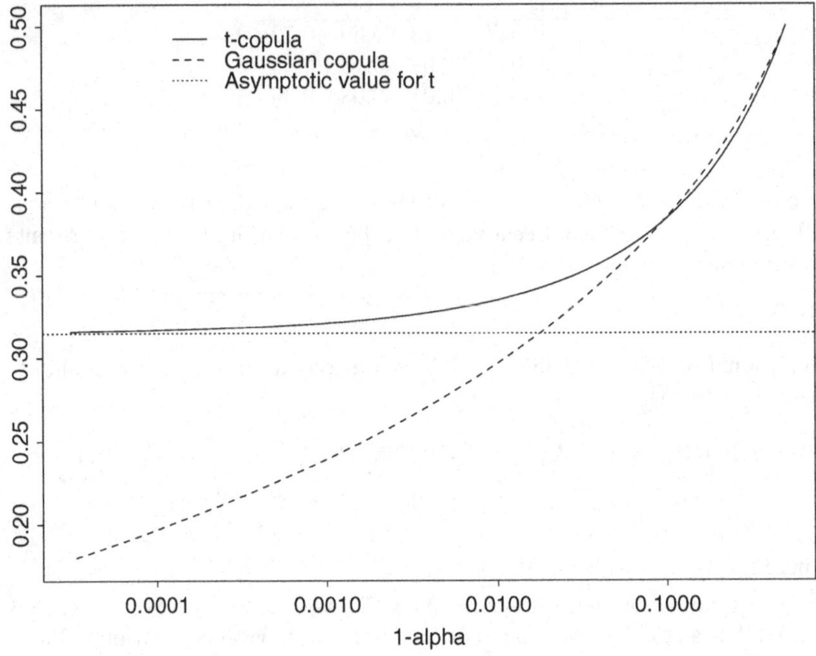

Figure 2: Exact values of the conditional probability $\mathbb{P}[Y > \mathrm{VaR}_\alpha(Y) \mid X = \mathrm{VaR}_\alpha(X)]$ for pairs of random variables $(X,Y)^t$ with the Gaussian and t-copulas, where the correlation parameter in both copulas is $\rho = 0.9$ and the degrees of freedom of the t-copula is $\nu = 4$.

and ρ. Perhaps surprisingly, even for negative and zero correlations, the t-copula gives asymptotic dependence in the upper tail. The strength of this dependence increases as ν decreases and the marginal distributions become heavier-tailed.

In Figure 2 we plot exact values of the conditional probability $\mathbb{P}[Y > \mathrm{VaR}_\alpha(Y) \mid X = \mathrm{VaR}_\alpha(X)]$ for pairs of random variables $(X,Y)^t$ with the Gaussian and t-copulas, where the correlation parameter of both copulas is $\rho = 0.9$ and the degrees of freedom of the t-copula is $\nu = 4$. For large values of α the conditional probabilities for the t-copula dominate those for the Gaussian copula. Moreover the former tend towards a non-zero asymptotic limit, whereas the limit in the Gaussian case is zero.

4.5 Concordance

In some situations we may be less concerned with measuring the strength of stochastic dependence between two random variables X and Y and we may wish simply to say whether they are *concordant* or *discordant*, that is, whether the dependence between X and Y is *positive* or *negative*. While it might seem natural to define X and Y to be positively dependent when $\rho(X,Y) > 0$ (or when $\rho_S(X,Y) > 0$ or $\rho_\tau(X,Y) > 0$), stronger conditions are generally used and we discuss two of these concepts in this section.

Definition 8. Two random variables X and Y are *positive quadrant dependent* (PQD), if

$$\mathbb{P}[X > x, Y > y] \geq \mathbb{P}[X > x]\mathbb{P}[Y > y] \quad \text{for all } x, y \in \mathbb{R}. \tag{4.7}$$

Since $\mathbb{P}[X > x, Y > y] = 1 - \mathbb{P}[X \leq x] + \mathbb{P}[Y \leq y] - \mathbb{P}[X \leq x, Y \leq y]$ it is obvious that (4.7) is equivalent to

$$\mathbb{P}[X \leq x, Y \leq y] \geq \mathbb{P}[X \leq x]\mathbb{P}[Y \leq y] \quad \text{for all } x, y \in \mathbb{R}.$$

Definition 9. Two random variables X and Y are *positively associated* (PA), if

$$\mathbb{E}[g_1(X,Y)g_2(X,Y)] \geq \mathbb{E}[g_1(X,Y)]\mathbb{E}[g_2(X,Y)] \tag{4.8}$$

for all real-valued, measurable functions g_1 und g_2 which are increasing in both components and for which the expectations involved are defined.

For further concepts of positive dependence see Chapter 2 of Joe (1997), where the relationships between the various concepts are also systematically explored. We note that PQD and PA are invariant under increasing transformations and we verify that the following chain of implications holds (cf. Definition 5):

$$\text{Comonotonicity} \Rightarrow \text{PA} \Rightarrow \text{PQD} \Rightarrow \rho(X,Y) \geq 0, \rho_S(X,Y) \geq 0, \rho_\tau(X,Y) \geq 0. \tag{4.9}$$

If X and Y are comonotonic, then from Theorem 2 we can conclude that $(X,Y) =_{\mathrm{d}} (F_1^{-1}(U), F_2^{-1}(U))$, where $U \sim \mathrm{U}(0,1)$. Thus the expectations in (4.8) can be written as

$$\mathbb{E}[g_1(X,Y)g_2(X,Y)] = \mathbb{E}[\widetilde{g}_1(U)\widetilde{g}_2(U)]$$

and

$$\mathbb{E}[g_1(X,Y)] = \mathbb{E}[\widetilde{g}_1(U)] \ , \mathbb{E}[g_2(X,Y)] = \mathbb{E}[\widetilde{g}_2(U)],$$

where $\widetilde{g}_1$ and $\widetilde{g}_2$ are increasing. Lemma 2.1 in Joe (1997) shows that

$$\mathbb{E}[\widetilde{g}_1(U)\widetilde{g}_2(U)] \geq \mathbb{E}[\widetilde{g}_1(U)]\mathbb{E}[\widetilde{g}_2(U)],$$

so that X and Y are PA. The second implication follows immediately by taking

$$\begin{aligned} g_1(u,v) &= \mathbf{1}_{\{u>x\}} \\ g_2(u,v) &= \mathbf{1}_{\{v>y\}}. \end{aligned}$$

The third implication PQD $\Rightarrow \rho(X,Y) \geq 0, \rho_S(X,Y) \geq 0$ follows from the identity (4.6) and the fact that PA and PQD are invariant under increasing transformations. PQD $\Rightarrow \rho_\tau(X,Y) \geq 0$ follows from Theorem 2.8 in Joe (1997).

In the sense of these implications (4.9), comonotonicity is the *strongest* type of concordance or positive dependence.

5 Fallacies

Where not otherwise stated, we consider bivariate distributions of the random vector $(X,Y)^t$.

Fallacy 1. Marginal distributions and correlation determine the joint distribution.

This is true if we restrict our attention to the multivariate normal distribution or the elliptical distributions. For example, if we know that $(X,Y)^t$ have a bivariate normal distribution, then the expectations and variances of X and Y and the correlation $\rho(X,Y)$ uniquely determine the joint distribution. However, if we only know the marginal distributions of X and Y and the correlation then there are many possible bivariate distributions for $(X,Y)^t$. The distribution of $(X,Y)^t$ is not uniquely determined by F_1, F_2 and $\rho(X,Y)$. We illustrate this with examples, interesting in their own right.

Example 1. Let X and Y have standard normal distributions and assume $\rho(X,Y) = \rho$. If $(X,Y)^t$ is *bivariate* normally distributed, then the distribution function F of $(X,Y)^t$ is given by

$$F(x,y) = C_\rho^{Ga}(\Phi(x), \Phi(y)).$$

We have represented this copula as a double integral earlier in (2.4). Any other copula $C \neq C_\rho^{Ga}$ gives a bivariate distribution with standard normal marginals which is *not* bivariate normal with correlation ρ. We construct a copula C of the type (2.6) by taking

$$\begin{aligned} f(x) &= \mathbf{1}_{\{(\gamma,1-\gamma)\}}(x) + \frac{2\gamma-1}{2\gamma}\mathbf{1}_{\{(\gamma,1-\gamma)^c\}}(x) \\ g(y) &= -\mathbf{1}_{\{(\gamma,1-\gamma)\}}(y) - \frac{2\gamma-1}{2\gamma}\mathbf{1}_{\{(\gamma,1-\gamma)^c\}}(y), \end{aligned}$$

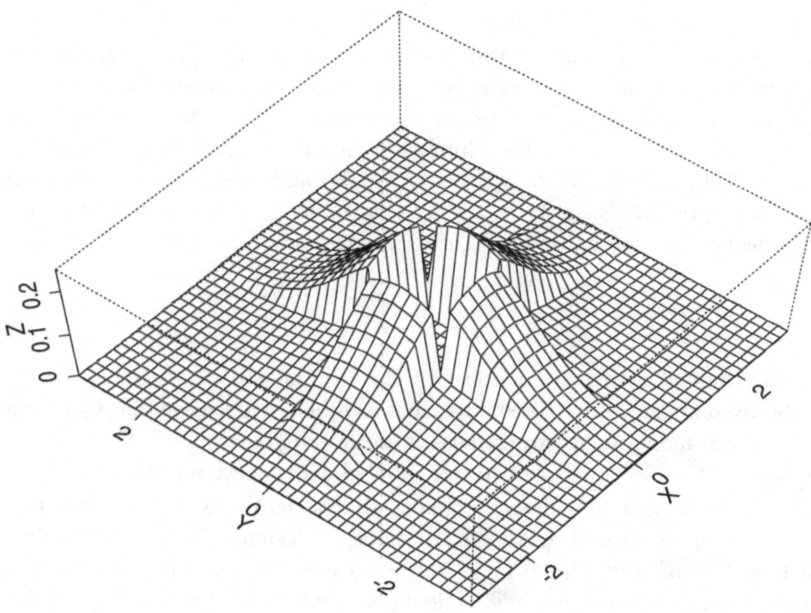

Figure 3: Density of a non-bivariate normal distribution which has standard normal marginals.

where $\frac{1}{4} \leq \gamma \leq \frac{1}{2}$. Since $h(x, y)$ disappears on the square $[\gamma, 1-\gamma]^2$ it is clear that C for $\gamma < \frac{1}{2}$ and $F(x, y) = C(\Phi(x), \Phi(y))$ is *never* bivariate normal; from symmetry considerations ($C(u, v) = v - C(1-u, v)$, $0 \leq u, v \leq 1$) the correlation irrespective of γ is zero. There are uncountably many bivariate distributions with standard normal marginals and correlation zero. In Figure 3 the density of F is shown for $\gamma = 0.3$; this is clearly very different from the joint density of the standard bivariate normal distribution with zero correlation.

Example 2. A more realistic example for risk management is the motivating example of the Introduction. We consider two bivariate distributions with Gamma(3,1) marginals (denoted $\mathrm{G}_{3,1}$) and the same correlation $\rho = 0.7$, but with different dependence structures, namely

$$\begin{aligned} F_{Ga}(x, y) &= C_{\tilde{\rho}}^{Ga}(\mathrm{G}(x), \mathrm{G}(y)), \\ F_{Gu}(x, y) &= C_{\alpha}^{Gu}(\mathrm{G}(x), \mathrm{G}(y)), \end{aligned}$$

where $C_{\tilde{\rho}}^{Ga}$ is the Gaussian dependence structure and C_{β}^{Gu} is the Gumbel copula introduced in (2.5). To obtain the desired linear correlation the parameter values were set to be $\tilde{\rho} = 0.71$ and $\beta = 0.54$ [7].

In Section 4.4 we showed that the two copulas have quite different tail dependence; the Gaussian copula is asymptotically independent if $\tilde{\rho} < 1$ and the Gumbel copula is asymptotically dependent if $\beta < 1$. At finite levels the greater tail dependence of the Gumbel copula is apparent in Figure 1. We fix $u = \mathrm{VaR}_{0.99}(X) = \mathrm{VaR}_{0.99}(Y) = \mathrm{G}_{3,1}^{-1}(0.99)$ and consider the conditional exceedance probability $\mathbb{P}[Y > u \mid X > u]$ under the two models. An easy empirical estimation based on Figure 1 yields

$$\begin{aligned}\widehat{\mathbb{P}}_{F_{Ga}}[Y > u \mid X > u] &= 3/9,\\ \widehat{\mathbb{P}}_{F_{Gu}}[Y > u \mid X > u] &= 12/16.\end{aligned}$$

In the Gumbel model exceedances of the threshold u in one margin tend to be accompanied by exceedances in the other, whereas in the Gaussian dependence model joint exceedances in both margins are rare. There is less "diversification" of large risks in the Gumbel dependence model.

Analytically it is difficult to provide results for the Value-at-Risk of the sum $X + Y$ under the two bivariate distributions,[8] but simulation studies confirm that $X + Y$ produces more large outcomes under the Gumbel dependence model than the Gaussian model. The difference between the two dependence structures might be particularly important if we were interested in losses which were triggered only by joint extreme values of X and Y.

Example 3. The Value-at-Risk of linear portfolios is certainly not uniquely determined by the marginal distributions and correlation of the constituent risks. Suppose $(X, Y)^t$ has a bivariate normal distribution with standard normal marginals and correlation ρ and denote the bivariate distribution function by F_ρ. Any mixture $F = \lambda F_{\rho_1} + (1 - \lambda) F_{\rho_2}$, $0 \le \lambda \le 1$, of bivariate normal distributions F_{ρ_1} and F_{ρ_2} also has standard normal marginals and correlation $\lambda\rho_1 + (1 - \lambda)\rho_2$. Suppose we fix $-1 < \rho < 1$ and choose $0 < \lambda < 1$ and $\rho_1 < \rho < \rho_2$ such that $\rho = \lambda\rho_1 + (1 - \lambda)\rho_2$. The sum $X + Y$ is longer tailed under F than under F_ρ. Since

$$\mathbb{P}_F[X + Y > z] = \lambda\overline{\Phi}\left(\frac{z}{2(1 + \rho_1)}\right) + (1 - \lambda)\overline{\Phi}\left(\frac{z}{2(1 + \rho_2)}\right),$$

and

$$\mathbb{P}_{F_\rho}[X + Y > z] = \overline{\Phi}\left(\frac{z}{2(1 + \rho)}\right),$$

we can use Mill's ratio

$$\overline{\Phi}(x) = 1 - \Phi(x) = \phi(x)\left(\frac{1}{x} + \mathrm{O}\left(\frac{1}{x^2}\right)\right)$$

[7]These numerical values were determined by stochastic simulation.

[8]See Müller & Bäuerle (1998) for related work on stop-loss risk measures applied to bivariate portfolios under various dependence models. A further reference is Albers (1999).

to show that

$$\lim_{z\to\infty} \frac{\mathbb{P}_F[X+Y>z]}{\mathbb{P}_{F_\rho}[X+Y>z]} = \infty.$$

Clearly as one goes further into the respective tails of the two distributions the Value-at-Risk for the mixture distribution F is larger than that of the original distribution F_ρ. By using the same technique as Embrechts, Klüppelberg & Mikosch (1997) (Example 3.3.29) we can show that, as $\alpha \to 1-$,

$$\begin{aligned}
\mathrm{VaR}_{\alpha,F}(X+Y) &\sim 2(1+\rho_2)\,(-2\log(1-\alpha))^{1/2} \\
\mathrm{VaR}_{\alpha,F_\rho}(X+Y) &\sim 2(1+\rho)\,(-2\log(1-\alpha))^{1/2},
\end{aligned}$$

so that

$$\lim_{\alpha\to 1-} \frac{\mathrm{VaR}_{\alpha,F}(X+Y)}{\mathrm{VaR}_{\alpha,F_\rho}(X+Y)} = \frac{1+\rho_2}{1+\rho} > 1,$$

irrespective of the choice of λ.

Fallacy 2. Given marginal distributions F_1 and F_2 for X and Y, all linear correlations between -1 and 1 can be attained through suitable specification of the joint distribution.

This statement is not true and it is simple to construct counterexamples.

Example 4. Let X and Y be random variables with support $[0,\infty)$, so that $F_1(x) = F_2(y) = 0$ for all $x, y < 0$. Let the right endpoints of F_1 and F_2 be infinite, $\sup_x\{x|F_1(x) < 1\} = \sup_y\{y|F_2(y) < 1\} = \infty$. Assume that $\rho(X,Y) = -1$, which implies $Y = aX + b$ a.s., with $a < 0$ and $b \in \mathbb{R}$. It follows that for all $y < 0$,

$$\begin{aligned}
F_2(y) &= \mathbb{P}[Y \le y] = \mathbb{P}[X \ge (y-b)/a] \ge \mathbb{P}[X > (y-b)/a] \\
&= 1 - F_1((y-b)/a) > 0,
\end{aligned}$$

which contradicts the assumption $F_2(y) = 0$.

The following theorem shows which correlations are possible for given marginal distributions.

Theorem 4. *[Höffding (1940) and Fréchet (1957)] Let $(X,Y)^t$ be a random vector with marginals F_1 and F_2 and unspecified dependence structure; assume $0 < \sigma^2[X], \sigma^2[Y] < \infty$. Then*

1. *The set of all possible correlations is a closed interval $[\rho_{\min}, \rho_{\max}]$ and for the extremal correlations $\rho_{\min} < 0 < \rho_{\max}$ holds.*

2. *The extremal correlation $\rho = \rho_{\min}$ is attained if and only if X and Y are countermonotonic; $\rho = \rho_{\max}$ is attained if and only if X and Y are comonotonic.*

3. *$\rho_{\min} = -1$ iff X and $-Y$ are of the same type; $\rho_{\max} = 1$ iff X and Y are of the same type.*

Proof. We make use of the identity (4.6) and observe that the Fréchet inequalities (4.1) imply

$$\max\{F_1(x) + F_2(y) - 1, 0\} \leq F(x,y) \leq \min\{F_1(x), F_2(y)\}.$$

The integrand in (4.6) is minimized pointwise if X and Y are countermonotonic, and maximized if X and Y are comonotonic. It is clear that $\rho_{\max} \geq 0$. However, if $\rho_{\max} = 0$ this would imply that $\min\{F_1(x), F_2(y)\} = F_1(x)F_2(y)$ for all x, y. This can only occur if F_1 or F_2 is degenerate, i.e. of the form $F_1(x) = \mathbf{1}_{\{x \geq x_0\}}$ or $F_2(y) = \mathbf{1}_{\{y \geq y_0\}}$, and this would imply $\sigma^2[X] = 0$ or $\sigma^2[Y] = 0$ so that the correlation between X and Y is undefined. Similarly we argue that $\rho_{\min} < 0$. If $F_\ell(x_1, x_2) = \max\{F_1(x) + F_2(y) - 1, 0\}$ and $F_u(x_1, x_2) = \min\{F_1(x), F_2(y)\}$ then the mixture $\lambda F_\ell + (1-\lambda)F_u$, $0 \leq \lambda \leq 1$ has correlation $\lambda\rho_{\min} + (1-\lambda)\rho_{\max}$. Using such mixtures we can construct joint distributions with marginals F_1 and F_2 and with arbitrary correlations $\rho \in [\rho_{\min}, \rho_{\max}]$. This will be used in Section 6 □

Example 5. Let $X \sim$ Lognormal$(0,1)$ and $Y \sim$ Lognormal$(0, \sigma^2)$, $\sigma > 0$. We wish to calculate $\rho_{\min}$ and $\rho_{\max}$ for these marginals. Note that X and Y are not of the same type although $\log X$ and $\log Y$ are. It is clear that $\rho_{\min} = \rho(e^Z, e^{-\sigma Z})$ and $\rho_{\max} = \rho(e^Z, e^{\sigma Z})$, where $Z \sim \mathcal{N}(0,1)$. This observation allows us to calculate $\rho_{\min}$ and $\rho_{\max}$ analytically:

$$\rho_{\min} = \frac{e^{-\sigma} - 1}{\sqrt{(e-1)(e^{\sigma^2} - 1)}},$$

$$\rho_{\max} = \frac{e^{\sigma} - 1}{\sqrt{(e-1)(e^{\sigma^2} - 1)}}.$$

These maximal and minimal correlations are shown graphically in Figure 4. We observe that $\lim_{\sigma\to\infty} \rho_{\min} = \lim_{\sigma\to\infty} \rho_{\max} = 0$.

This example shows it is possible to have a random vector $(X,Y)^t$ where the correlation is almost zero, even though X and Y are comonotonic or countermonotonic and thus have the strongest kind of dependence possible. This seems to contradict our intuition about probability and shows that small correlations cannot be interpreted as implying weak dependence between random variables.

Fallacy 3. The worst case VaR (quantile) for a linear portfolio $X+Y$ occurs when $\rho(X,Y)$ is maximal, i.e. X and Y are comonotonic

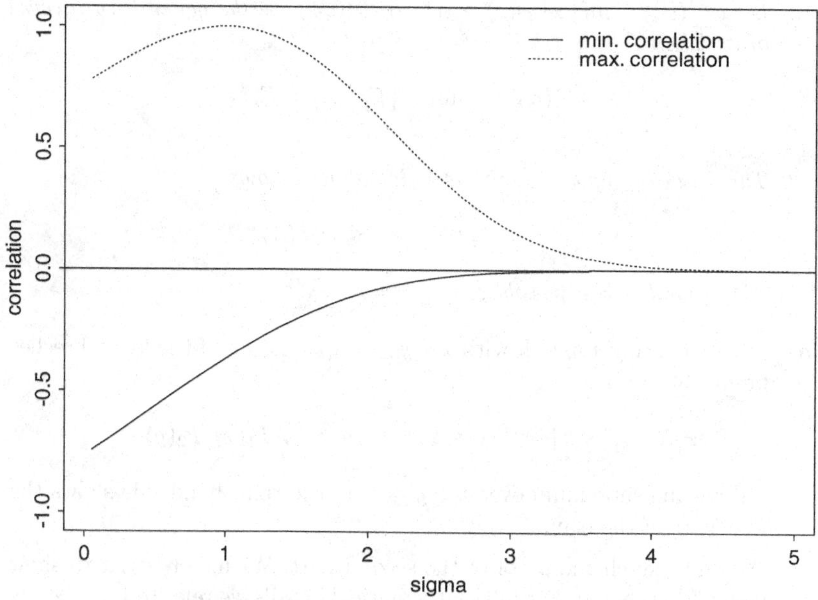

Figure 4: $\rho_{\min}$ and $\rho_{\max}$ graphed against σ.

As we have discussed in Section 3.3 it is common to consider variance as a measure of risk in insurance and financial mathematics and, whilst it is true that the variance of a linear portfolio, $\sigma^2(X+Y) = \sigma^2(X) + \sigma^2(Y) + 2\rho(X,Y)\sigma(X)\sigma(Y)$, is maximal when the correlation is maximal, it is in general not correct to conclude that the Value-at-Risk is also maximal. For elliptical distributions it is true, but generally it is false.

Suppose two random variables X and Y have distribution functions F_1 and F_2 but that their dependence structure (copula) is unspecified. In the following theorem we give an upper bound for $\mathrm{VaR}_\alpha(X+Y)$.

Theorem 5. [*Makarov (1981) and Frank, Nelsen & Schweizer (1987)*]

1. *For all* $z \in \mathbb{R}$,

$$\mathbb{P}[X+Y \le z] \ge \sup_{x+y=z} C_\ell(F_1(x), F_2(y)) =: \psi(z).$$

This bound is sharp in the sense that, setting $t = \psi(z-) = \lim_{u \to z-} \psi(u)$, *there exists a copula, which we denote by* $C^{(t)}$, *such that under the distribution with distribution function* $F(x,y) = C^{(t)}(F_1(x), F_2(y))$ *we have that* $\mathbb{P}[X+Y < z] = \psi(z-)$.[9]

[9] In general there is no copula such that $\mathbb{P}[X+Y \le z] = \psi(z)$, not even if F_1 and F_2 are both continuous; see Nelsen (1999).

2. *Let* $\psi^{-1}(\alpha) := \inf\{z \mid \psi(z) \geq \alpha\}$, $\alpha \in (0,1)$, *be the generalized inverse of* ψ. *Then*

$$\psi^{-1}(\alpha) = \inf_{C_\ell(u,v)=\alpha} \{F_1^{-1}(u) + F_2^{-1}(v)\}.$$

3. *The following upper bound for Value-at-Risk holds:*

$$VaR_\alpha(X+Y) \leq \psi^{-1}(\alpha).$$

This bound is best-possible.

Proof. 1. For any $x, y \in \mathbb{R}$ with $x + y = z$ application of the lower Fréchet bound (4.1) yields

$$\mathbb{P}[X+Y \leq z] \geq \mathbb{P}[X \leq x, Y \leq y] \geq C_\ell(F_1(x), F_2(y)).$$

Taking the supremum over $x + y = z$ on the right hand side shows the first part of the claim.

We only sketch the proof of the second part. We merely want to show how $C^{(t)}$ is chosen. For full mathematical details we refer to Frank *et al.* (1987). We restrict ourselves to continuous distribution functions F_1 and F_2. Since copulas are distributions with uniform marginals we transform the problem onto the unit square by defining $A = \{(F_1(x), F_2(y)) | x + y \geq z\}$ the boundary of which is $s = \{(F_1(x), F_2(y)) | x + y = z\}$. We need to find a copula $C^{(t)}$ such that $\iint_A dC^{(t)} = 1 - t$. Since F_1 and F_2 are continuous, we have that $\psi(z-) = \psi(z)$ and therefore $t \geq u + v - 1$ for all $(u, v) \in s$. Thus the line $u + v - 1 = t$ can be considered as a tangent to s and it becomes clear how one can choose $C^{(t)}$. $C^{(t)}$ belongs to the distribution which is uniform on the line segments $\overline{(0,0)(t,t)}$ and $\overline{(t,1)(1,t)}$. Therefore

$$C^{(t)}(u,v) = \begin{cases} \max\{u+v-1, t\} & (u,v) \in [t,1] \times [t,1], \\ \min\{u,v\} & \text{otherwise.} \end{cases} \tag{5.1}$$

Since the set $\overline{(t,1)(1,t)} \subset A$ has probability mass $1-t$ we have under $C^{(t)}$ that $\mathbb{P}[X+Y \geq z] = \iint_A dC^{(t)} \geq 1-t$ and therefore $\mathbb{P}[X+Y < z] \leq t$. But since t is a lower bound for $\mathbb{P}[X+Y < z]$ it is necessary that $\mathbb{P}[X+Y < z] = t$.

2. This follows from the duality theorems in Frank & Schweizer (1979).

3. Let $\epsilon > 0$. Then we have

$$\mathbb{P}[X+Y \leq \psi^{-1}(\alpha) + \epsilon] \geq \psi(\psi^{-1}(\alpha) + \epsilon) \geq \alpha.$$

Taking the limit as $\epsilon \to 0+$ yields $\mathbb{P}[X+Y \le \psi^{-1}(\alpha)] \ge \alpha$ and therefore $\mathrm{VaR}_\alpha(X+Y) \le \psi^{-1}(\alpha)$. This upper bound cannot be improved. Again, take $\epsilon > 0$. Then if $(X,Y)^t$ has copula $C^{(\psi^{-1}(\alpha)-\epsilon/2)}$ one has

$$\begin{aligned}\mathbb{P}[X+Y \le \psi^{-1}(\alpha)-\epsilon] &\le \mathbb{P}[X+Y < \psi^{-1}(\alpha)-\epsilon/2] \\ &= \psi((\psi^{-1}(\alpha)-\epsilon/2)-) \le \psi(\psi^{-1}(\alpha)-\epsilon/2) < \alpha\end{aligned}$$

and therefore $\mathrm{VaR}_\alpha(X+Y) > \psi^{-1}(\alpha)-\epsilon$.

□

Remark 2. The results in Frank *et al.* (1987) are more general than Theorem 5 in this article. Frank *et al.* (1987) give lower *and* upper bounds for $\mathbb{P}[L(X,Y) \le z]$ where $L(\cdot,\cdot)$ is continuous and increasing in each coordinate. Therefore a best-possible *lower* bound for $\mathrm{VaR}_\alpha(X+Y)$ also exists. Numerical evaluation methods of ψ^{-1} are described in Williamson & Downs (1990). These two authors also treat the case where we restrict attention to particular subsets of copulas. By considering the sets of copulas $\mathcal{D} = \{C \mid C(u,v) \ge uv,\ 0 \le u,v \le 1\}$, which has minimal copula $C_{ind}(u,v) = uv$, we can derive bounds of $\mathbb{P}[X+Y \le z]$ under positive dependence (PQD as defined in Definition 8). Multivariate generalizations of Theorem 5 can be found in Li, Scarsini & Shaked (1996).

In Figure 5 the upper bound $\psi^{-1}(\alpha)$ is shown for $X \sim \mathrm{Gamma}(3,1)$ and $Y \sim \mathrm{Gamma}(3,1)$, for various values of α. Notice that $\psi^{-1}(\alpha)$ can easily be analytically computed analytically for this case since for α sufficiently large

$$\psi^{-1}(\alpha) = \inf_{u+v-1=\alpha}\{F_1^{-1}(u)+F_2^{-1}(v)\} = F_1^{-1}((\alpha+1)/2) + F_1^{-1}((\alpha+1)/2).$$

This is because $F_1 = F_2$ and the density of Gamma(3, 1) is unimodal, see also Example 6. For comparative purposes $\mathrm{VaR}_\alpha(X+Y)$ is also shown for the case where X, Y are independent and the case where they are comonotonic. The latter is computed by addition of the univariate quantiles since under comonotonicity $\mathrm{VaR}_\alpha(X+Y) = \mathrm{VaR}_\alpha(X)+\mathrm{VaR}_\alpha(Y)$. [10] The example shows that for a fixed $\alpha \in (0,1)$ the maximal value of $\mathrm{VaR}_\alpha(X+Y)$ is considerably larger than the value obtained in the case of comonotonicity. This is not surprising since we know that VaR is not a subadditive risk measure (Artzner *et al.* 1999) and there are situations where $\mathrm{VaR}_\alpha(X+Y) > \mathrm{VaR}_\alpha(X) + \mathrm{VaR}_\alpha(Y)$. In a sense, the difference $\psi^{-1}(\alpha)-(\mathrm{VaR}_\alpha(X)+\mathrm{VaR}_\alpha(Y))$ quantifies the amount by which VaR *fails* to be subadditive for particular marginals and a particular α. For a coherent risk measure ϱ, we must have that $\varrho(X+Y)$

[10] This is also true when X or Y do not have continuous distributions. Using Proposition 4.5 in Denneberg (1994) we deduce that for comonotonic random variables $X+Y = (u+v)(Z)$ where u and v are *continuous* increasing functions and $Z = X+Y$. Remark 1 then shows that $\mathrm{VaR}_\alpha(X+Y) = (u+v)(\mathrm{VaR}_\alpha(Z)) = u(\mathrm{VaR}_\alpha(Z)) + v(\mathrm{VaR}_\alpha(Z)) = \mathrm{VaR}_\alpha(X)+\mathrm{VaR}_\alpha(Y)$.

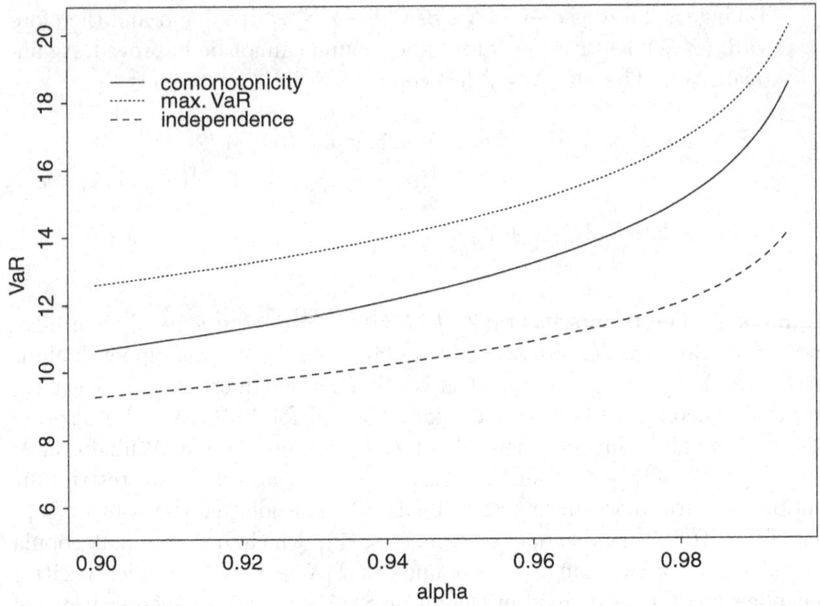

Figure 5: $\psi^{-1}(\alpha)$(max. VaR) graphed against α.

attains its maximal value in the case of comonotonicity and that this value is $\varrho(X) + \varrho(Y)$ (Delbaen 1999). The fact that there are situations which are worse than comonotonicity as far as VaR is concerned, is another way of showing that VaR is not a coherent measure of risk.

Suppose we define a measure of diversification by

$$D = (\mathrm{VaR}_\alpha(X) + \mathrm{VaR}_\alpha(Y)) - \mathrm{VaR}_\alpha(X + Y),$$

the idea being that comonotonic risks are undiversifiable ($D = 0$) but that risks with weaker dependence *should* be diversifiable ($D > 0$). Unfortunately, Theorem 5 makes it clear that we can always find distributions with linear correlation strictly less than the (comonotonic) maximal correlation (see Theorem 4) that give negative diversification ($D < 0$). This weakens standard diversification arguments, which say that "low correlation means high diversification". As an example Table 2 gives the numerical values of the correlations of the distributions yielding maximal $\mathrm{VaR}_\alpha(X+Y)$ for $X, Y \sim \mathrm{Gamma}(3, 1)$.

It might be supposed that VaR is in some sense asymptotically subadditive, so that negative diversification disappears as we let α tend to one, and comonotonicity becomes the worst case. The following two examples show that this is also wrong.

α	0.25	0.5	0.75	0.8	0.85	0.9	0.95	0.99
ρ	−0.09	0.38	0.734	0.795	0.852	0.901	0.956	0.992

Table 2: Correlations of the distributions giving maximal $\mathrm{VaR}_\alpha(X+Y)$.

Example 6. The quotient $\mathrm{VaR}_\alpha(X+Y)/(\mathrm{VaR}_\alpha(X)+\mathrm{VaR}_\alpha(Y))$ can be made arbitrarily large. In general we do *not* have $\lim_{\alpha\to 1-}\psi^{-1}(\alpha)/(\mathrm{VaR}_\alpha(X)+\mathrm{VaR}_\alpha(Y))=1$. To see this consider Pareto marginals $F_1(x)=F_2(x)=1-x^{-\beta}$, $x\geq 1$, where $\beta>0$. We have to determine $\inf_{u+v-1=\alpha}\{F_1^{-1}(u)+F_2^{-1}(v)\}$. Since $F_1=F_2$, the function

$$g:(\alpha,1)\to\mathbb{R}_+,\ u\mapsto F_1^{-1}(u)+F_2^{-1}(\alpha+1-u)$$

is symmetrical with respect to $(\alpha+1)/2$. Since the Pareto density is decreasing, the function g is decreasing on $(\alpha,(\alpha+1)/2]$ and increasing on $[(\alpha+1)/2,1)$; hence $g((\alpha+1)/2)=2F_1^{-1}((\alpha+1)/2)$ is the minimum of g and $\psi^{-1}(\alpha)=2F_1^{-1}((\alpha+1)/2)$. Therefore

$$\begin{aligned}\frac{\mathrm{VaR}_\alpha(X+Y)}{\mathrm{VaR}_\alpha(X)+\mathrm{VaR}_\alpha(Y)} &\leq \frac{\psi^{-1}(\alpha)}{\mathrm{VaR}_\alpha(X)+\mathrm{VaR}_\alpha(Y)}\\ &= \frac{F_1^{-1}((\alpha+1)/2)}{F_1^{-1}(\alpha)}=\frac{(1-\frac{\alpha+1}{2})^{-1/\beta}}{(1-\alpha)^{-1/\beta}}=2^{1/\beta}.\end{aligned}$$

The upper bound $2^{1/\beta}$, which is independent of α, can be reached.

Example 7. Let X and Y be independent random variables with identical distribution $F_1(x)=1-x^{-1/2}$, $x\geq 1$. This distribution is extremely heavy-tailed with no finite mean. Consider the risks $X+Y$ and $2X$, the latter being the sum of comonotonic risks. We can calculate

$$\mathbb{P}[X+Y\leq z]=1-\frac{2\sqrt{z-1}}{z}<\mathbb{P}[2X\leq z],$$

for $z>2$. It follows that

$$\mathrm{VaR}_\alpha(X+Y)>\mathrm{VaR}_\alpha(2X)=\mathrm{VaR}_\alpha(X)+\mathrm{VaR}_\alpha(Y)$$

for $\alpha\in(0,1)$, so that, from the point of view of VaR, independence is worse than perfect dependence no matter how large we choose α. VaR is not subadditive for this rather extreme choice of distribution and diversification arguments do not hold; one is better off taking one risk and doubling it than taking two independent risks. Diversifiability of two risks is not only dependent on their dependence structure but also on the choice of marginal distribution. In fact, for distributions with $F_1(x)=F_2(x)=1-x^{-\kappa}$, $\kappa>0$, we do have asymptotic subadditivity in the case $\kappa>1$. That means

$\text{VaR}_\alpha(X+Y) < \text{VaR}_\alpha(X) + \text{VaR}_\alpha(Y)$ if α large enough. To see this use Lemma 1.3.1 of Embrechts *et al.* (1997) and the fact that $1 - F_1$ is regularly varying of index $-\kappa$. (For an introduction to regular variation theory see the appendix of the same reference.)

6 Simulation of Random Vectors

There are various situations in practice where we might wish to simulate dependent random vectors $(X_1, \ldots, X_n)^t$. In finance we might wish to simulate the future development of the values of assets in a portfolio, where we know these assets to be dependent in some way. In insurance we might be interested in multiline products, where payouts are triggered by the occurrence of losses in one or more dependent business lines, and wish to simulate typical losses. The latter is particularly important within DFA. It is very tempting to approach the problem in the following way:

1. Estimate marginal distributions $F_1, \ldots, F_n$,
2. Estimate the matrix of pairwise correlations $\rho_{ij} = \rho(X_i, X_j), i \neq j$,
3. Combine this information in some simulation procedure.

Unfortunately, we now know that Step 3 represents an attempt to solve an ill-posed problem. There are two main dangers. Given the marginal distributions the correlation matrix is subject to certain restrictions. For example, each ρ_{ij} must lie in an interval $[\rho_{\min}(F_i, F_j), \rho_{\max}(F_i, F_j)]$ bounded by the minimal and maximal attainable correlations for marginals F_i and F_j. It is possible that the estimated correlations are not consistent with the estimated marginals so that no corresponding multivariate distribution for the random vector exists. In the case where a multivariate distribution exists it is often not unique.

The approach described above is highly questionable. Instead of considering marginals and correlations separately it would be more satisfactory to attempt a direct estimation of the multivariate distribution. It might also be sensible to consider whether the question of interest permits the estimation problem to be reduced to a one-dimensional one. For example, if we are really interested in the behaviour of the sum $X_1 + \cdots + X_n$ we might consider directly estimating the univariate distribution of the sum.

6.1 Given marginals and linear correlations

Suppose, however, we are required to construct a multivariate distribution F in $\mathbb{R}^n$ which is consistent with given marginals distributions $F_1, \ldots, F_n$ and a linear correlation matrix ρ. We assume that ρ is a *proper* linear correlation matrix, by which we mean in the remainder of the article that it is a symmetric, positive semi-definite matrix with $-1 \le \rho_{ij} \le 1,\ i,j = 1,\dots,n$ and $\rho_{ii} = 1,\ i = 1,\dots,n$. Such a matrix will always be the linear correlation matrix of *some* random vector in $\mathbb{R}^n$ but we must check that it is compatible with the given marginals. Our problem is to find a multivariate distribution F such that if $(X_1,\dots,X_n)^t$ has distribution F then the following conditions are satisfied:

$$X_i \sim F_i, \quad i = 1,\dots,n, \tag{6.1}$$

$$\rho(X_i, X_j) = \rho_{ij}, \quad i,j = 1,\dots,n. \tag{6.2}$$

In the bivariate case, provided the prespecified correlation is attainable, the construction is simple and relies on the following.

Theorem 6. *Let F_1 and F_2 be two univariate distributions and $\rho_{\min}$ and $\rho_{\max}$ the corresponding minimal and maximal linear correlations. Let $\rho \in [\rho_{\min}, \rho_{\max}]$. Then the bivariate mixture distribution given by*

$$F(x_1, x_2) = \lambda F_\ell(x_1, x_2) + (1-\lambda) F_u(x_1, x_2), \tag{6.3}$$

where $\lambda = (\rho_{\max} - \rho)/(\rho_{\max} - \rho_{\min})$, $F_\ell(x_1, x_2) = \max\{F_1(x_1) + F_2(x_2) - 1, 0\}$ and $F_u(x_1, x_2) = \min\{F_1(x_1), F_2(x_2)\}$, has marginals F_1 and F_2 and linear correlation ρ.

Proof. Follows easily from Theorem 4. □

Remark 3. A similar result to the above holds for rank correlations when $\rho_{\min}$ and $\rho_{\max}$ are replaced by -1 and 1 respectively.

Remark 4. Also note that the mixture distribution is not the unique distribution satisfying our conditions. If $\rho \ge 0$ the distribution

$$F(x_1, x_2) = \lambda F_1(x_1) F_2(x_2) + (1-\lambda) F_u(x_1, x_2), \tag{6.4}$$

with $\lambda = (\rho_{\max} - \rho)/\rho_{\max}$ also has marginals F_1 and F_2 and correlation ρ. Many other mixture distributions (e.g. mixtures of distributions with Gumbel copulas) are possible.

Simulation of one random variate from the mixture distribution in Theorem 6 is achieved with the following algorithm:

1. Simulate U_1, U_2 independently from the standard uniform distribution,
2. If $U_1 \le \lambda$ take $(X_1, X_2)^t = (F_1^{-1}(U_2), F_2^{-1}(1 - U_2))^t$,
3. If $U_1 > \lambda$ take $(X_1, X_2)^t = (F_1^{-1}(U_2), F_2^{-1}(U_2))^t$.

Constructing a multivariate distribution in the case $n \geq 3$ is more difficult. For the existence of a solution it is certainly necessary that $\rho_{\min}(F_i, F_j) \leq \rho_{ij} \leq \rho_{\max}(F_i, F_j)$, $i \neq j$, so that the pairwise constraints are satisfied. In the bivariate case this is sufficient for the existence of a solution to the problem described by (6.1) and (6.2), but in the general case it is not sufficient as the following example shows.

Example 8. Let F_1, F_2 and F_3 be Lognormal(0, 1) distributions. Suppose that ρ is such that ρ_{ij} is equal to the minimum attainable correlation for a pair of Lognormal(0, 1) random variables (≈ -0.368) if $i \neq j$ and $\rho_{ij} = 1$ if $i = j$. This is both a proper correlation matrix and a correlation matrix satisfying the pairwise constraints for lognormal random variables. However, since ρ_{12}, ρ_{13} and ρ_{23} are all minimum attainable correlations, Theorem 4 implies that X_1, X_2 and X_3 are pairwise countermonotonic random variables. Such a situation is unfortunately impossible as is is clear from the following proposition.

Proposition 4. *Let X, Y and Z be random variables with joint distribution F and continuous marginals F_1, F_2 and F_3.*

1. *If (X, Y) and (Y, Z) are comonotonic then (X, Z) is also comonotonic and $F(x, y, z) = \min\{F_1(x), F_2(y), F_3(z)\}$.*
2. *If (X, Y) is comonotonic and (Y, Z) is countermonotonic then (X, Z) is countermonotonic and $F(x, y, z) = \max\{0, \min\{F_1(x), F_2(y)\} + F_3(z) - 1\}$.*
3. *If (X, Y) and (Y, Z) are countermonotonic then (X, Z) is comonotonic and $F(x, y, z) = \max\{0, \min\{F_1(x), F_3(z)\} + F_2(y) - 1\}$*

Proof. We show only the first part of the proposition, the proofs of the other parts being similar. Using (4.3) we know that $Y = S(X)$ a.s. and $Z = T(Y)$ a.s. where $S, T : \mathbb{R} \to \mathbb{R}$ are increasing functions. It is clear that $Z = T \circ S(X)$ a.s. with $T \circ S$ increasing, so that X and Z are comonotonic. Now let $x, y, z \in \mathbb{R}$ and because also (X, Z) is comonotonic we may without loss of generality assume that $F_1(x) \leq F_2(y) \leq F_3(z)$. Assume for simplicity, but without loss of generality, that $Y = S(X)$ and $Z = T(Y)$ (i.e. ignore almost surely). It follows that $\{X \leq x\} \subseteq \{Y \leq y\}$ and $\{Y \leq y\} \subseteq \{Z \leq z\}$ so that

$$F(x, y, z) = \mathbb{P}[X \leq x] = F_1(x).$$

□

Example 9. Continuity of the marginals is an essential assumption in this proposition. It does not necessarily hold for discrete distributions as the next counterexample shows. Consider the multivariate two-point distributions given by

$$\begin{aligned}\mathbb{P}[(X,Y,Z)^t = (0,0,1)^t] &= 0.5,\\ \mathbb{P}[(X,Y,Z)^t = (1,0,0)^t] &= 0.5.\end{aligned}$$

(X,Y) and (Y,Z) are comonotonic but (X,Z) is countermonotonic.

Proposition 4 permits us now to state a result concerning existence and uniqueness of solutions to our problem in the special case where random variables are either pairwise comonotonic or countermonotonic.

Theorem 7. [*Tiit (1996)*] *Let $F_1, \dots, F_n$, $n \geq 3$, be continuous distributions and let ρ be a (proper) correlation matrix satisfying the following conditions for all $i \neq j$, $i \neq k$ and $j \neq k$:*

- $\rho_{ij} \in \{\rho_{\min}(F_i, F_j), \rho_{\max}(F_i, F_j)\}$,
- *If $\rho_{ij} = \rho_{\max}(F_i, F_j)$ and $\rho_{ik} = \rho_{\max}(F_i, F_k)$ then $\rho_{jk} = \rho_{\max}(F_j, F_k)$,*
- *If $\rho_{ij} = \rho_{\max}(F_i, F_j)$ and $\rho_{ik} = \rho_{\min}(F_i, F_k)$ then $\rho_{jk} = \rho_{\min}(F_j, F_k)$,*
- *If $\rho_{ij} = \rho_{\min}(F_i, F_j)$ and $\rho_{ik} = \rho_{\min}(F_i, F_k)$ then $\rho_{jk} = \rho_{\max}(F_j, F_k)$.*

Then there exists a unique distribution with marginals $F_1, \dots, F_n$ and correlation matrix ρ. This distribution is known as an extremal distribution. In $\mathbb{R}^n$ there are 2^{n-1} possible extremal distributions.

Proof. Without loss of generality suppose

$$\rho_{1j} = \begin{cases} \rho_{\max}(F_1, F_j) & \text{for } 2 \leq j \leq m \leq n, \\ \rho_{\min}(F_1, F_j) & \text{for } m < j \leq n, \end{cases}$$

for some $2 \leq m \leq n$. The conditions of the theorem ensure that the pairwise relationship of any two margins is determined by their pairwise relationship to the first margin. The margins for which ρ_{1j} takes a maximal value form an equivalence class, as do the margins for which ρ_{1j} takes a minimal value. The joint distribution must be such that $(X_1, \dots, X_m)$ are pairwise comonotonic, $(X_{m+1}, \dots, X_n)$ are pairwise comonotonic, but two random variables taken from different groups are countermonotonic. Let $U \sim U(0,1)$. Then the random vector

$$(F_1^{-1}(U), F_2^{-1}(U), \dots, F_m^{-1}(U), F_{m+1}^{-1}(1-U), \dots, F_n^{-1}(1-U))^t,$$

has the required joint distribution. We use a similar argument to that of Proposition 4 and assume, without loss of generality, that

$$\min_{1 \leq i \leq m} \{F_i(x_i)\} = F_1(x_1), \quad \min_{m < i \leq n} \{F_i(x_i)\} = F_{m+1}(x_1).$$

It is clear that the distribution function is

$$\begin{aligned} F(x_1, \dots, x_n) &= \mathbb{P}[X_1 \le x_1, X_{m+1} \le x_{m+1}] \\ &= \max\{0, \min_{1\le i\le m} \{F_i(x_i)\} + \min_{m\le i\le n} \{F_i(x_i)\} - 1\}, \end{aligned}$$

which in addition shows uniqueness of distributions with pairwise extremal correlations. □

Let $G_j, j = 1, \dots, 2^{n-1}$ be the extremal distributions with marginals $F_1, \dots, F_n$ and correlation matrix ρ_j. Convex combinations

$$G = \sum_{j=1}^{2^{n-1}} \lambda_j G_j,\ \lambda_j \ge 0,\ \sum_{j=1}^{2^{n-1}} \lambda_j = 1,$$

also have the same marginals and correlation matrix given by $\rho = \sum_{j=1}^{2^{n-1}} \lambda_j \rho_j$. If we can decompose an arbitrary correlation matrix ρ in this way, then we can use a convex combination of extremal distributions to construct a distribution which solves our problem. In Tiit (1996) this idea is extended to quasi-extremal distributions. Quasi-extremal random vectors contain sub-vectors which are extremal as well as sub-vectors which are independent.

A disadvantage of the extremal (and quasi-extremal) distributions is the fact that they have no density, since they place all their mass on edges in $\mathbb{R}^n$. However, one can certainly think of practical examples where such distributions might still be highly relevant.

Example 10. Consider two portfolios of credit risks. In the first portfolio we have risks from country A, in the second risks from country B. Portfolio A has a profit-and-loss distribution F_1 and portfolio B a profit-and-loss distribution F_2. With probability p the results move in the same direction (comonotonicity); with probability $(1-p)$ they move in opposite directions (countermonotonicity). This situation can be modelled with the distribution

$$F(x_1, x_2) = p \cdot \min\{F_1(x_1), F_2(x_2)\} + (1-p) \cdot \max\{F_1(x_1) + F_2(x_2) - 1, 0\},$$

and of course generalizes to more than two portfolios.

6.2 Given marginals and Spearman's rank correlations

This problem has been considered in Iman & Conover (1982) and their algorithm forms the basis of the @RISK computer program (Palisade 1997).

It is clear that a Spearman's rank correlation matrix is also a linear correlation matrix (Spearman's rank being defined as the linear correlation of ranks). It is not known to us whether a linear correlation matrix is necessarily a Spearman's rank correlation matrix. That is, given an arbitrary symmetric, positive semi-definite matrix with unit elements on the diagonal and off-diagonal elements in the interval $[-1, 1]$, can we necessarily find a random vector with continuous marginals for which this is the rank correlation matrix, or alternatively a multivariate distribution for which this is the linear correlation matrix of the copula? If we estimate a rank correlation matrix from data, is it guaranteed that the estimate is itself a rank correlation matrix? A necessary condition is certainly that the estimate is a linear correlation matrix, but we do not know if this is sufficient.

If the given matrix is a true rank correlation matrix, then the problem of the existence of a multivariate distribution with prescribed marginals is solved. The choice of marginals is in fact irrelevant and imposes no extra consistency conditions on the matrix.

Iman & Conover (1982) do not attempt to find a multivariate distribution which has exactly the given rank correlation matrix ρ. They simulate a standard multivariate normal variate $(X_1, \ldots, X_n)^t$ with linear correlation matrix ρ and then transform the marginals to obtain $(Y_1, \ldots, Y_n)^t = (F_1^{-1}(\Phi(X_i)), \ldots, F_n^{-1}(\Phi(X_n)))^t$.The rank correlation matrix of $\mathbf{Y}$ is identical to that of $\mathbf{X}$. Now because of (3.2)

$$\rho_S(Y_i, Y_j) = \rho_S(X_i, X_j) = \frac{6}{\pi} \arcsin \frac{\rho(X_i, X_j)}{2} \approx \rho(X_i, X_j),$$

and, in view of the bounds for the absolute error,

$$\left| \frac{6}{\pi} \arcsin \frac{\rho}{2} - \rho \right| \leq 0.0181, \ \rho \in [-1, 1],$$

and for the relative error,

$$\frac{\left|\frac{6}{\pi} \arcsin \frac{\rho}{2} - \rho\right|}{|\rho|} \leq \frac{\pi - 3}{\pi},$$

the rank correlation matrix of $\mathbf{Y}$ is very close to that which we desire. In the case when the given matrix belongs to an extremal distribution (i.e. comprises only elements 1 and -1) then the error disappears entirely and we have constructed the unique solution of our problem.

This suggests how we can find a sufficient condition for ρ to be a Spearman's rank correlation matrix and how, when this condition holds, we can construct a distribution that has the required marginals and exactly this rank correlation matrix. We define the matrix $\tilde{\rho}$ by

$$\tilde{\rho}_{ij} = 2 \sin \frac{\pi \rho_{ij}}{6}, \tag{6.5}$$

and check whether this is a proper linear correlation matrix. If so, then the vector $(Y_1, \ldots, Y_n)^t = (F_1^{-1}(\Phi(X_i)), \ldots, F_n^{-1}(\Phi(X_n)))^t$ has rank correlation matrix ρ, where $(X_1, \ldots, X_n)^t$ is a standard multivariate normal variate with linear correlation matrix $\tilde{\rho}$.

In summary, a necessary condition for ρ to be a rank correlation matrix is that it is a linear correlation matrix and a sufficient condition is that $\tilde{\rho}$ given by (6.5) is a linear correlation matrix. We are not aware at present of a necessary and sufficient condition.

A further problem with the approach described above is that we only ever construct distributions which have the dependence structure of the multivariate normal distribution. This dependence structure is limited as we observed in Example 2; it does not permit asymptotic dependence between random variables.

6.3 Given marginals and copula

Whenever marginal distributions $F_1, \ldots, F_n$ and a copula $C(u_1, \ldots, u_n)$ are specified, a unique multivariate distribution with distribution function $F(x_1, \ldots, x_n) = C(F_1(x_1), \ldots, F_n(x_n))$ satisfying these specifications can be found. The problem of simulating from this distribution is no longer the theoretical one of whether a solution exists, but rather the technical one of how to perform the simulation. We assume the copula is given in the form of a parametric function which the modeller has chosen; we do not consider the problem of how copulas might be estimated from data, which is certainly more difficult than estimating linear or rank correlations.

Once we have simulated a random vector $(U_1, \ldots, U_n)^t$ from C, then the random vector $(F_1^{-1}(U_1), \ldots, F_n^{-1}(U_n))^t$ has distribution F. We assume that efficient univariate simulation presents no problem and refer to Ripley (1987),Gentle (1998) or Devroye (1986) for more on this subject. The major technical difficulty lies now in simulating realisations from the copula.

Where possible a transformation method can be applied; that is, we make use of multivariate distributions with the required copula for which a multivariate simulation method is already known. For example, to simulate from the bivariate Gaussian copula it is trivial to simulate $(Z_1, Z_2)^t$ from the standard bivariate normal distribution with correlation ρ and then to transform the marginals with the univariate distribution function so that $(\Phi(Z_1), \Phi(Z_2))^t$ is distributed according to the desired copula. For the bivariate Gumbel copula a similar approach can be taken.

Example 11. Consider the Weibull distribution having survivor function $\overline{F}_1(x) = 1 - F_1(x) = \exp\left(-x^\beta\right)$ for $\beta > 0,\ x \geq 0$. If we apply the Gumbel copula to this survivor function (not to the distribution function) we get a bivariate distribution with Weibull marginals and survivor function

$$\overline{F}(z_1, z_2) = \mathbb{P}[Z_1 > z_1, Z_2 > z_2] = C(\overline{F}_1(z_1), \overline{F}_1(z_2)) = \exp\left[-(z_1 + z_2)^\beta\right].$$

Lee (1979) describes a method for simulating from this distribution. We take $(Z_1, Z_2)^t = (US^{1/\beta}, (1-U)S^{1/\beta})^t$ where U is standard uniform and S is a mixture of Gamma distributions with density $h(s) = (1 - \beta + \beta s)\exp(-s)$ for $s \geq 0$. Then $(\overline{F}_1(Z_1), \overline{F}_1(Z_2))^t$ will have the desired copula distribution.

Where the transformation method cannot easily be applied, another possible method involves recursive simulation using univariate conditional distributions. We consider the general case $n > 2$ and introduce the notation

$$C_i(u_1, \ldots, u_i) = C(u_1, \ldots, u_i, 1, \ldots, 1),\ i = 2, \ldots, n-1$$

to represent i–dimensional marginal distributions of $C(u_1, \ldots, u_n)$. We write $C_1(u_1) = u_1$ and $C_n(u_1, \ldots, u_n) = C(u_1, \ldots, u_n)$. Let us suppose now that $(U_1, \ldots, U_n)^t \sim C$; the conditional distribution of U_i given the values of the first $i-1$ components of $(U_1, \ldots, U_n)^t$ can be written in terms of derivatives and densities of the i–dimensional marginals

$$\begin{aligned} C_i(u_i \mid u_1, \ldots, u_{i-1}) &= \mathbb{P}[U_i \leq u_i \mid U_1 = u_1, \ldots, U_{i-1} = u_{i-1}] \\ &= \frac{\partial^{i-1} C_i(u_1, \ldots, u_i)}{\partial u_1 \cdots \partial u_{i-1}} \Big/ \frac{\partial^{i-1} C_{i-1}(u_1, \ldots, u_{i-1})}{\partial u_1 \cdots \partial u_{i-1}}, \end{aligned}$$

provided both numerator and denominator exist. This suggests that in the case where we can calculate these conditional distributions we use the algorithm:

- Simulate a value u_1 from $U(0,1)$,
- Simulate a value u_2 from $C_2(u_2 \mid u_1)$,
- Continue in this way,
- Simulate a value u_n from $C_n(u_n \mid u_1, \ldots, u_{n-1})$.

To simulate a value from $C_i(u_i \mid u_1, \ldots, u_{i-1})$ we would in general simulate u from $U(0,1)$ and then calculate $C_i^{-1}(u \mid u_1, \ldots, u_{i-1})$, if necessary by numerical root finding.

7 Conclusions

In this article we have shown some of the problems that can arise when the concept of linear correlation is used with non-elliptical multivariate distributions. In the world of elliptical distributions correlation is a natural and elegant summary of dependence, which lends itself to algebraic manipulation and the standard approaches of risk management dating back to Markowitz. In the non-elliptical world our intuition about correlation breaks down and leads to a number of fallacies. The first aim of this article has been to suggest that practitioners of risk management must be aware of these pitfalls and must appreciate that a deeper understanding of dependence is needed to model the risks of the real world.

The second main aim of this article has been to address the problem of simulating dependent data with given marginal distributions. This question arises naturally when one contemplates a Monte Carlo approach to determining the risk capital required to cover dependent risks. We have shown that the ideal situation is when the multivariate dependence structure (in the form of a copula) is fully specified by the modeller. Failing this, it is preferable to be given a matrix of rank correlations rather than a matrix of linear correlations, since rank correlations are defined at a copula level, and we need not worry about their consistency with the chosen marginals. Both correlations are, however, scalar-valued dependence measures and if there is a multivariate distribution which solves the simulation problem, it will not be the unique solution. The example of the Introduction showed that two distributions with the same correlation can have qualitatively very different dependence structures and, ideally, we should consider the whole dependence structure which seems appropriate for the risks we wish to model.

Acknowledgements

Daniel Staumann was supported by RiskLab Switzerland. Alexander McNeil would like to thank Swiss Re for financial support. We would like to thank Eduardo Vilela and Rüdiger Frey for fruitful discussions.

References

Albers, W. (1999), 'Stop–loss premiums under dependence', *Insurance: Mathematics and Economics* **24**, 173–185.

Artzner, P., Delbaen, F., Eber, J.-M. & Heath, D. (1999), 'Coherent measures of risk', *Mathematical Finance* **9**(3), 203–228. Reprinted in this volume.

Boyer, B. H., Gibson, M. S. & Loretan, M. (1999), Pitfalls in tests for changes in correlations. International Finance Papers No. 597, Board of Governors of the Federal Reserve System.

Campbell, J. Y., Lo, A. W. & MacKinlay, A. (1997), *The Econometrics of Financial Markets*, Princeton University Press, Princeton.

Capéraà, P., Fougères, A.-L. & Genest, C. (1997), 'A nonparametric estimation procedure for bivariate extreme value copulas', *Biometrika* **84**(3), 567–577.

CAS (1997), *CAS Forum Summer 1997: DFA Call Papers.*

Delbaen, F. (1999), Coherent risk measures on general probability spaces. Preprint ETH Zürich.

Denneberg, D. (1994), *Non-additive Measure and Integral*, Kluwer Academic Publishers, Dordrecht.

Devroye, L. (1986), *Non-uniform Random Variate Generation*, Springer, New York.

Dhaene, J. & Goovaerts, M. J. (1996), 'Dependency of risks and stop-loss orders', *ASTIN Bulletin* **26**(2), 201–212.

Embrechts, P., Klüppelberg, C. & Mikosch, T. (1997), *Modelling Extremal Events for Insurance and Finance*, Springer, Berlin.

Fang, K.-T., Kotz, S. & Ng, K.-W. (1987), *Symmetric Multivariate and Related Distributions*, Chapman & Hall, London.

Frank, M., Nelsen, R. B. & Schweizer, B. (1987), 'Best–possible bounds for the distribution of a sum — a problem of Kolmogorov', *Probability Theory and Related Fields* **74**, 199–211.

Frank, M. & Schweizer, B. (1979), 'On the duality of generalized infimal and supremal convolutions', *Rendiconti di Matematica* **12**(1), 1–23.

Fréchet, M. (1957), 'Les tableaux de corrélation dont les marges sont données', *Annales de l'Université de Lyon, Sciences Mathématiques et Astronomie, Série A* **4**, 13–31.

Frees, E. W. & Valdez, E. (1998), 'Understanding relationships using copulas', *North American Actuarial Journal* **2**(1), 1–25.

Galambos, J. (1987), *The Asymptotic Theory of Extreme Order Statistics*, Kreiger Publishing Co., Melbourne, FL.

Genest, C., Ghoudi, K. & Rivest, L.-P. (1995), 'A semiparametric estimation procedure of dependence parameters in multivariate families of distributions', *Biometrika* **82**(3), 543–552.

Genest, C. & Rivest, L.-P. (1993), 'Statistical inference procedures for bivariate Archimedean copulas', *Journal of the American Statistical Association* **88**(423), 1034–1043.

Gentle, J. E. (1998), *Random Number Generation and Monte Carlo Methods*, Springer, New York.

Gibbons, J. D. (1988), *Nonparametric Statistical Inference*, Dekker, New York.

Harlow, W. (1991), 'Asset allocation in a downside-risk framework', *Financial Analysts Journal* **47**(5), 28–40.

Höffding, W. (1940), 'Massstabinvariante Korrelationstheorie', *Schriften des Mathematischen Seminars und des Instituts für Angewandte Mathematik der Universität Berlin* **5**, 181–233.

Hutchinson, T. P. & Lai, C. D. (1990), *Continuous Bivariate Distributions, Emphasizing Applications*, Rumsby Scientific Publishing, Adelaide.

Iman, R. L. & Conover, W. (1982), 'A distribution–free approach to inducing rank correlation among input variables', *Communications in Statistics — Simulation and Computation* **11**, 311–334.

Joag-dev, K. (1984), 'Measures of dependence', in *Handbook of Statistics*, Vol. 4, P. R. Krishnaiah, ed., ' North–Holland/Elsevier, New York, pp. 79–88.

Joe, H. (1997), *Multivariate Models and Dependence Concepts*, Chapman & Hall, London.

Kelker, D. (1970), 'Distribution theory of spherical distributions and a location–scale parameter generalization', *Sankhiă A* **32**, 419–430.

Kendall, M. & Stuart, A. (1979), *Handbook of Statistics*, Griffin & Company, London.

Kimeldorf, G. & Sampson, A. R. (1978), 'Monotone dependence', *Annals of Statistics* **6**, 895–903.

Klugman, S. A. & Parsa, R. (1999), 'Fitting bivariate loss distributions with copulas', *Insurance: Mathematics & Economics* **24**, 139–148.

Lee, L. (1979), 'Multivariate distributions having Weibull properties', *Journal of Multivariate Analysis* **9**, 267–277.

Li, H., Scarsini, M. & Shaked, M. (1996), 'Bounds for the distribution of a multivariate sum', in *Distributions with Fixed Marginals and Related Topics*, L. Rüschendorff, B. Schweizer & M. D. Taylor, eds, Institute of Mathematical Statistics, Hayward, CA, pp. 198–212.

Lowe, S. P. & Stanard, J. N. (1997), 'An integrated dynamic financial analysis and decision system for a property catastrophe insurer', *ASTIN Bulletin* **27**(2), 339–371.

Makarov, G. (1981), 'Estimates for the distribution function of a sum of two random variables when the marginal distributions are fixed', *Theory of Probability and its Applications* **26**, 803–806.

Marshall, A. W. (1996), 'Copulas, marginals and joint distributions', in *Distributions with Fixed Marginals and Related Topics*, L. Rüschendorff, B. Schweizer & M. D. Taylor, eds, Institute of Mathematical Statistics, Hayward, CA, pp. 213–222.

Mikusinski, P., Sherwood, H. & Taylor, M. (1992), 'The Fréchet bounds revisited', *Real Analysis Exchange* **17**, 759–764.

Müller, A. & Bäuerle, N. (1998), 'Modelling and comparing dependencies in multivariate risk portfolios', *ASTIN Bulletin* **28**(1), 59–76.

Nelsen, R. B. (1999), *An Introduction to Copulas*, Springer, New York.

Palisade (1997), *Manual for @RISK*, Palisade Corporation, Newfield, NY.

Resnick, S. I. (1987), *Extreme Values, Regular Variation and Point Processes*, Springer, New York.

Ripley, B. D. (1987), *Stochastic Simulation*, Wiley, New York.

Schweizer, B. & Sklar, A. (1983), *Probabilistic Metric Spaces*, North–Holland/Elsevier, New York.

Schweizer, B. & Wolff, E. (1981), 'On nonparametric measures of dependence for random variables', *Annals of Statistics* **9**, 879–885.

Sibuya, M. (1961), 'Bivariate extreme statistics', *Annals of Mathematical Statistics* **11**, 195–210.

Tiit, E. (1996), 'Mixtures of multivariate quasi-extremal distributions having given marginals', in *Distributions with Fixed Marginals and Related Topics*, L. Rüschendorff, B. Schweizer & M. D. Taylor, eds, Institute of Mathematical Statistics, Hayward, CA, pp. 337–357.

Tjøstheim, D. (1996), 'Measures of dependence and tests of independence', *Statistics* **28**, 249–284.

Wang, S. (1997), Aggregation of correlated risk portfolios: Models and algorithms. Preprint Casualty Actuarial Society (CAS).

Wang, S. & Dhaene, J. (1998), 'Comonoticity, correlation order and premium principles', *Insurance: Mathematics and Economics* **22**, 235–242.

Williamson, R. C. & Downs, T. (1990), 'Probabilistic arithmetic: Numerical methods for calculating convolutions and dependency bounds', *Journal of Approximate Reasoning* **4**, 89–158.

Yaari, M. (1987), 'The dual theory of choice under risk', *Econometrica* **55**, 95–115.

Measuring Risk with Extreme Value Theory

Richard L. Smith

1 Introduction

As financial trading systems have become more sophisticated, there has been increased awareness of the dangers of very large losses. This awareness has been heightened by a number of highly publicised catastrophic incidents:

- *Barings.* In February 1995, the Singapore subsidiary of this long-established British bank lost about \$1.3 billion because of the illegal activity of a single trader, Nick Leeson. As a result the bank collapsed, and was subsequently sold for one pound.
- *Orange County.* Bob Citron, the Treasurer of Orange County, had invested much of the county's assets in a series of derivative instruments tied to interest rates. In 1994, interest rates rose, and Orange County went bankrupt, losing \$1.7 billion.
- *Daiwa Bank.* A single trader, Toshihide Iguchi, lost \$1.1 billion of the bank's money over a period of 11 years, the losses only coming to light when Iguchi confessed to his managers in July 1995.
- *Long Term Capital Management.* In the most spectacular example to date, this highly-regarded hedge fund nearly collaped in September 1998. LTCM was trading a complex mixture of derivatives which, according to some estimates, gave it an exposure to market risk as high as \$200 billion. Things started to go wrong after the collapse of the Russian economy in the summer of 1998, and to avoid a total collapse of the company, 15 major banks contributed to a \$3.75 billion rescue package.

These and other examples have increased awareness of the need to quantify probabilities of large losses, and for risk management systems to control such events. The most widely used tool is *Value at Risk* (henceforth, VaR). Originally started as an internal management tool by a number of banks, it gained a higher profile in 1994 when J.P. Morgan published its RiskMetrics system[1]. Subsequent books aimed at financial academics and traders (Jorion 1996, Dowd 1998) explained the statistical basis behind VaR. Despite

[1] `http://www.riskmetrics.com/rm/index.html`

the complexity of financial data management that these systems need, the statistical principles behind them are quite simple.

According to the most usual definition, we have to fix a time horizon T and a failure probability α. A common value for T is ten trading days, while α is often set to be .05 or .01. The VaR is then defined to be the largest number x such that the probability of a loss as large as x over the time horizon T is no more than α. Since it is widely accepted that, conditionally on the current volatility σ, the daily log returns ($Y_t = 100\log(X_t/X_{t-1})$ where X_t is the price on day t) are independent normally distributed with standard deviation σ, the VaR becomes a routine calculation of normal probabilities. When the joint behaviour of a large number of assets is considered, as is needed to calculate the VaR of a portfolio, it is usual to adopt a multivariate normal distribution, though much work goes into the computation of the variances and covariances required. For instance, it is common to use some variant of either principal components analysis or factor analysis to reduce the dimensionality of the statistical estimation problem.

What has been outlined is the simplest approach to VaR estimation. There are at least three competing approaches, none of them so reliant on distributional assumptions. The *historical data* approach uses historical market movements to determine loss probabilities in a statistically nonparametric way. The disadvantage of this is that historical data may not adequately represent current market conditions, or may not be available in sufficient quantity to allow reliable risk calculations to be made. The *stress testing* approach puts much less emphasis on the assessment of small probabilities, instead relying on computing losses under various scenarios of unlikely but plausible market conditions. Finally there is the approach discussed in the present article, using *Extreme Value Theory* (EVT) to characterise the lower tail behaviour of the distribution of returns without tying the analysis down to a single parametric family fitted to the whole distribution.

The use of EVT in financial market calculations is a fairly recent innovation, but there is a much longer history of its use in the insurance industry. The excellent recent book by Embrechts *et al.* (1997) surveys the mathematical theory of EVT and discusses its applications to both financial and insurance risk management. In Section 2 of the current article, I outline some of the statistical techniques used in EVT and illustrate them with a recent example of insurance data. However, I also highlight some aspects of financial data — specifically, the presence of variable volatility — that makes direct application of such methods to financial data inappropriate.

In subsequent sections, I outline some current areas of theoretical development that have strong potential for applicability in the insurance and financial industries —

- Bayesian methods (Section 3) as a device for taking account of model uncertainty in extreme risk calculations,

- Multivariate EVT (Section 4) as an alternative approach to risk assessment in high-dimensional systems,
- A random changepoint model (Section 5) as one approach to long-term stochastic volatility.

The overall message of the article is that EVT contains rich possibilities for application to finance and insurance risk management, but that these areas of application also pose many new challenges to the methodology.

2 Outline of Extreme Value Theory

The mathematical foundation of EVT is the class of extreme value limit laws, first derived heuristically by Fisher and Tippett (1928) and later from a rigorous standpoint by Gnedenko (1943). Suppose $X_1, X_2, \ldots,$ are independent random variables with common distribution function $F(x) = \Pr\{X \le x\}$ and let $M_n = \max\{X_1, \ldots, X_n\}$. For suitable normalising constants $a_n > 0$ and b_n, we seek a limit law G satisfying

$$\Pr\left\{\frac{M_n - b_n}{a_n} \le x\right\} = F^n(a_n x + b_n) \to G(x) \tag{2.1}$$

for every x. The key result of Fisher–Tippett and Gnedenko is that there are only three fundamental *types* of extreme value limit laws [2]. These are

$$\begin{aligned}
\text{Type I}: \quad & \Lambda(x) &=& \exp(-e^{-x}), \quad -\infty < x < \infty, \\
\text{Type II}: \quad & \Phi_\alpha(x) &=& \begin{cases} 0, & x \le 0, \\ \exp(-x^{-\alpha}), & x > 0, \end{cases} \\
\text{Type III}: \quad & \Psi_\alpha(x) &=& \begin{cases} \exp(-(-x)^{-\alpha}), & x \le 0, \\ 1, & x > 0. \end{cases}
\end{aligned}$$

In Types II and III, α is a positive parameter. The three types may also be combined into a single *generalised extreme value* distribution, first proposed by von Mises (1936), of form

$$G(x) = \exp\left\{-\left(1 + \xi\frac{x-\mu}{\psi}\right)_+^{-1/\xi}\right\}, \tag{2.2}$$

where $y_+ = \max(y, 0)$, $\psi > 0$ and μ and ξ are arbitrary real parameters. The case $\xi > 0$ corresponds to Type II with $\alpha = 1/\xi$, $\xi < 0$ to Type III with $\alpha = -1/\xi$, and the limit $\xi \to 0$ to Type I.

[2] Two probability distributions G_1 and G_2 are said to be *of the same type* if they may be related by a location-scale transformation, $G_1(y) = G_2(Ay + B)$ for some $A > 0$ and B. Thus, in saying that there are only three types, we mean that any extreme value limit law may be reduced to one of the three given forms by a location-scale transformation.

Classical EVT is sometimes applied directly, for example by fitting one of the extreme value limit laws to the annual maxima of a series, and much historical work was devoted to this approach (Gumbel 1958). From a modern viewpoint, however, the classical approach is too narrow to be applied to a wide range of problems.

An alternative approach is based on *exceedances over thresholds* (Smith 1989, Davison and Smith 1990, Leadbetter 1991). According to this approach, we fix some high threshold u and look at all *exceedances* of u. The distribution of *excess values* is given by

$$F_u(y) = \Pr\{X \le u + y \mid X > u\} = \frac{F(u+y) - F(u)}{1 - F(u)}, \quad y > 0. \tag{2.3}$$

By analogy with classical EVT, there is a theory about the asymptotic form of $F_u(y)$, first given by Pickands (1975). According to this, if the underlying distribution function F is such that a classical extreme value distribution (2.1) exists, then there are constants $c_u > 0$ such that as $u \to \omega_F$ [3] such that

$$F_u(c_u z) \to H(z), \tag{2.4}$$

where

$$H(z) = \begin{cases} 1 - \left(1 + \frac{\xi z}{\sigma}\right)_+^{-1/\xi}, & \xi \neq 0, \\ 1 - e^{-z/\sigma}, & \xi = 0, \end{cases} \tag{2.5}$$

where $\sigma > 0$ and $-\infty < \xi < \infty$. This is known as the *generalised Pareto distribution* (GPD). There is a close analogy between (2.5) and (2.2), because ξ is the same and there are also mathematical relations among μ, ψ and σ (Davison and Smith 1990).

The threshold approach is most usually applied by fitting the GPD to the observed excesses over the threshold. One advantage of this method over the annual maximum approach is that since each exceedance is associated with a specific event, it is possible to make the parameters σ and ξ depend on covariates. This has been done, for instance, in assessing the probability of a high-level exceedance in the tropospheric ozone record as a function of meteorology (Smith and Shively 1995). Other aspects of the method are the selection of a suitable threshold, and treatment of time series dependence. In environmental applications, the latter aspect is often dealt with by the simple procedure of restricting attention to *peaks* within clusters of high exceedances (Davison and Smith 1990), though as we shall see, such a simple-minded approach does not appear to work for handling stochastic volatility in financial time series.

There are other approaches to extreme value modelling, based on variants of the theoretical results already discussed. One approach extends the annual

[3] $\omega_F = \sup\{x : F(x) < 1\}$, the right-hand endpoint of the distribution, usually but not necessarily assumed to be $+\infty$.

maximum approach to the joint distribution of the k largest or smallest order statistics in each year – this was first developed statistically by Smith (1986) and Tawn (1988), though the underlying probability theory is much older (see, for example, Section 2.3 of Leadbetter *et al.* 1983). This method is not used much, but we shall see an example of it in Section 3.

A more substantial variant is to take the point-process viewpoint of high-level exceedances, which again has been very well developed as a probabilistic technique (e.g. the books by Leadbetter *et al.* (1983) and Resnick (1987) both use it, though from quite different viewpoints) and was developed as a statistical technique by Smith (1989). According to this viewpoint, the exceedance times and excess values of a high threshold are viewed as a two-dimensional point process (Fig. 1). If the process is stationary and satisfies a condition that there are asymptotically no clusters among the high-level exceedances, then its limiting form is non-homogeneous Poisson and the intensity measure of a set A of the form $(t_1, t_2) \times (y, \infty)$ (see Fig. 1) may be expressed in the form

$$\Lambda(A) = (t_2 - t_1) \cdot \left(1 + \xi \frac{y - \mu}{\psi}\right)_+^{-1/\xi}. \tag{2.6}$$

Here, the interpretation of the parameters μ, ψ and ξ is exactly the same as in (2.2) – indeed, if the time scale in (2.6) is measured in years then the corresponding version of (2.2) is precisely the probability that a set $A = (t_1, t_1 + 1) \times (y, \infty)$ is empty, or in other words, that the annual maximum is $\leq y$. However, one can also derive the GPD as a consequence of (2.6) and hence tie this view of the theory with the peaks over threshold analysis.

A more general form of the model allows for time-dependent behaviour by replacing the fixed parameters μ, ψ, ξ with functions μ_t, ψ_t, ξ_t where t denotes time. In particular, we consider models of this form in Section 5, and equation (5.1) gives the generalisation of (2.6) in this case. In this way, dependence on covariates or other time-dependent phenomena may be incorporated into the model.

A number of diagnostic techniques have been devised to test whether these assumptions are satisfied in practice. Among these are the mean excess plot (Davison and Smith, 1990), which is a plot of the mean of all excess values over a threshold u against u itself. This is based on the following identity: if Y is a random variable with distribution function (2.5), provided $\xi < 1$, then for $u > 0$,

$$\mathrm{E}\{Y - u \mid Y > u\} = \frac{\sigma + \xi u}{1 - \xi}.$$

Thus, a sample plot of mean excess against threshold should be approximately a straight line with slope $\xi/(1 - \xi)$. This is a useful tool in selecting the threshold.

In practice, the plot can be hard to interpret because for large u there are few exceedances and hence very high variability in the mean, but its

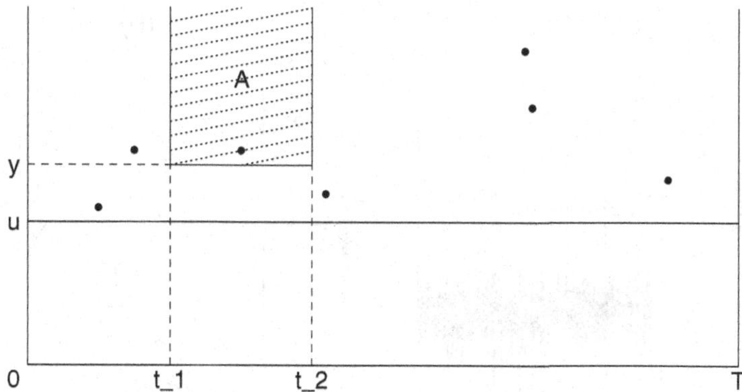

Figure 1: Illustration of high-level exceedances represented as a two-dimensional point process.

real purpose is to detect significant shifts in slope at lower thresholds. As an example, Fig. 2(a) shows the negative log daily returns for Standard and Poors index (S&P 500), 1947–1987. The values are negated because our interest in this discussion is in the possibility of very large losses, so the values of interest appear as large positive values in the plot. In particular, the spike at the right hand end of the plot is the October 19, 1987 value. A mean excess plot (Fig. 2(b)) shows an apparent 'kink' near $y = 3.8$, so it would seem unwise to include values below that threshold. (In fact this discussion is too simple because we have not taken variable volatility into account, but we return to that point later.)

In contrast, Fig. 3(a) shows 15 years of insurance claims data from a well-known multinational company (Smith and Goodman 2000), and the corresponding mean excess plot in Fig. 3(b). In this case the series is dominated by two very large claims in the middle of the plot, which together account for 35% of all claims in the series, but in spite of this apparent evidence of outliers, the mean excess plot is surprisingly stable. Repeated fitting of the model (2.6), to a variety of thresholds (Table 1), shows comparatively little variation in the parameters μ, ψ and ξ, which is another indication that the model is a good fit.

Other diagnostics may be derived from the fitted point process. For example, under the model (2.6), the one-dimensional point process of exceedance times of a fixed threshold u is a nonhomogeneous Poisson with parameter $\lambda = \{1 + \xi(u - \mu)/\psi\}^{-1/\xi}$. As noted already following (2.6), in general we may permit the parameters μ, ψ, ξ to be functions of time and in that case the constant λ is replaced by a time-dependent intensity λ_t. For this model, with constant λ as a special case, if the observations begin at time T_0 and the

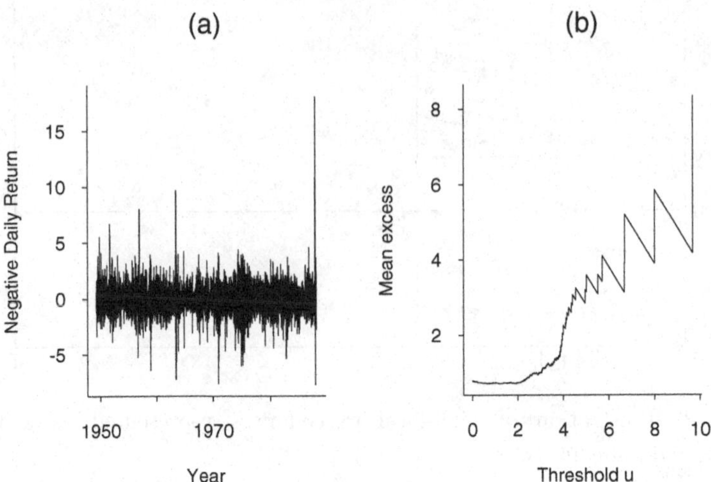

Figure 2: Negative daily returns from the S&P 500 (a), and a Mean Excess Plot based on these data (b).

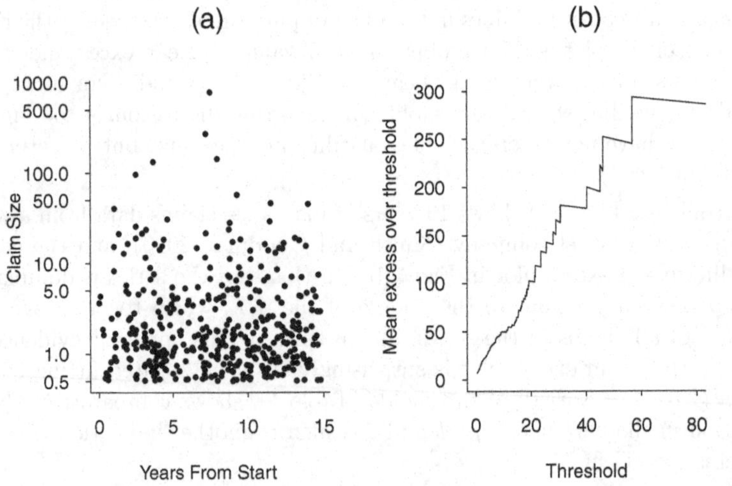

Figure 3: Scatterplot of large insurance claims against time (a), and a Mean Excess Plot based on these data (b).

successive exceedance times are at $T_1, T_2, \ldots$, the variables

$$Z_k = \int_{T_{k-1}}^{T_k} \lambda_t dt, \quad k \geq 1, \tag{2.7}$$

Table 1: Parameter estimates for the insurance claims data based on a variety of thresholds.

Threshold	Num. of exceedances	μ	$\log\psi$	ξ
0.5	393	26.5	3.30	1.00
2.5	132	26.3	3.22	0.91
5	73	26.8	3.25	0.89
10	42	27.2	3.22	0.84
15	31	22.3	2.79	1.44
20	17	22.7	3.13	1.10
25	13	20.5	3.39	0.93

should be independent exponentially distributed random variables with mean 1. This may be tested graphically, for example, via a QQ [4] plot of observed order statistics versus their expected values under the independent exponential assumption.

We can also test the marginal distribution of excesses in similar fashion. In this case the appropriate test statistic is

$$W_k = \frac{1}{\xi_{T_k}} \log\left[1 + \xi_{T_k}\left\{\frac{Y_{T_k} - u}{\psi_{T_k} + \xi_{T_k}(u - \mu_{T_k})}\right\}\right], \tag{2.8}$$

Y_{T_k} being the observed value of the process at time T_k and the notation indicating that the parameters μ, ψ and ξ are all dependent on time in the most general form of the model. Once again, if the assumed model is correct then the $\{W_k\}$ are independent exponentially distributed random variables with mean 1, and this may be tested in various ways, for example, through a QQ plot of the order statistics. The plots based on the Z and W statistics were first suggested by Smith and Shively (1995).

As an example, Fig. 4 shows the Z and W plots for the insurance data of Fig. 3, in the case that the extreme values parameters μ, ψ, ξ are assumed constants independent of time. In this case, both plots look quite close to a straight line of unit slope, indicating an acceptable fit to the model. As a standard for later comparison, we calculate the R^2 for regression $(1 - \sum(y_i - x_i)^2/\sum(y_i - \bar{y})^2$ where (x_i, y_i) are the coordinates of the i'th point in the plot). The R^2 values in this example are .992 for the Z plot and .980 for the W plot.

Fig. 5, based on the S&P index data, is more problematic. The W plot ($R^2 = 0.971$) is a good fit except for the very largest value, which is of

[4] quantile-quantile

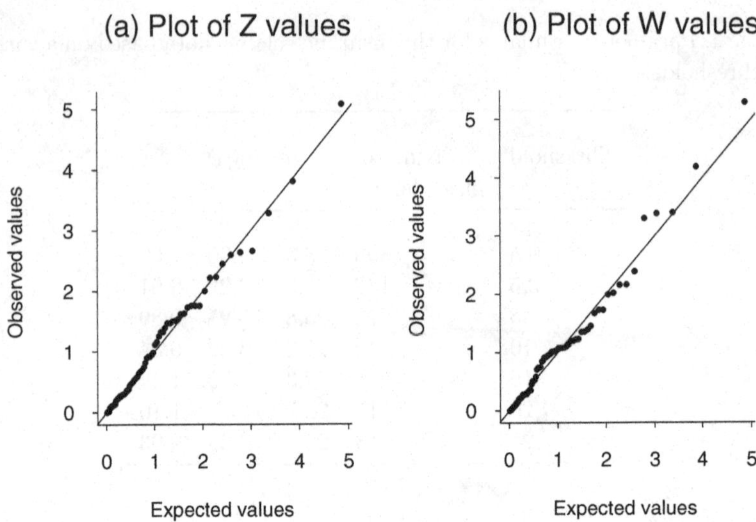

Figure 4: Z-plot and W-plot for the insurance data, all exceedances over threshold 5.

course the October 19, 1987 market crash, so this is easily understood if not so easily explained. However, the Z plot ($R^2 = 0.747$) is completely wrong, and no obvious variant on the methodology (such as changing the threshold, transforming the response variable, or adding simple time trends to the model) will do anything to correct this. The explanation is that variation in volatility results in substantial variation in the mean time between exceedances over the threshold, and no simple modification of the model can account for this.

3 Bayesian Statistics for Risk Assessment

So far, our statistical viewpoint has implicitly been classical frequentist, including maximum likelihood estimation for the parameters of the extreme value model. In this section, I argue that there may be substantial advantages of application and interpretation by taking a Bayesian approach to the analysis. The approach taken is not specifically tied to the subjective viewpoint of probability theory, since there may also be substantial advantages to the proposed approach from a frequentist viewpoint, though the evidence on the latter point is still unclear at the present time.

To illustrate the ideas, I take an example from a quite different field to the ones discussed so far. Fig. 6(a) shows the five best performances by different athletes in the women's 3000 metre track race for each year from 1972 to 1992, together with the remarkable new world record established in 1993 by the Chinese athlete Wang Junxia. Many questions have been raised about

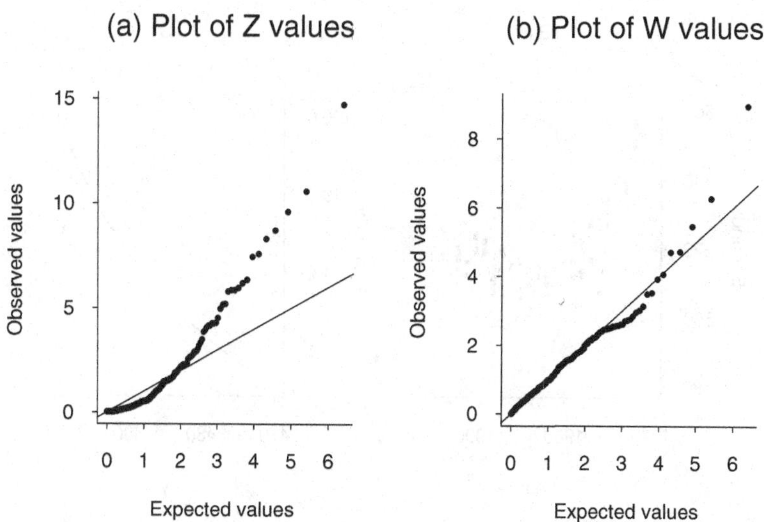

Figure 5: Z-plot and W-plot for the S&P 500, all exceedances over threshold 2.

Wang's performance, including the possibility that it may have been assisted by drugs, though no direct evidence of that was ever found. The present discussion is based on analysis by Robinson and Tawn (1995) and Smith (1997a).

The natural extreme value model for this problem is a Type III or Weibull distribution, which implies a finite lower bound β on the distribution of running times. The basic model for the distribution function of an annual maximum is (2.2). This is applied to running times multiplied by -1, so as to convert minima into maxima. When $\xi < 0$, the distribution function (2.2) has a finite upper bound (for which $G(x) = 1$) at $x = \mu - \psi/\xi$. Thus when this is applied to -1 times the running times, there is a finite *minimum* running time at $\beta = -\mu + \psi/\xi$. We therefore concentrate on estimation of this parameter, finding the maximum likelihood estimate and likelihood-based confidence intervals for β, based on the data up to 1992. If Wang's actual performance lay outside this confidence interval, that could be interpreted as evidence that something untoward had taken place. As noted briefly in Section 2, the actual estimate was based on the k best performances in each year, where in this analysis, $k = 5$.

In one analysis, Smith (1997a) analysed the data from 1980, ignoring the trend in the early part of the series, and obtained a 95% confidence interval for β of (481.9, 502.4) (seconds). Wang's actual record was 486.1, so while this lies towards the lower end of the confidence interval, the analysis does not definitively establish that there was anything wrong. Earlier, Robinson

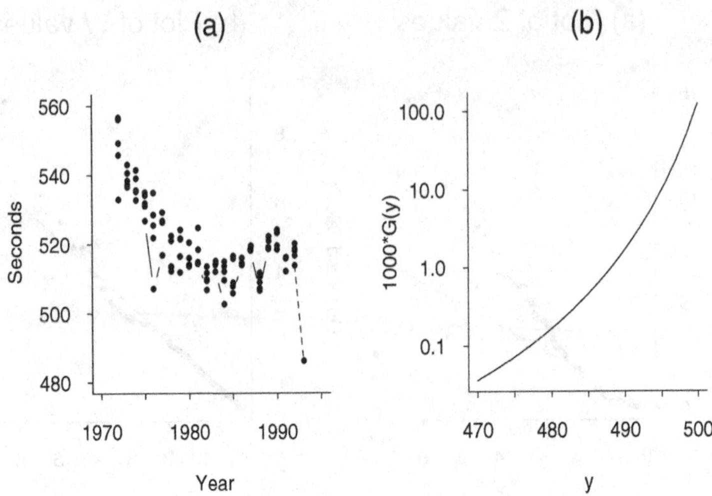

Figure 6: (a) Plot of best five performances by different athletes in each year from 1972–1992, together with Wang Junxia's performance from 1993. (b) Plot of predictive conditional probability distribution given all data up to 1992.

and Tawn (1995) gave a number of alternative analyses based on various interpretations of the trend in Fig. 6(a), but all led to the same conclusion, that Wang's record lay within a 95% confidence interval for β.

However, Smith (1997a) went on to argue that obtaining a confidence interval for β was solving the wrong problem. Consider the situation as it appeared at the end of 1992. A natural question to ask is: what is the probability distribution of the best performance that will be observed in 1993? This is a question about the *predictive* distribution of an as yet unobserved random variable. As a partial protection against the obvious selection bias associated with the choice of year, the paper proposed that the predictive probability be calculated conditionally on the event that a new world record be set.

There is no known frequentist solution to this problem that adequately takes account of the fact that the model parameters are unknown[5], but a Bayesian solution is straightforward. If the required conditional predictive distribution is denoted $G(y;\theta)$, this being the probability that the best per-

[5] A naïve solution is to substitute a point estimator of the unknown parameters, such as the maximum likelihood solution, into the predictive distribution: in the notation of (3.1), $\tilde{G}(y) = G(y;\hat{\theta})$ where $\hat{\theta}$ is the MLE. In the present example, the MLE $\hat{\beta}$ based on the data up to 1992 is greater than the actual time run by Wang, so such an approach would automatically lead to the value 0 for the predictive probability. However we can see this this approach is too simplistic, because as has already been pointed out, a 95% confidence interval for β includes Wang's record.

formance in 1993 will be smaller than y, given that it is better than the existing record, as a function of model parameters θ, then the Bayesian solution is based on the estimate

$$\tilde{G}(y) = \int G(y;\theta)\pi(\theta \mid \mathbf{X})d\theta, \tag{3.1}$$

where $\pi(\theta \mid \mathbf{X})$ denotes the posterior density of the parameters θ given past data $\mathbf{X}$. Writing $\theta = (\mu, \sigma, \xi)$ and x in place of $-y$, $G(y;\theta)$ is given by (2.2). As already noted, the transformation from x to $-y$ was made to convert minima into maxima.

Using a vague prior for θ and a Monte Carlo integration to evaluate (3.1), a predictive probability of 0.00047 (since slightly revised to 0.0006) was attached to the actual record run by Wang. The complete curve of $\tilde{G}(y)$ is shown in Fig. 6(b). This calculation seems definitively to establish that her performance was inconsistent with previous performances in the event. It does not, of course, provide any direct evidence of drug abuse.

The relevance of this example to risk assessment in finance and insurance is threefold:

(1) There is a clear distinction between *inference* about unknown *parameters* and *predictive distributions* about future *variables*. Many risk applications, including VaR itself, revolve around questions of the form "What is the probability that I will lose a certain amount of money over a certain period of time?" These are questions about prediction, not inference.

(2) In evaluating predictive distributions, account must be taken of the fact that model parameters are unknown.

(3) Bayesian methods provide an operational solution to the problem of calculating predictive distributions in the presence of unknown parameters. There are pure frequentist solutions based on asymptotic theory (for example, Barndorff-Nielsen and Cox (1996)), and it remains an open question just how well Bayesian solutions to these kinds of problems perform from a frequentist point of view, but the evidence currently available is encouraging, provided proper account is taken of the loss function in a decision-theoretic formulation of the problem (Smith 1997b, 1999).

As an example of the possible application of these ideas to risk assessment problems, suppose we want to calculate the predictive distribution of the largest loss over a future one-year time period, based on the data in Figure 3 and assuming a constant distribution. Fig. 7 shows a plot of the Bayes posterior median (solid curve) of the probability of exceeding a given level y, for each of a series of y values represented on the vertical axis. In this

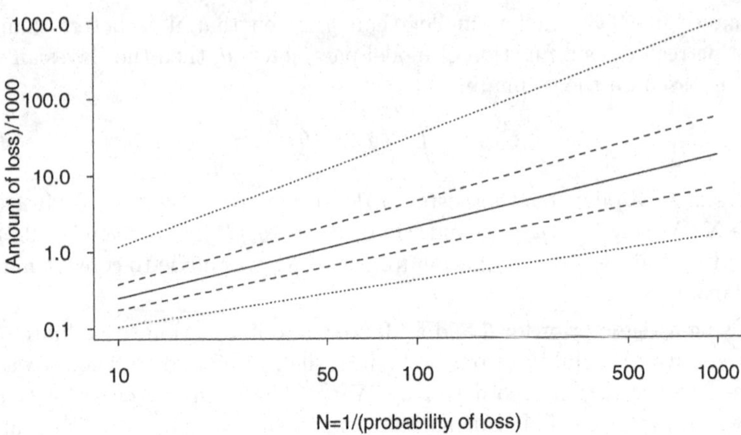

Figure 7: Median posterior loss curve with 50% and 95% probability bounds for insurance data, one-year losses, based on all exceedances over threshold 5.

plot we represent the probability of exceedance as $1/N$, and the value of N is represented on the horizontal axis. Also shown on the plot are 50% and 95% posterior probability intervals for the probability of exceedance, defined by the dashed lines and the dotted lines respectively. In the more detailed analysis of this data set, Smith and Goodman (1999) have provided a number of alternative analyses taking account of alternative features of the data. In particular, the data included a "type of claim" indicator, and when this is taken into account, the predictive distribution changes substantially, but that lies beyond the scope of the present discussion.

4 Multivariate Extremes

So far, our discussion has been entirely about extreme value theory for a single variable. However, it is more usual for VaR calculations to be made about a portfolio of assets rather than a single asset. In this context, a portfolio is simply a linear combination of individual asset prices. If the composition of the portfolio is held fixed, then it may be possible to assess the risk using univariate EVT, by simply treating the portfolio price as the variable of interest. However, the real rationale for doing this is often to help design the portfolio — for example, one may want to do this to maximise the expected return subject to some constraint on the VaR of the portfolio. To solve a problem of this nature, in which the weights on the different assets have to be determined, it is essential to consider the joint distribution of the asset prices. Conventional VaR theory is based on an assumption of multivariate normality for the joint distribution of log returns, but it is highly questionable whether such an assumption is appropriate for the calculation of extreme tail probabilities.

One approach to this problem is through multivariate EVT. Limiting relations such as (2.1) and (2.4) may be generalised to vector-valued processes, and for any $p \geq 1$, lead to a class of p-dimensional multivariate extreme value distributions (MVEVDs) and their threshold equivalents. There are numerous mathematically equivalent representations of MVEVDs, but one convenient form, due to Pickands (1981), is as follows. We may without loss of generality assume that all the marginal distributions have been transformed to the 'unit Fréchet' distribution function $e^{-1/x}$, $0 < x < \infty$; the joint distribution function of $x = (x_1, \ldots, x_p)$ is then of the form

$$G(x) = \exp\left\{-\int_{S_p} \max_{1 \leq j \leq p} \left(\frac{w_j}{x_j}\right) dH(w)\right\}, \tag{4.1}$$

where $S_p = \{(w_1, \ldots, w_p) : w_1 \geq 0, \ldots, w_p \geq 0, \sum w_j = 1\}$ is the unit simplex in p dimensions and H is some non-negative measure on S_p satisfying

$$\int_{S_p} w_j dH(w) = 1, \quad j = 1, \ldots, p. \tag{4.2}$$

Resnick (1987) is an excellent source of information about MVEVDs. The difficulty for statistical applications is that when $p > 1$, the class of MVEVDs does not reduce to a finite-dimensional parametric family, so there is potential explosion in the class of models to be considered. Most approaches to date have focussed either on simple parametric subfamilies, or on semiparametric approaches combining univariate EVT for the marginal distributions with nonparametric estimation of the measure H. Some example papers representing both approaches are Coles and Tawn (1991, 1994), Smith (1994) and de Haan and Ronde (1998). Recently, it has even been suggested that multivariate EVT may not be a rich enough theory to encompass all the kinds of behaviour one would like to be able to handle, and alternative measures of tail dependence have been developed. The main proponents of this approach so far have been Ledford and Tawn (1996, 1997, 1998); the last paper, in particular, contains an application to foreign exchange rates.

As I see it, the main difficulty with the application of this approach to VaR is in how to extend the methodology from the joint extremes of a small number of processes to the very large number of assets in a typical portfolio. Most of the papers just cited are for $p = 2$; some have considered extensions to $p = 3, 4, 5, \ldots$ but the model complexity increases greatly with p and there seems to be no hope of applying multivariate EVT directly to large portfolios in which p may be of the order of hundreds.

Recently, Smith and Weissman (1999) have proposed some alternative representations of extreme value processes aimed at characterising the joint distribution of extremes in multivariate time series of the form $\{X_{ij}, \ i =$

$1, 2, \ldots, 1 \le j \le p\}$. As in the preceding discussion, there is no loss of generality in assuming that unit Fréchet marginal distributions apply in the tail, because we may use univariate EVT to estimate the marginal tail distributions and then apply a probability integral transformation to each component. Smith and Weissman then defined a class of *multivariate maxima of moving maxima* (M_4 processes for short) by the equation

$$X_{ij} = \max_{\ell \ge 1} \max_{-\infty < k < \infty} a_{\ell k j} Z_{\ell, i-k}. \tag{4.3}$$

where $\{Z_{\ell,i}\}$ are a two-dimensional array of independent unit Fréchet random variables and the constants $\{a_{\ell k j}\}$ satisfy

$$a_{\ell k j} \ge 0, \quad \sum_{\ell} \sum_{k} a_{\ell k j} = 1 \quad \text{for all } j = 1, \ldots, p. \tag{4.4}$$

The main focus of the paper by Smith and Weissman is to argue that under fairly general conditions, extremal properties of a wide class of multivariate time series may be calculated by approximating the process by one of M_4 form. The fundamental ideas behind representations of this form are due to Deheuvels (1978, 1983), and they can be regarded as an alternative approach to those based on the representation (4.1).

In principle, (4.3) is simpler to handle than (4.1). Moreover it is a more general result, dealing directly with the case of multivariate time series and not just of independent multivariate observations. Another feature which makes (4.3) more directly interpretable for financial time series is that it represents the process in terms of an independent series of 'shocks' — in essence, large values among the $\{Z_{\ell,i}\}$ (the shocks) determine the pattern of extremes among the $\{X_{ij}\}$ and this has an obvious interpretation for the financial markets. On the other hand, estimating a three-dimensional array of unknown constants is a challenging problem in itself, and it is likely that some restrictions to specific classes will be necessary before this is feasible. Another difficulty with models of this form is that they suffer from degeneracies — the joint density of a set of random variables defined by (4.3) is typically singular with respect to Lebesgue measure and this causes problems for maximum likelihood techniques. However, this difficulty can be avoided by adding some additional noise to the observations and research is continuing into ways in which this might be done.

5 A Changepoint Model for Stochastic Volatility

We have seen that the standard extreme value methods do not appear to apply to the S&P 500 data. The explanation is non-constant volatility: it is apparent from simple inspection of the data in Fig. 2(a) that the variance of the series is much bigger in some years than in others, and consequently there is substantial variation in the rate in which any high threshold is exceeded. This problem is near-universal in financial time series: every other example which I have tried has exhibited problems similar to those with the Z-plot in Fig. 5.

There is by now a rich literature of models for financial time series taking into account changes in volatility. These divide broadly into two categories: models of the GARCH family, in which the variance of the process at time t, usually denoted σ_t, is expressed deterministically as a function of past values σ_s, $s < t$, and of the observations themselves; and models in which the volatility is treated as a stochastic process estimated by some form of state space model analysis. An excellent review of developments in both approaches is the paper by Shephard (1996).

It therefore seems worthwhile to develop extensions of the extreme value statistical methodology to take into account variable volatility. So far, very few attempts have been made to do this. McNeil and Frey (2000) have taken an approach built around the standard GARCH model, but in which the innovations, instead of being normally distributed as in the usual GARCH approach, are allowed to be long-tailed and estimated by methods similar to those presented earlier in this article, but taking account of the variation in σ_t estimated for the GARCH process. In another recent paper, Tsay (1999) has used methods similar to those of the present article, but allowing the extreme value parameters to depend on daily interest rates.

The aim of the present section is to suggest an alternative approach which is not tied to GARCH or to any particular model of volatility, but which simply assumes that the extreme value parameters change from one period to another according to a random changepoints process. Only an outline will be presented here; a fuller description is being prepared for publication elsewhere.

To describe the basic model, we first generalise (2.6) to

$$\Lambda(A) = \int_{t_1}^{t_2} \left(1 + \xi_t \frac{y - \mu_t}{\psi_t}\right)_+^{-1/\xi_t} dt \tag{5.1}$$

in which the notation explicitly reflects that the parameters μ, ψ and ξ are time-dependent.

The model is of hierarchical Bayesian structure, and is defined as follows. We assume that the process is observed over a time period $[0, T^*]$.

Level I At the top level of the hierarchy, we define hyperparameters m_μ, s_μ^2, m_ψ, s_ψ^2, m_ξ, s_ξ^2 with a prior distribution (to be specified later).

Level II Conditionally on the parameters of level I, let the number of changepoints K have a Poisson distribution with mean νT^*. Conditionally on K, let the individual changepoints $C_1, \ldots, C_K$ be independent uniform on $[0, T^*]$, and then ordered so that $0 < C_1 < \cdots < C_K < T^*$. (An equivalent description is that the random changepoints form a realisation of a homogeneous Poisson process with intensity ν.) For convenience we also write $C_0 = 0$, $C_{K+1} = T^*$. Also, let $\mu_1, \ldots, \mu_{K+1}$ be independently drawn from the $N(m_\mu, s_\mu^2)$ distribution, $\log \psi_1, \ldots, \log \psi_{K+1}$ independently drawn from $N(m_\psi, s_\psi^2)$ and $\xi_1, \ldots, \xi_{K+1}$ independently drawn from $N(m_\xi, s_\xi^2)$.

Level III Conditionally on the parameters in Level II, suppose that for each k between 1 and $K+1$, the exceedance times and values over a threshold u on the time interval $C_{k-1} < t \le C_k$ are defined by the Poisson process with cumulative intensity given by (5.1), in which $\mu^{(t)} = \mu_k$, $\psi^{(t)} = \psi_k$, $\xi^{(t)} = \xi_k$.

For the prior distributions at level I, we assume that $(m_\mu,\ s_\mu^2)$ are of 'gamma-normal' type: let τ_μ be drawn from the gamma distribution with density proportional to $\tau_\mu^{\alpha-1} \exp(-\beta\tau_\mu)$, $0 < \tau_\mu < \infty$, and then define $s_\mu^2 = 1/\tau_\mu$, $m_\mu \sim N(\eta, \frac{1}{\kappa\tau_\mu})$. This model may be summarised by the notation $(m_\mu, s_\mu^2) \sim GN(\alpha, \beta, \eta, \kappa)$. Similarly, we assume the pairs (m_ψ, s_ψ^2), (m_ξ, s_ξ^2) are independently drawn from the same distribution. We fix $\alpha = \beta = \kappa = 0.001$ and $\eta = 0$ to represent a proper but very diffuse prior distribution.

The treatment of the prior parameter ν is somewhat problematic in this set-up. It might be thought desirable to put a vague hyperprior on ν, but this is not possible because an improper prior leads to an improper posterior (and, in the practical implementation of the algorithm, the number of changepoints grows to ∞). Instead, therefore, I have specified a value for ν. In different runs, the values $\nu = 20$, 25 and 30 have all been tried, with some differences in the posterior distribution of the number of changepoints (see Fig. 8(a)) but fortunately these were not too great.

The actual algorithm uses the reversible jump Markov chain Monte Carlo sampler in manner very similar to that of Green (1995): indeed, my whole approach was very much motivated by Green's treatment of a famous data set concerned with coal mining disasters. However, in the present article I omit all details of the algorithm, which essentially consists of iteratively updating all the parameters of the models using a reversible jump sampler to take care of the fact that the number of changepoints, and hence the dimension of the model to be estimated, are *a priori* unknown.

Fig. 8 shows the outcome of one run of this analysis, based on a total of 100,000 iterations of the reversible jump sampler with every 100th iteration recorded and used to construct the plots. Thus, for example, the histogram in Fig. 8(a) is based on 1,000 sampled values of the number of changepoints. The posterior distribution of the number of changepoints has (in this run) a mean of 23.9 and a standard deviation of 2.5. Figs. 8(b) and 8(c) show the Z and W plots computed from the posterior means in the changepoint model;

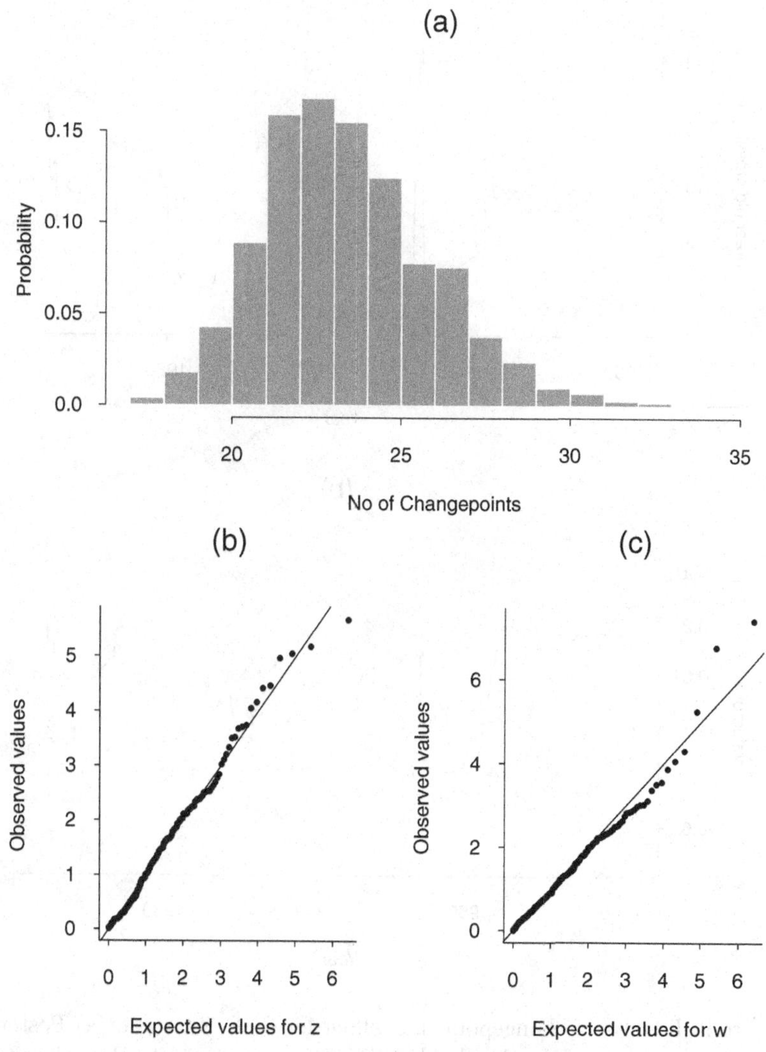

Figure 8: Results of changepoint modelling for S&P 500 data. (a) Posterior distribution for number of changepoints. (b) Z plot. (c) W plot. Based on threshold 2, prior mean number of changepoints $\nu = 25$.

in this case, $R^2 = .992$ for the Z plot and .981 for the W plot. There is still some concern about the very largest values in the W plot but otherwise the fit in the model seems much better than in the earlier discussion of Fig. 5.

Fig. 9(a) shows the posterior mean estimate of the crossing rate of the threshold 2, as it varies across time. This shows very clearly the effects of

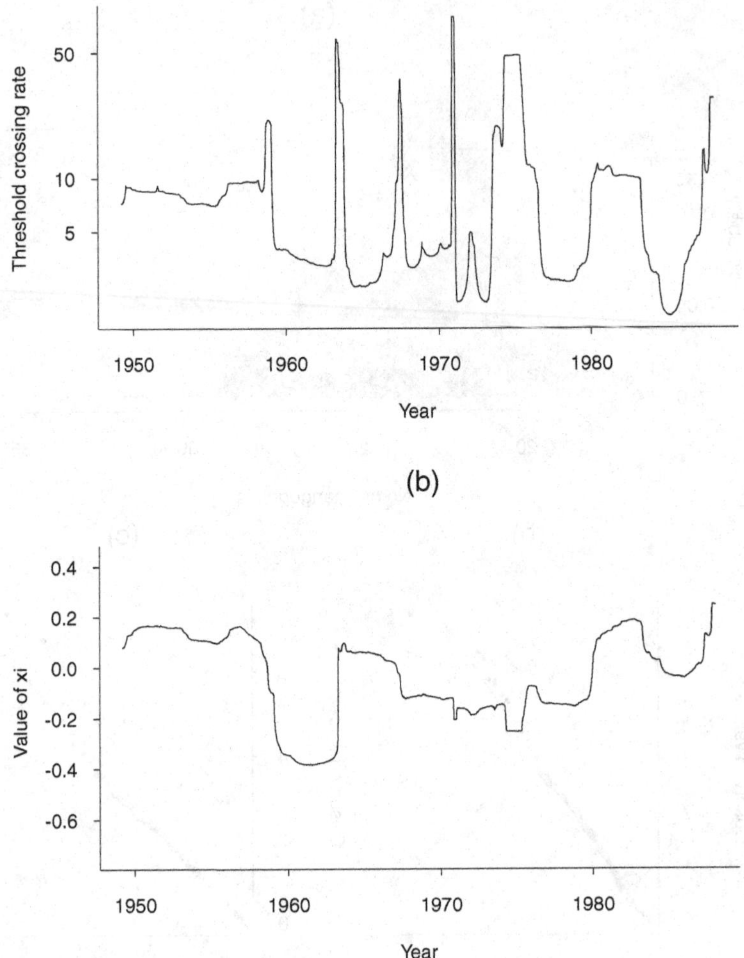

Figure 9: Results of changepoint modelling for S&P 500 data. (a) Posterior mean crossing rate of the threshold as a function of time t. (b) Posterior mean of ξ_t.

stochastic volatility, with periods when there is a high probability of crossing the threshold such as around 1971, during 1973–1974 or the early 1980s, interspersed with periods when the probability of crossing the threshold is much lower, such as the late 1970s or mid 1980s. Fig. 9(b) shows a similar plot for the posterior mean of the ξ_t parameter as t ranges over the data set. Many, though not all, of the rises and falls in this plot match the rises and falls in the threshold crossing rate.

Finally, we consider the consequences of the changepoint model for the estimated extreme value parameters. Table 2 shows the estimated posterior parameters and standard errors for (a) December 1987, (b) January 1978 (chosen to represent a quiet period), (c) an overall average over all days of the series (this was calculated by sampling from the total Monte Carlo output) and (d) based on the maximum likelihood estimates for a single homogeneous model fitted to the whole series, as in Section 2. Perhaps the most interesting parameter here is ξ, which represents the overall shape of the tail. For December 1987, the posterior mean is $\hat{\xi} = .21$, representing a fairly long-tailed case (but not excessively so — values of ξ in the range 0.5 to 1 often occur in insurance applications, including the one mentioned earlier in this article). For January 1978, the posterior mean is –.02, insignificantly different from 0, which is an exponential tail (in other words, short-tailed). The *overall* average over the whole series is –.05, which seems to reflect that the typical behaviour is short-tailed with mainly the high-volatility periods being long-tailed. However, the maximum likelihood estimates based on a homogeneous model imply $\hat{\xi} = .22$ with a standard error of .06. This seems completely misleading, implying that long-tailed behaviour is a feature of the whole series rather than just of short high-volatility periods of it. The interpretation, I believe, is that the effect of mixing over inhomogenous periods has inflated the apparent value of ξ and has made the distribution seem more long-tailed that it really is most of the time. A similar phenomenon has also been observed for the insurance data of Section 2 (Smith and Goodman 1999), though in that case the mixing was over different types of insurance claim rather than inhomogeneous periods of time.

6 Conclusions

The interaction between extreme value theory and the assessment of financial risk poses many exciting possibilities. Many of these seem to require new techniques. In this article I have presented three areas in which new methodology seems to be required. Bayesian statistics is a valuable tool for the assessment of predictive distributions which is very often the real question of interest, rather than inference for unknown parameters. The possibility of applying VaR analysis to large portfolios implies the need for multivariate extreme value techniques in high dimensions, in contrast with most of the multivariate extreme value theory developed to date which has concentrated on low-dimensional problems. Finally, the last section proposed one way of dealing with the stochastic volatility problem, via a changepoint model for the extreme value parameters. However this in itself is a tentative approach; there is ample scope for exploration of alternative approaches for combining extreme value theory and stochastic volatility.

Table 2: Bayes posterior means of model parameters (posterior standard deviations in parentheses) for specific time periods. Row 1: December 1987. Row 2: January 1978. Row 3: Averaged over all time periods in the data. Row 4: Maximum likelihood estimates and standard errors based on a single homogeneous model fitted to the whole series.

Time	μ	$\log\psi$	ξ
December 1987	5.12 (.97)	0.24 (.44)	0.21 (.18)
January 1978	3.03 (1.04)	–0.61 (.43)	–0.02 (.27)
Averaged over time	3.27 (1.16)	–0.51 (.56)	–0.05 (.32)
Homogeneous model	3.56 (.10)	–0.09 (.09)	0.22 (.06)

Acknowledgments

The work reported here was carried out primarily during a visit to the Isaac Newton Institute of Cambridge University, July–December 1998, supported in part by a Visiting Fellowship of the EPSRC, grant number GR K99015, by a grant from the Tsunami Initiative, and by NSF grant DMS-9705166.

References

Barndorff-Nielsen, O.E. and Cox, D.R. (1996), 'Prediction and asymptotics'. *Bernoulli* **2**, 319–340.

Coles, S.G. and Tawn, J.A. (1991), 'Modelling extreme multivariate events'. *J. R. Statist. Soc. B* **53**, 377–392.

Coles, S.G. and Tawn, J.A. (1994), 'Statistical methods for multivariate extremes: an application to structural design (with discussion)'. *Applied Statistics* **43**, 1–48.

Davison, A.C. and Smith, R.L. (1990), 'Models for exceedances over high thresholds (with discussion)'. *J.R. Statist. Soc. B* **52**, 393–442.

Deheuvels, P. (1978). 'Caractérisation complète des lois extrèmes multivariées et de la convergence des types extrèmes'. *Publ. Inst. Statist. Univ. Paris* **23**, 1–36.

Deheuvels, P. (1983). 'Point processes and multivariate extreme values'. *J. Multivar. Anal.* **13**, 257–272.

Dowd, K. (1998), *Beyond Value at Risk: the New Science of Risk Management.* Wiley.

Embrechts, P., Klüppelberg, C. and Mikosch, T. (1997), *Modelling Extremal Events for Insurance and Finance.* Springer.

Fisher, R.A. and Tippett, L.H.C. (1928), 'Limiting forms of the frequency distributions of the largest or smallest member of a sample'. *Proc. Camb. Phil. Soc.* **24**, 180–190.

Gnedenko, B.V. (1943), 'Sur la distribution limite du terme maximum d'une série aléatoire'. *Ann. Math.* **44**, 423–453.

Green, P.J. (1995), 'Reversible jump Markov chain Monte Carlo computation and Bayesian model determination'. *Biometrika* **82**, 711–732.

Gumbel, E.J. (1958), *Statistics of Extremes.* Columbia University Press.

Haan, L. de and Ronde, J de (1998), 'Sea and wind: Multivariate extremes at work'. *Extremes* **1**, 7–45.

Jorion, P. (1996), *Value at Risk: the New Benchmark for Controlling Market Risk.* Irwin.

Leadbetter, M.R. (1991), 'On a basis for 'peaks over thresholds' modeling'. *Statistics and Probability Letters* **12**, 357–362.

Leadbetter, M.R., Lindgren, G. and Rootzén, H. (1983), *Extremes and Related Properties of Random Sequences and Series.* Springer Verlag.

Ledford, A.W. and Tawn, J.A. (1996), 'Statistics for near independence in multivariate extreme values'. *Biometrika* **83**, 169–187.

Ledford, A.W. and Tawn, J.A. (1997), 'Modelling dependence within joint tail regions'. *J.R. Statist. Soc. B* **59**, 475–499.

Ledford, A.W. and Tawn, J.A. (1998), 'Diagnostics for dependence within time-series extremes'. Preprint, Department of Mathematics and Statistics, University of Surrey.

McNeil, A.J. and Frey, R. (2000), 'Estimation of tail-related risk measures from heteroscedastic financial time series: an extreme value approach'. *J. Empirical Finance* **7**, 271–300.

Mises, R. von (1936), 'La distribution de la plus grande de *n* valeurs'. Reprinted in *Selected Papers II*, Amer. Math. Soc., Providence, RI (1954), 271–294.

Pickands, J. (1975), 'Statistical inference using extreme order statistics'. *Ann. Statist.* **3**, 119–131.

Pickands, J. (1981), 'Multivariate extreme value distributions'. *Bull. I.S.I.* **XLIX** (Book 2), 859–878.

Resnick, S. (1987), *Extreme Values, Point Processes and Regular Variation.* Springer Verlag.

Robinson, M.E. and Tawn, J.A. (1995), 'Statistics for exceptional athletics records'. *Applied Statistics* **44**, 499–511.

Shephard, N. (1996), 'Statistical aspects of ARCH and stochastic volatility'. In *Time Series Models: in Econometrics, Finance and other Fields*, edited by D.R. Cox, D.V. Hinkley and O.E. Barndorff-Nielsen. Chapman and Hall, pp. 1–67.

Smith, R.L. (1986), 'Extreme value theory based on the r largest annual events'. *J. Hydrology* **86**, 27–43.

Smith, R.L. (1989), 'Extreme value analysis of environmental time series: an application to trend detection in ground-level ozone (with discussion)'. *Statistical Science* **4**, 367–393.

Smith, R.L. (1994), 'Multivariate threshold methods'. In *Extreme Value Theory and Applications*, edited by J. Galambos, J. Lechner and E. Simiu. Kluwer Academic Publishers, pp. 225–248.

Smith, R.L. (1997a), 'Statistics for exceptional athletics records'. Letter to the editor, *Applied Statistics* **46**, 123–127.

Smith, R.L. (1997b), 'Predictive inference, rare events and hierarchical models'. Preprint, University of North Carolina.

Smith, R.L. (1999), 'Bayesian and frequentist approaches to parametric predictive inference (with discussion)'. In *Bayesian Statistics 6*, edited by J.M. Bernardo, J.O. Berger, A.P. Dawid and A.F.M. Smith. Oxford University Press, 589–612.

Smith, R.L. and Goodman, D.J. (2000), 'Bayesian risk assessment'. In *Extremes and Integrated Risk Management*, edited by P. Embrechts. Risk Books, pp. 235–251.

Smith, R.L. and Shively, T.S. (1995), 'A point process approach to modeling trends in tropospheric ozone'. *Atmospheric Environment* **29**, 3489–3499.

Smith, R.L. and Weissman, I. (1999), 'Characterization and estimation of the multivariate extremal index'. Tentatively accepted for *Extremes*.

Tawn, J.A. (1988), 'An extreme value theory model for dependent observations'. *J. Hydrology* **101**, 227–250.

Tsay, R.S. (1999), 'Extreme value analysis of financial data'. Preprint, Graduate School of Business, University of Chicago.

Extremes in Operational Risk Management

E.A. Medova and M.N. Kyriacou

Abstract

Operational risk is defined as a consequence of critical contingencies, most of which are quantitative in nature, and many questions regarding economic capital allocation for operational risk continue to be open. Existing quantitative models that compute the value at risk for market and credit risk do not take into account operational risk. They also make various assumptions about 'normality' and so exclude extreme and rare events. In this paper we formalize the definition of operational risk and apply extreme value theory for the purpose of calculating the economic capital requirement against unexpected operational losses.

1 Introduction

Highly publicized events such as those at LTCM, Barings and Sumitomo have all involved mismanagement leading to extraordinary losses and raising concerns about financial instability at international levels. As a result, along with the established capital charges for market and credit risks, the Basle Committee on Banking Supervision is proposing an explicit capital charge to guard the banks against operational risks. The response from the banks has been an increasing number of operational risk management initiatives with corresponding efforts to formulate a framework for capital allocation for operational risk. This paper contains a model for calculating the economic capital against extreme risks which is our contribution to the quantification of operational risk.

One of the first definitions of operational risk (British Bankers' Association, 1997) was specified by a list of possible causes [4]:

> *The risks associated with human error, inadequate procedures and control, fraudulent and criminal activities;*
>
> *the risks caused by technological shortcomings, system breakdowns;*
>
> *all risks which are not 'banking' and arising from business decisions as competitive action, pricing, etc;*

legal risk and risk to business relationships, failure to meet regulatory requirements or an adverse impact on the bank's reputation;

'external factors' include: natural disasters, terrorist attacks and fraudulent activity, etc.

After four years of intensive debate on what constitutes an operational risk the current Basle proposal defines operational risk as [2]:

Operational risk is the risk of direct or indirect loss resulting from inadequate or failed internal processes, people and systems or from external events.

Strategic and reputational risks are not included in this new definition, but as before it focuses on the causes of operational risk and is open to endless discussion about the detailed definition of each loss category. The 'semantic Wild West' of operational risk [15] is still with us and the view of operational risk as 'everything not covered by exposure to credit and market risk' remains the one most often used by practitioners.

Our own operational risk study started with a search for a definition suitable for quantitative modelling. The resulting modelling approach [19] is presented in Section 2. According to Basle: 'A capital charge for operational risk should cover unexpected losses. Provisions should cover expected losses'. The Committee clarifies the complex issues of risk management by adopting a 'three-pillared' approach. The first pillar is concerned with capital allocation, the second pillar with supervision and controls and the third with transparency and consistency of risk management procedures. With the view that statistical analysis of loss data and consistency of modelling techniques may be considered respectively as parts of Pillars 2 and 3, we adopt the 'practitioners' definition of operational risk and propose a model for the capital allocation of Pillar 1. We also assume that provisions and improvements in management control (Pillars 2 and 3) will cover low value frequently occurring losses and we concentrate here on extreme and rare operational risks. A definition of operational risk suitable for quantitative modelling and our framework for economic capital allocation are presented in Section 2. This stochastic model is based on results from extreme value theory and in Section 3 we review key results on stable distributions and the classical theory of extremes. In Section 4 we detail our model [19] and discuss related implementation issues. A Bayesian hierarchical simulation method is applied to the parameter estimation of extreme distributions from small-sized samples. The method also provides a more transparent assessment of risk by taking into account data on losses due to different risk factors or business units. We illustrate our operational risk framework on an example of an anonymous European bank during the period of the Russian Crisis in Section 5 and draw conclusions and sketch future directions in Section 6.

2 Firm-wide operational risk management

Market or credit risk definitions came naturally from specific businesses, effectively market trading, lending or investment, with the corresponding consistent probabilistic definition of the *value at risk* (VaR). Operational risk definitions on the other hand are based on an identification of causes whose consequences are often not measurable. Such differences in defining types of risk result in segregated capital allocation rules for operational risk. Yet the importance of integration of all forms of risk is obvious.

Recall that VaR provides a measure of the market risk of a portfolio due to adverse market movements under *normal* market conditions and is expressed here in return terms as

$$\mathrm{P}(\mathbf{R} < \mathrm{VaR}) = \int_{-\infty}^{\mathrm{VaR}} N(R)\, dR = \pi, \tag{1}$$

where the return $\mathbf{R}$ is the normalised portfolio value change over a specified time horizon, N denotes a suitable normal density and π is a probability corresponding to a one-sided confidence level (typically 5% or 1%). More generally, N is replaced by an appropriate return density $f_{\mathbf{R}}$, for example one which is obtained by simulation.

Similarly, credit ratings correspond to normal credit conditions, for example with default corresponding to a rating below CCC. In credit modelling the *default point* threshold is difficult to formalize as it depends on the evolution of the institution's assets (for a discussion, see M. Ong [21]). The 'value of the firm' framework as implemented by CreditMetrics defines a series of levels of the firm's assets which determine the credit rating of the firm. In Ong's interpretation: 'Assuming that asset returns denoted by the symbol $\mathbf{R}$ are normally distributed with mean μ and standard deviation σ, the generalization concerning the firm's credit quality immediately translates to the slicing of the asset returns distribution into distinct bands. Each band, representing the different threshold levels of asset returns, can be mapped one-to-one to the credit migration frequencies in the transition matrix'. Thus the firm's default probability expressed in terms of its asset return distribution is given by

$$\mathrm{P}(\mathbf{R} < \mathrm{CVaR}) = \int_{-\infty}^{\mathrm{CVaR}} N(R)\, dR = \rho < \pi. \tag{2}$$

Again, more general or empirical (historical) distributions might be substituted for the Gaussian in (2) to represent 'normal' market conditions more accurately.

One might thus naturally ask how the definition of 'normality' relates to operational risk and to the problem of internal bank controls and external supervision. These questions are critical when a specific loss event happens, particularly when it is related to extreme losses. As market, credit and operational risks become entangled at the time of occurrence of large losses, it is important that an operational risk analyst deals with both market and credit risk management without double-counting. While risk capital is generally understood as a way of protecting a bank against 'unexpected' losses – expected losses are covered by business-level reserves – it is not clear as to what degree risk capital should be used to cover the most extreme risks. In an attempt to answer these questions we construct a framework that allows the allocation of capital against extreme operational losses while identifying the roles of credit and market risks in their occurrence.

Let us assume that a bank's market and credit risk management is informed by quantitative models that compute the value at risk for market risk and credit risk and that allocate economic capital to these risks. It is clear that such capital allocation is not sufficient to cover unexpected losses due to natural disasters, fraudulent activities and human errors. Currently used models do not take into account operational risks. For example, VaR models allocate capital 'under normal market conditions' and so exclude extreme or rare events such as natural disasters and major social or political events. As a consequence, inadequate models contribute to operational losses as a part of an 'inadequate internal process'.

The first step in operational risk management should be a careful analysis of all available data to identify the statistical patterns of losses related to identifiable risk factors. Ideally, this analysis would form part of the financial surveillance system of the bank. In the future perhaps such an analysis might also form part of the duties of bank supervisors. In other words, at a conceptual level such an analysis relates to the third of the Basle Committee's three pillars. The important point is that this surveillance is concerned with the identification of the 'normality' of business processes. In statistical terms it means a fundamental justification of the Gaussian or *normal* model to describe the central part of the distribution which does not allow for large fluctuations in data. The identification of market and credit risk models suitable for the *tail* events forms a natural part of an operational risk assessment. It allows an analyst to classify a bank's losses into two categories:

(1) *significant* in value but *rare*, corresponding to *extreme* loss event distributions;

(2) *low* value but *frequently* occurring, corresponding to '*normal*' loss event distributions.

Thus an analysis of profit and loss data and the verification or rejection of the assumption of normality may both be considered as the part of the (usually internal) risk supervisory process. We take the view that over time control procedures will be developed by a financial institution for the reduction of the low value/frequent losses and for their illumination and disclosure – the second pillar of the Basle approach. These control procedures, and any continuing expected level of losses, should be accounted for in the operational budget.

Any deviation from the normality assumed, or increased volatility in the markets, will tend to underestimate market value at risk. Similarly, under normal conditions for credit risk, which corresponds to credit ratings higher than BBB, credit models provide measures for credit risk. This allows us to assume that only losses of large magnitude need be considered for operational risks. With the view that control procedures verify the assumptions of internal market and credit models, and that losses within the limits of market and credit value at risk can be accommodated, we assume that only losses of larger magnitude need be considered for operational risk capital provision. Hence we adopt the accepted practice of defining operational risk as 'everything which is not market or credit risk' and *assume operational losses to be in the category of losses which are larger than those due to market or credit risks under normal market conditions.*

As all forms of risk are driven by the same fundamental market conditions, capital allocation for market, credit risks and operational risk must be derived from the same profit and loss distribution simultaneously[1]. Therefore for integrated profit and loss data at the firm- or business unit-level the following thresholds for losses are obtained from market and credit risk models as:

- the unexpected loss level due to market risk, denoted by $u_{\mathrm{VaR}\pi}$, which is exceeded with probability π
- the level of loss due to both credit and market risks, denoted by $u_{\mathrm{CVaR}\rho}$, which is exceeded with probability $\rho \leq \pi$, so that $u_{\mathrm{CVaR}\rho} \leq u_{\mathrm{VaR}\pi}$.

Losses beyond the $u_{\mathrm{CVaR}\rho}$ level, or so called *unexpected losses*, are assumed to belong to the operational risk category. Therefore extreme operational losses are modelled as excesses over both market and credit losses on the P&L distribution as shown in Figure 1, with the risk measures corresponding to the appropriate approximating distribution. The required capital allocation for operational risk will be derived from the parameters of the *asymptotic* distribution of *extremes* of profit and loss.

For the purpose of operational risk management we obtain an *unexpected loss threshold* u obtained from the operational risk model to be developed (see Section 4). We shall suppose that the $u_{\mathrm{CVaR}\rho}$ level approximately equals this threshold u. Relations between the thresholds for market and credit risk may be obtained by variety of methods as implemented by internal models. These levels should be re-examined in this context with respect to the overall implementation of risk management procedures according to the definitions of 'expected' and 'unexpected' losses.

[1]This conceptual view of total risk modelling does not necessarily mean simultaneous implementation of market, credit and operational risk model components.

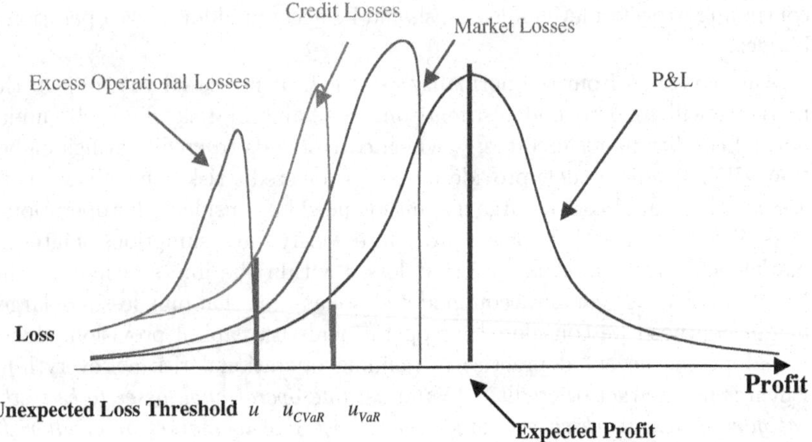

Figure 1: Decomposition of the loss-tail of a profit & loss distribution into its three loss-types (market, credit and operational losses) and definition of the threshold for extreme operational losses.

3 Stable random variables and extreme value theory

Our formalism in defining operational risk focuses on tail events. But consistency in estimation of profit and loss distributions at different levels of a financial institution and at different time scales is difficult to achieve and any successful implementation would rely on approximation and heuristics. The asymptotic theories of sums and maxima of random variables are thus of crucial importance for risk management. Here we recall some definitions and principal results used in our proposed procedure for operational risk capital allocation.

The summary effects of daily fluctuations in price return or of a portfolio is well captured by a limiting *normal* distribution for data whose underlying distribution has finite variance, but this normal limit is often inadequate for highly variable data. *Stable* distributions approximate the distribution of sums of *independent identically distributed* (i.i.d.) random variables with infinite variance and include the Gaussian as a special case. There are many famous monographs on asymptotic theory for sums dating from the 1950s: Gnedenko and Kolmogorov (1954) [11], Feller(1966) [9], Mandelbrot (1982) [18], Samorodnitsky and Taqqu (1990) [23].

Many results of the *asymptotic theory for sums* (or *central limit theory*) have their complements in the *asymptotic theory of extreme order statistics* known as *extreme value theory* (EVT). EVT has been applied in engineering, hydrology, insurance and currently applies to financial risk management. Some of most useful references are: Galambos (1978) [10], Leadbetter *et al.* (1983) [16], Du Mouchel (1983) [7], Castillo (1988) [5], Embrechts, Kluppelberg & Mikosch (1997) [8], R. Smith (1985, 1990, 1996) [25–29], Danielson and de Vries (1997) [6] and McNeil and Saladin (1997) [20].

One of the fundamental problems of risk management is identification of the functional form of a profit and loss distribution. Simulation methods will 'construct' such a distribution without requiring an analytic form, but this usually involves a complex implementation and considerable computing time.

Every random profit/loss $\mathbf{X}$ has associated with it a distribution function with four basic parameters that have physical or geometric meaning. These are the *location* μ, the *scale* σ, the *tail index* α, or equivalently the *shape* $\xi := 1/\alpha$, and the *skewness* β.

Stable distributions have a number of equivalent definitions in terms of the 'stability' property, the domain of attraction, or as a special subclass of the infinitely divisible distributions. Most important for applications is the fact that any α-stable random variable can be expressed as a convergent sum of random variables indexed by the arrival times of a Poisson process (for definitions, see [23]).

A random variable $\mathbf{X}$ is said to have an *α-stable* distribution if for any $n \geq 2$ there is a positive number c_n and a real number d_n such that

$$\mathbf{X}_1 + \mathbf{X}_2 + \cdots + \mathbf{X}_n \stackrel{\mathrm{d}}{=} c_n \mathbf{X} + d_n, \tag{3}$$

where $\mathbf{X}_1$, $\mathbf{X}_2$, $\ldots$, $\mathbf{X}_n$ are independent copies of $\mathbf{X}$ and $c_n = n^{1/\alpha}$ for some number α, $0 \leq \alpha \leq 2$, called the *index of stability.*

Stable distributions are suitable for modelling a wide class of empirical distributions. In fitting such distributions to heavy-tailed samples, the parameter α measures the thickness of tails and the finiteness of moments of the distribution of $\mathbf{X}$. The distribution functions of stable random variables are often not available in a closed form with the exception of a few special cases. Feller [9] describes stable distributions analytically by specifying their characteristic function given by

$$\begin{aligned}\phi_x(t; \alpha, \beta, \mu, \sigma) &:= \mathrm{E}[\exp(it\mathbf{X}) \mid \alpha, \beta, \mu, \sigma] \\ &= \begin{cases} \exp\left(i\mu t - \sigma^\alpha |t|^\alpha \left(1 - i\beta \mathrm{sgn}(t) \tan\left(\frac{\pi\alpha}{2}\right)\right)\right) & \text{if } \alpha \neq 1 \\ \exp\left(i\mu t - \sigma|t| \left(1 + i\beta \mathrm{sgn}(t) \frac{2}{\pi} \tan \log|t|\right)\right) & \text{if } \alpha = 1 \end{cases}\end{aligned} \tag{4}$$

for $-\infty < t < \infty$, $0 \leq \alpha \leq 2$, $\sigma \geq 0$, $-1 \leq \beta \leq 1$ and μ real, where $\mathrm{E}[\,\cdot\,]$ denotes expectation.

A r.v. $\mathbf{X}$ is has a stable distribution if, and only if, it has a *domain of attraction,* i.e. if there is a sequence of independent identically distributed (i.i.d.) random variables $\mathbf{Y}_1, \mathbf{Y}_2, \ldots$ and sequences of positive numbers $\{d_n\}$ and real numbers $\{a_n\}$ such that

$$\frac{\mathbf{Y}_1 + \mathbf{Y}_2 + \cdots + \mathbf{Y}_n}{d_n} + a_n \stackrel{\mathrm{d}}{\Longrightarrow} \mathbf{X}, \tag{5}$$

where $\stackrel{\mathrm{d}}{\Longrightarrow}$ denotes convergence in distribution as $n \to \infty$. In general $d_n := n^{1/\alpha} h(n)$, where $h(x)$, $x > 0$, is a *slowly (or regularly) varying function at infinity*, i.e. for sufficiently large $u > 0$

$$\lim_{x \to \infty} \frac{h(ux)}{h(x)} = 1. \tag{6}$$

When $\mathbf{X}$ is Gaussian, i.e. $\alpha = 2$, and $\mathbf{Y}_1, \mathbf{Y}_2, \ldots$ are i.i.d. with finite variance, then (5) is the statement of the *Central Limit Theorem* (CLT). Generalizations of the CLT involve *infinitely divisible* random variables [11]. The family of *infinitely divisible distributions* includes the stable distributions. A random variable is *infinitely divisible* if, and only if, for every natural number n it can be represented as the sum

$$\mathbf{X} = \mathbf{X}_{n_1} + \mathbf{X}_{n_2} + \cdots + \mathbf{X}_{n_n} \tag{7}$$

of n i.i.d. random variables.

Equivalently, for every natural number n there exists a characteristic function given by $\phi_X(t)$ whose nth power is equal to the characteristic function ϕ_{X_n} of $\mathbf{X}$, i.e.

$$\phi_X - (\phi_{X_n})^n. \tag{8}$$

In terms of distribution functions, the distribution function F of $\mathbf{X}$ is given by a convolution of corresponding F_n's as

$$F = F_n^{n^*} := F_n * F_n * \cdots * F_n. \tag{9}$$

Let $\mathbf{X}_1, \ldots, \mathbf{X}_n$ represent i.i.d. random variables with distribution function F and define their *partial sum* by $\mathbf{S}_n = \mathbf{X}_1 + \mathbf{X}_2 + \cdots + \mathbf{X}_n$ and their *maximum* by $\mathbf{M}_n = \max(\mathbf{X}_1, \mathbf{X}_2, \ldots, \mathbf{X}_n)$.

It can be shown [8, 9,12] that regular variation in the tails (6) and infinite divisibility (7) together imply *subexponentiality* of a distribution, i.e. for $n \geq 2$

$$\lim_{x \to \infty} \frac{\overline{F}^{n^*}(x)}{\overline{F}(x)} = n, \tag{10}$$

where, for example, $\overline{F} := 1 - F$ denotes the *survivor function* corresponding to F. It follows that

$$\mathrm{P}(\mathbf{S}_n > x) \approx \mathrm{P}(\mathbf{M}_n > x) \quad \text{as} \quad x \to \infty. \tag{11}$$

Thus behaviour of the distribution for a sum in its tail may be explained by that of its maximum term, leading to many complementary results to those of central limit theory for the 'max-stable' distributions studied in extreme value theory.

The possible limiting distributions for the maximum $\mathbf{M}_n$ of n i.i.d. random variables are identified as the class of *max-stable* distributions, the maximum domain of attraction is analogous to the domain of attraction and the Poisson representation mentioned above is the main theoretical tool for studying the process of exceedances of a specified level.

The current theoretical foundations of EVT are given in Embrecht, Kluppelberg and Mikosch's book [8]. Since [8] and R. Smith's papers [25–29] focus on applications to insurance and risk management, we will only state here results required for modelling operational risk.

The Fisher–Tippett theorem proves the convergence of the sample maxima to the non-degenerate limit distribution $H_{\xi;\mu,\sigma}$ under some linear rescaling such that for $c_n > 0$ and d_n real, $c_n^{-1}(\mathbf{M}_n - d_n) \overset{\mathrm{d}}{\Longrightarrow} H_{\xi;\mu,\sigma}$, as the *sample size* n increases, i.e. for $-\infty < x < \infty$

$$P\left[\left(\frac{\mathbf{M}_n - d_n}{c_n}\right) \leq x\right] \to H_{\xi;\mu,\sigma} \quad \text{as} \quad n \to \infty. \tag{12}$$

Three classical extreme value distributions of normalised sample maxima which are included in this representation are the *Gumbel*, *Frechet* and *Weibull* distributions. The *generalised extreme value* (GEV) *distribution* $H_{\xi;\mu,\sigma}$ provides a representation for the non-degenerate limit distribution of normalised maxima with *shape parameter* ξ

$$H_{\xi;\mu,\sigma}(x) = \begin{cases} \exp\left[-\left(1+\xi\dfrac{x-\mu}{\sigma}\right)^{-1/\xi}\right] & \text{if } \xi \neq 0,\ 1+\xi\dfrac{x-\mu}{\sigma} > 0 \\ \exp\left[-\exp\left(-\dfrac{x-\mu}{\sigma}\right)\right] & \text{if } \xi = 0. \end{cases} \tag{13}$$

For the case of α-max-stable distributions, the shape parameter ξ satisfies $1/2 \leq \xi = 1/\alpha < \infty$ and determines the existence of moments. For the Gaussian case $\alpha = 1/\xi = 2$, while for $\xi > 1$ the distribution has no moments finite.

Modelling worst case losses will involve fitting an extreme value distribution. This can be done by grouping the data into epochs (month, years, etc) and using its maximum (minimum) over an epoch as one representative of a GEV. However the longer the epoch the larger the loss of data with this approach. The central idea of a method based on *exceedances* is to avoid such a loss of information and to consider all data which lie above a given *threshold* value [16, 17, 23, 25].

Given an i.i.d. sequence of random variables $\mathbf{X}_1, \ldots, \mathbf{X}_\mathbf{n}$ drawn from an underlying distribution F, we are interested in the distribution of *excesses*

$\mathbf{Y} := \mathbf{X} - u$ over a high *threshold* u. We define an *exceedance* of the level u if in the event $\mathbf{X} = x$ we have $x > u$. The distribution of *excesses* is given by the conditional distribution function in terms of the tail of the underlying distribution F as

$$F_u(y) := \mathrm{P}(\mathbf{X} - u \leq y \mid \mathbf{X} > u) = \frac{F(u+y) - F(u)}{1 - F(u)} \text{ for } 0 \leq y \leq \infty. \quad (14)$$

The limiting distribution $G_{\xi,\beta}(y)$ of excesses as $u \to \infty$ is known as the *generalised Pareto distribution* (GPD) with *shape parameter* ξ and *scale parameter* β given by

$$G_{\xi,\beta}(y) = \begin{cases} 1 - \left(1 + \xi \dfrac{y}{\beta}\right)^{-1/\xi} & \xi \neq 0 \\ 1 - \exp\left(-\dfrac{y}{\beta}\right) & \xi = 0 \end{cases} \quad \text{where } y \in \begin{cases} [0, \xi] & \xi \geq 0 \\ [0, -\beta/\xi] & \xi < 0. \end{cases} \quad (15)$$

Pickands [21] has shown that the GPD is a good approximation of F_u in that

$$\lim_{u \to x_F} \sup_{0 \leq y \leq y_F} |F_u(y) - G_{\xi,\beta}(y)| = 0, \quad (16)$$

where x_F (possibly infinite) is the right hand end point of the support of the distribution given by F and $y_F := x_F - u$ for some positive measurable function of the threshold u given by $\beta(u)$, provided that this distribution is in the max-domain of attraction of the generalized extreme value distribution.

For $\xi > 0$, the tail of the density corresponding to F decays slowly like a *power* function and F belongs to the family of *heavy-tailed* distributions that includes among others the Pareto, log-gamma, Cauchy and t-distributions. Such distributions may not possess moments. Indeed, for the GPD with $\xi > 0$, $\mathrm{E}[\mathbf{Y}^k]$ is infinite for $k > 1/\xi$, so that for $\xi > 1$ the GPD has no mean and for $\xi > 1/2$ it has infinite variance. For $0 \leq \xi \leq 1/2$, the tail of F decreases *exponentially fast* and F belongs to the class of *medium-tailed* distributions with two moments finite comprising the normal, exponential, gamma and log-normal distributions. Finally, for $\xi < 0$ the underlying distribution F is characterised by a *finite* right endpoint and such *short-tailed* distributions as the uniform and beta.

Financial losses and operational losses in particular are often such that underlying extremes tend to increase without bound over time rather than clustering towards a well-defined upper limit. This suggests that the shape parameter for the GPD estimated from such data can be expected to be non-negative.

Equation (14) may be re-written in terms of survivor functions as

$$\overline{F}(u+y) = \overline{F}(u)\overline{F}_u(y). \quad (17)$$

The survivor function $\overline{F}(u)$ may be estimated empirically by simply calculating the proportion of the sample exceeding the threshold u, i.e. $\overline{F}(u) = N_u/n$. The corresponding q-quantiles of the underlying distribution F are then given by

$$\begin{aligned} x_q &= u + \frac{\beta}{\xi}\left[\left(\frac{n}{n_u}(1-p)\right)^{-\xi} - 1\right] \quad \xi \geq 0 \\ x_q &= u - \frac{\beta}{\xi} \quad\quad\quad\quad\quad\quad\quad\quad\quad \xi < 0 \end{aligned} \tag{18}$$

and the mean of the GPD or *expected excess* function equals

$$\mathrm{E}(\mathbf{X} - u \mid \mathbf{X} > 0) = \frac{\beta + \xi u}{1 - \xi} \text{ for } \xi < 1,\ u > 0. \tag{19}$$

These may be estimated by replacing the shape and scale parameters by their sample estimates [20, 27].

4 Stochastic model for measuring of operational risk

Occurrences of extreme losses over time may be viewed as a point process $\mathbf{N}_u$ of exceedances which converges weakly to a Poisson limit [7, 17, 20]. The GPD provides a model for the excesses over an appropriate threshold u, while the Poisson limit approximation helps to make inferences about the *intensity* of their occurrence. The resulting asymptotic model is known as the *peaks over threshold* (POT) model [8, 16, 17].

For u fixed the parameters of the POT model are the *shape* ξ and the *scale* β_u parameters of the GPD and the Poisson *exceedance rate* λ_u. In terms of these parameters, the alternative *location* μ and *scale* σ parameters are given respectively by

$$\mu = u + \frac{\beta}{\xi}(\lambda^{\xi} - 1) \tag{20}$$

$$\sigma = \beta\lambda^{\xi}. \tag{21}$$

Conversely, the location and alternative scale parameters determine the scale parameter and exceedance rate respectively as

$$\beta_u = \sigma + \xi(u - \mu) \tag{22}$$

$$\lambda_u := \left(1 + \xi\frac{(u - \mu)}{\sigma}\right)^{-1/\xi}. \tag{23}$$

The POT model captures both aspects of operational risk measures – severity and frequency of loss – in terms of excess sizes and corresponding exceedance times. The choice of threshold must satisfy the asymptotic convergence conditions in (11) and (16), i.e. be large enough for a valid approximation, but when u is too high classical parameter estimators for ξ and β_u may have too high a variance due to the small size of exceedances. In the literature [6–8, 20, 25–29] various techniques have been proposed for a statistically reliable choice of threshold. We will assume that the chosen threshold u satisfies a 'bias versus variance trade-off' optimality condition. In our operational risk framework such a u may be termed an *unexpected loss threshold.* Since in this threshold method all excess data is used for parameter estimation, the intensity is measured in the same time units as the given underlying profit and loss data.

Justified by the presented theoretical results from the asymptotic theory of extremes and based upon the point process representation of exceedances given by the POT model, we are now in a position to quantify operational risk. In summary, the operational risk measures are the expected severity and intensity of losses over a suitably chosen threshold u for this model estimated from appropriate profit and loss data.

- Severity of the losses is modelled by the GPD. The expectation of excess loss distribution, i.e. *expected severity* is our coherent risk measure [1] given by

$$\mathrm{E}(\mathbf{X} - u \mid \mathbf{X} > u) = \frac{\beta_u + \xi u}{1 - \xi} \text{ with } \beta := \sigma + \xi(u - \mu). \tag{24}$$

- The number of exceedances $\mathbf{N}_u$ over the threshold u and the corresponding exceedance times are modelled by a Poisson point process with *intensity* (frequency per unit time) given by

$$\lambda_u := \left(1 + \xi \frac{(u - \mu)}{\sigma}\right)^{-1/\xi}. \tag{25}$$

- *Extra* capital provision for operational risk over the *unexpected loss threshold* u is estimated as the *expectation of the excess loss* distribution (expected severity) scaled by the *intensity* λ_u of the Poisson process, *viz.*

$$\lambda_u \mathrm{E}(\mathbf{X} - u \mid \mathbf{X} > u) = \lambda_u \frac{\beta_u + \xi u}{1 - \xi}, \tag{26}$$

  where u, β, ξ and λ are the parameters of the POT model and time is measured in the same units as data collection frequency, e.g. hours, days, weeks, etc. (Note that usually β_u and λ_u will be expressed in terms of the μ and σ as in (24) and (25).)

- The *total* amount of capital provided against extreme operational risks for the time period T will then be calculated by

$$u_T + \lambda_u T \mathrm{E}(\mathbf{X} - u \mid \mathbf{X} > u) = u_T + \lambda T \frac{\beta + \xi u}{1 - \xi}, \qquad (27)$$

  where u_T may in the first instance be considered to be equal to u under the assumption of max-stability.

In general this threshold value u_T over a long horizon T should be adjusted with respect to the time horizon appropriate to integrated risk management and to the thresholds obtained from market and credit models. This is a topic of our current research. The accuracy of our economic capital allocation (26) depends of course on both the correct choice of threshold and accurate estimates of the GPD parameters.

Extreme losses are rare by definition and consequently the issue of small data sets becomes of crucial importance to the accuracy of the resulting risk measures. In addition, operational risk data sets are not homogeneous and are often classified into several subsamples, each associated with a different risk factor or business unit. The conventional *maximum likelihood* (ML) estimation method performs unstably when it is applied to small or even moderate sample sizes, i.e. less than fifty observations. *Bayesian simulation* methods for parameter estimates allow one to overcome problems associated with lack of data through intensive computation.

The *Bayesian hierarchical Markov Chain Monte Carlo* (MCMC) *simulation* model [3, 24] treats uncertainties about parameters by considering them to be random variables (Bayesian view) and generates (simulates) an empirical parameter distribution approximating the conditional posterior parameter distribution given the available loss data. A Bayesian hierarchy is used to link the posterior parameters of interest through the use of prior distribution hyperparameters – in our case estimates of the parameters are linked through the data on different risk types. Our computational procedures were built on R. Smith's statistical procedures and algorithms for GPD assumption verification and corresponding threshold choice [25] using the special library for extreme value statistics of Splus software. Stability of parameter estimation in the presence of small samples is achieved by taking as estimates the *medians* of the possibly disperse empirical marginal posterior parameter distributions.

Operational loss data may be organized into a matrix according to loss type and to business unit as in Table 1 (in which for simplicity only a single cell entry is shown).

The simulated values of the parameters of the POT model are used for calculation of capital provision according to formulas (25) and (26). For overall capital allocation at the top level of the bank, we hope to reduce the overall assessed capital allocation due to portfolio diversification effects and to identify the high-risk factors for specific business units of the firm.

Table 1: Firm-wide matrix of operational losses.

Business unit **Loss factor**	1	...	j	...	N	**Firm-wide**
Technology failure	X_1^1		X_1^j		X_1^N	$X_1^1, X_1^2,\ldots, X_1^N$
Fraud	X_2^1		X_2^j		X_2^N	$X_2^1, X_2^2,\ldots,X_2^N$
...		...		...		...
External event	X_n^1		X_n^j		X_n^N	$X_n^1, X_n^2,\ldots, X_n^N$
Total	$X_1^1, X_1^2, \ldots, X_1^N$	...	$X_2^1, X_2^2, \ldots, X_2^N$	...	$X_n^1, X_n^2, \ldots, X_n^N$	$X^1, X^2,\ldots,X^N$

The procedure can be applied to one business unit across different loss types. Alternatively, it may be applied to one type of loss across all business units as will be demonstrated below in Section 6. Conceptually, both loss factor and business unit dimensions can be simultaneously accommodated at the cost of increased complexity – a topic of our current research. Essentially, the technique is to apply computational power to substitute for insufficient amounts of data, but its empirical estimation efficiency when back-tested on large data sets is surprisingly good.

5 Simulation of peaks over threshold model parameters by MCMC

Bayesian parameter estimation treats uncertainties about parameters by considering parameters to be random variables possessing probability density functions. If the *prior density* $f_{\boldsymbol{\theta}|\boldsymbol{\psi}}$ of the random parameter vector $\boldsymbol{\theta}$ is parametric, given a vector of random *hyperparameters* $\boldsymbol{\psi}$, and of a mathematical form such that the calculated *posterior density* $f_{\boldsymbol{\theta}|X_1,\ldots,X_n,\psi} := f_{\boldsymbol{\theta}|\psi^*}$ is of the same form with new hyperparameters ψ^+ determined by ψ and the observations $X_1,\ldots,X_n$, then we say that $f_{\boldsymbol{\theta}|\psi}$ is a parametric family of densities *conjugate prior* to the *sampling density* $f_{\mathbf{X}|\boldsymbol{\theta}}$.

The *Bayesian hierarchical model* provides a transparent risk assessment by taking into account the possible classification of the profit and loss sample according to loss data subtypes or *classes*, i.e. risk factors or business units, as well as the aggregate. In this model the prior density for the hyperparameters $\boldsymbol{\psi}$ is *common* to all loss subtype prior densities for the parameters $\boldsymbol{\theta}$. The *hyper-hyper parameters* φ are chosen to generate a vague conjugate prior indicating a lack of information on the hyper-parameters' prior distribution before the excess loss data is seen. Thus we have a Bayesian hierarchical decomposition of the posterior parameter density $f_{\boldsymbol{\theta}|\mathbf{X},\psi}$ given the observations and the *initial* hyper-hyper-parameters φ as

$$\begin{aligned} f_{\boldsymbol{\theta}|\mathbf{X},\psi} &\propto f_{\mathbf{X}|\boldsymbol{\theta}}(X \mid \theta) f_{\boldsymbol{\theta}|\psi}(\theta \mid \psi) f_{\psi}(\psi \mid \varphi) \\ &\propto f_{\mathbf{X}|\boldsymbol{\theta}}(X \mid \theta) f_{\psi|\boldsymbol{\theta}}(\psi \mid \theta, \varphi) \\ &\propto f_{\mathbf{X}|\boldsymbol{\theta}}(X \mid \theta) f_{\psi}(\psi \mid \varphi^{+}), \end{aligned} \tag{27}$$

where $\propto$ denotes proportionality (up to a positive constant). We may thus perform the Bayesian update of the prior parameter density $f_{\boldsymbol{\theta}} \propto f_{\boldsymbol{\theta}|\psi} f_{\psi}$ in two stages – first updating the hyper-hyper-parameters φ to φ^{+} conditional on a given value of θ and then computing the value of the corresponding posterior density for this θ given the observations X. Figure 2 depicts schematically the relationships between the 3 parameter levels and the excess loss observations for each risk class. Note that even though the prior specification of parameters for individual risk classes is as an independent sample from the *same* hyper-parameter Gaussian prior distribution, their posterior multivariate Gaussian specification will *not* maintain this independence given observations which are statistically dependent.

The Bayesian *posterior density* $f_{\boldsymbol{\theta}|\mathbf{X},\psi}$ may be computed via *Markov chain Monte Carlo* (MCMC) simulation [24, 28, 29]. The idea, which goes back to Metropolis, Teller *et al.* and the hydrogen bomb project, is to simulate sample paths of a Markov chain. The states of the chain are the values of the parameter vector θ and its visited states converge to a stationary distribution which is the Bayesian joint posterior parameter distribution $f_{\boldsymbol{\theta}|\mathbf{X},\psi}$ (termed the *target* distribution) given the loss data X and a vector ψ of hyperparameters as discussed above. In this context, a *Markov chain* is a discrete time continuous state stochastic process whose next random state depends statistically only on its current state and not on the past history of the process. Its random dynamics are specified by the corresponding state transition probability density. In this application the parameter vector state space of the chain is discretised for computation in order to create a parameter histogram approximation to the required multivariate posterior parameter distribution.

For our application, the parameter vector θ represents the generalized Pareto distribution (GPD) parameters of interest $\{\mu_j, \log\sigma_j, \xi_j : j = 1, 2 \ldots, J\}$ for the $j = 1, \ldots, J$ data classes (business units or risk factors) and the hyperparameter vector ψ consists of $\{m_\mu, s_\mu^2, m_{\log\sigma}, s_{\log\sigma}^2, m_\xi, s_\xi^2\}$ which are the parameters of a common (across all business units) multivariate Gaussian prior distribution of the GPD parameters. To implement the strategy, *Gibbs sampling* and the *Metropolis-Hastings* algorithm [3] are used to construct the Markov chain possessing our specific target posterior distribution as its stationary distribution. This target distribution is defined by standard Bayesian calculations in terms of the peaks over threshold likelihood function and appropriate prior distributions. Running the Markov chain for very many

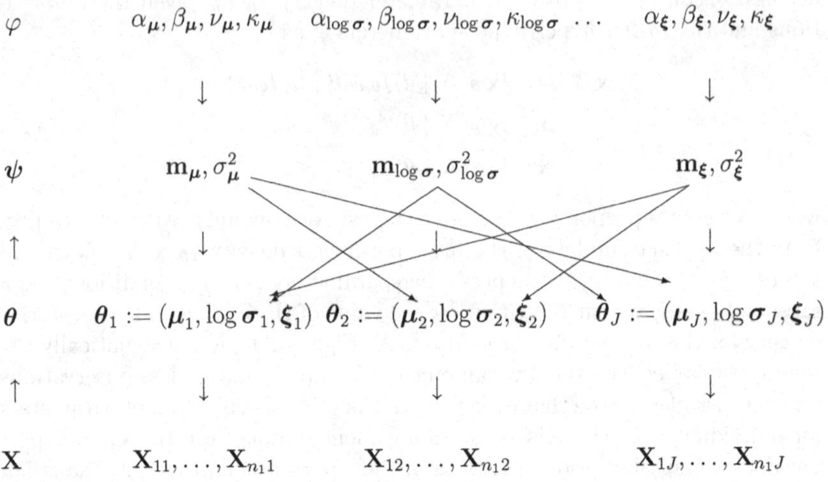

Figure 2: Hierarchical Bayesian model parameter and observation dependencies conditional on their hyperparameters.

transitions (about 1M) produces an empirical parameter distribution that is used to estimate the posterior density $f_{\boldsymbol{\theta}|\mathbf{X},\psi}$.

These MCMC dynamical methods generate the sequence $\{\theta_j^0, \theta_j^1, \theta_j^2, \ldots\}$ of parameter estimates $\theta_j = \{\mu_j, \log\sigma_j\}$, $j = 1, 2, \ldots, J$ for each data class with θ_j^{t+1} (for time $t \geq 0$) depending solely upon θ_j^t. This process represents the traditional exchange of computational intensity for low data availability. After sufficient iterations the Markov chain will forget its initial state and converge to the stationary required posterior distribution $f_{\boldsymbol{\theta}|\mathbf{X},\psi}$ not depending on the initial state θ_j^0 or time t. By discarding the first k (=10k) states of the chain, constituting the *burn-in period*, the remainder of the Markov chain output may be taken to be a parameter sample drawn from the high-dimensional target parameter posterior distribution.

In summary, the MCMC simulation is used to generate an empirical parameter distribution approximating the conditional posterior multivariate parameter distribution given the available loss data. A *Bayesian hierarchical* model is used to link the posterior parameters of interest through the use of common prior distribution hyperparameters. The simulation is implemented using hybrid methods and parameter estimates are taken as median values of the generated empirical marginal parameter distributions.

Table 2: Summary statistics for daily aggregated P&L data. Losses are positive and profits are negative.

Min:	–1532.394960	Mean:	–92.353455
Max:	2214.319020		
1st Qu.:	–320.839980	Median:	–119.276080
3rd Qu.:	68.261120		
Sample size:	296		
Std Dev.:	463.733057		
Excess Kurtosis:	5.047392		

6 Example: bank trading losses analysis through the Russian crisis

We apply the framework set out above to analyse the losses of the trading activities of a major European investment bank during the period 1 October 1997 to 31 December 1998. Financial turmoil in the summer of 1998 caused by the Russian government's domestic bond default on 24 August caused losses which can be seen as *external* to the bank's normal operating conditions – possibly in the category of unexpected large losses. In financial crises the separation of financial risks into various types (market, credit etc.) proves to be difficult and the Russian crisis is no exception. To reduce bank exposure to the consequences of such events a correct model for risk evaluation and capital provision should be identified, with the corresponding unexpected threshold level given by current historical loss data. In what follows the necessary diagnostics to test and verify the POT model assumptions for *aggregated* P&L data are first performed. Next we back-test the predictive power of the POT model in terms of the proposed capital provision estimation rule and then study its breakdown by business unit. Our data (rescaled for confidentiality reasons) contains daily P&L reports from four business unit/trading desks. Daily events are aggregated across the four desks. The *aggregated* P&L data consists of $n = 296$ profits or losses with a net profit figure of 27,337 monetary units. They range from a 2,214 loss to a 1,532 profit; see Table 2 for summary statistics and Figure 3 for a time-series plot and histogram of aggregated P&L data.

In Figure 4 we plot the empirical excesses of the aggregated P&L data for an increasing sequence of thresholds. The positive steep slope above a threshold of about 500 indicates a heavy loss tail. The shape parameter plot, based on maximum likelihood estimation of ξ, seems to have stable standard

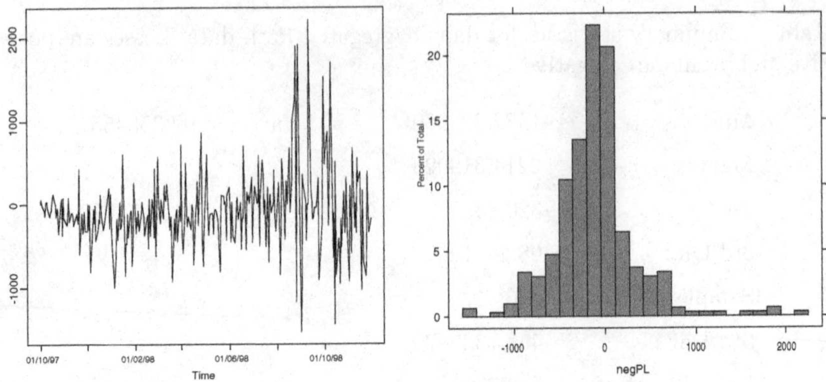

Figure 3: Daily P&L data aggregated over the four trading desks: time-series plot (left) and histogram (right). Note that losses are positive and profits are negative.

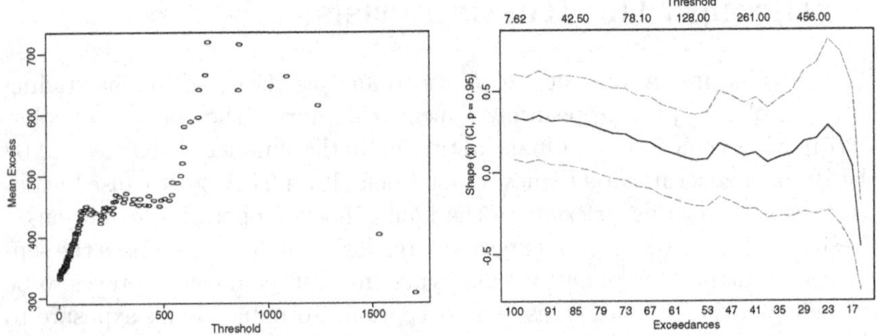

Figure 4: Empirical mean excess plot and shape parameter ξ ML estimates for an increasing sequence of thresholds in aggregated P&L data. Dotted lines represent estimated 95% confidence intervals of ξ ML estimates.

deviation 0.15 up to a minimum of $n_u = 55$ exceedances. Samples of size less than $n_u = 55$ exceedances (or equivalently thresholds higher than $u = 150$) yield ML ξ estimates with significantly large estimated 95% confidence intervals. Hence a realistic threshold should not be set higher than $u = 150$ when fitting the POT model with the ML approach.

Figure 5 shows the empirical quantiles versus a standard normal distribution and a GPD with scale parameter $\beta = 1$ and shape parameter $\xi = 0.25$, which represents the best Q–Q plot against the GPD for various values of ξ. These Q–Q plots verify earlier observations that the loss tail is heavier than that to be expected from a normal distribution.

As noted above, the choice of threshold should guarantee the stability of the ML estimate of the shape parameter ξ when using maximum likelihood

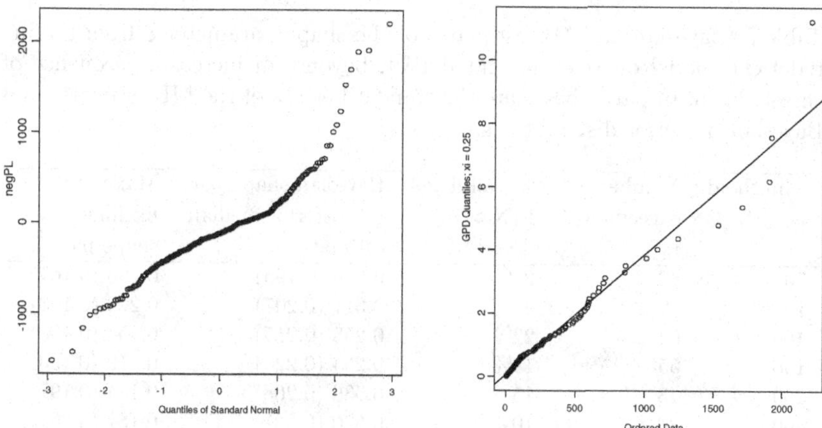

Figure 5: Q–Q plots of aggregated P&L data. (a) Comparing the empirical quantiles (vertical axis) with the quantiles expected from a standard normal distribution (horizontal axis); losses are positive. (b) Comparing the empirical quantiles (horizontal axis) with the quantiles expected from a GPD ($\beta = 1$, $\xi = 0.25$).

estimation. The ML estimates of ξ together with their standard errors for an increasing sequence of thresholds, $50 \leq u \leq 1534$, and the corresponding MCMC Bayesian estimates of ξ based on the medians and standard deviations of the marginal posterior distributions for the same thresholds are shown in Table 3. The Bayesian estimates of the shape parameter ξ from the MCMC algorithm show relative stability in the range $u = 250$ to 1534 (corresponding to the 15% to 1% tails of the underlying P&L empirical distribution) at a value about $\hat{\xi} = 0.53$, indicating an α-stable distribution with only a single moment finite. Moreover the Bayesian method allows estimation of the shape parameter from smaller-sized samples, less than $n_u = 20$ exceedances, whereas the corresponding ML estimates become totally unreliable for such small samples. For example, the ML shape parameter estimate for $u = 600$ (i.e. $n_u = 16$ exceedances) is negative, which is not at all representative of its true value. By contrast, the Bayesian shape parameter estimates are stable even for just $n_u = 4$ exceedances (i.e. $u = 1534$), although, as shown in Table 3, the corresponding posterior distribution in this case is rather dispersed.

The estimated standard errors of the ML estimates are merely an indication of accuracy which in fact deteriorates dangerously for higher thresholds (or, equivalently, lower tail probabilities and samples of smaller size). However, by calculating the posterior distributions of μ, σ and ξ by the Bayesian MCMC method, statistics – such as standard deviation or quantiles – based on the entire distribution can be considered in addition to the median point estimates corresponding to absolute parameter error loss functions.

Table 3: Bayesian and ML estimates of the shape parameter ξ from the fitted POT model on the aggregated P&L beyond an increasing sequence of thresholds u. In parentheses are the standard errors of the ML estimates and Bayesian posterior distributions.

Threshold u	Number of exceedances n_u	% Tail fitted $P(\mathbf{X} > u)$	Bayesian shape par. $\hat{\xi}$ (posterior median estimate	Maximum likelihood shape par. $\hat{\xi}$
50	82	28%	0.396 (0.195)	0.296 (0.167)
75	72	25%	0.311 (0.207)	0.220 (0.163)
100	64	22%	0.258 (0.215)	0.154 (0.158)
150	55	18%	0.254 (0.226)	0.119 (0.163)
250	43	15%	0.536 (0.268)	0.144 (0.197)
400	30	10%	0.520 (0.221)	0.181 (0.261)
600	16	5%	0.573 (0.325)	–0.228 (0.527)
1000	8	2.7%	0.524 (0.422)	NA*
1534	4	1%	0.527 (0.662)	NA

*NA: not available

Such an EVT analysis can assist in model evaluation by more robustly identifying the heavy-tail distributions. In our example, the Bayesian estimates of the shape parameter for the aggregated data suggest that only the first moment (i.e. the mean) is finite.

Prediction of actual losses by the economic loss capital provision at firm level

To test the capital allocation rule consider five 'event' dates: 17th, 21st, 25th, 28th August 1998 and 11th September 1998. Two events are before and three after the Russian government's GKO default on 24th August 1998, *cf.* Figure 3. The fifth event-date (11th September) is selected so that the subsample includes the maximum historic loss as its last observation. (Note that losses are treated as positive unless stated otherwise.) For a fixed loss threshold $u = 150$, we fit to data both the normal distribution and the POT model using both maximum likelihood and Bayesian estimation. With the threshold set at $u = 150$ the number of exceedances for all five data sets and the full sample are equal to $n_u = 27, 29, 31, 33, 36$ and 55 respectively. The results are illustrated in Figure 6 where the dots represent the empirical distribution function based on the full aggregated P&L data. There is a marked difference between the suggested GPD model and the normal distribution in all six experiments. The GPD approximates the excess loss distribution F_u significantly better using the Bayesian posterior median estimates of ξ, μ and σ (see Figure 7, p. 271). No maximum likelihood estimates are available for the first data set (to 17th August 1998). Hosking and Wallis [13,14] show empirically that no ML estimates exist for $n_u < 50$. Our data supports this for $n_u = 27$. The Bayesian method yields a posterior distribution for the shape parameter with median estimate $\hat{\xi} = 0.22$. Prediction results are improved by 21st August 1998 with the Bayesian estimates still performing better than the maximum likelihood estimates. For data up to 28 August 1998 both estimation techniques start to yield comparable fits. This is so for the data up to the 11th September 1998 and indeed for the full sample. When this experiment is repeated for the threshold $u = 600$ corresponding to the 5% tail of the empirical loss distribution only Bayesian estimates (based on 16 exceedances in the full sample) are reliable.

For the five dates selected the results of the Bayesian calculations of the operational risk capital allocation (using (25)) are given in Table 4. All estimates are based on the medians of the corresponding posterior distributions. Table 4A corresponds to the statistically fit threshold $u = 150$, while Table 4B corresponds to the more theoretically reliable threshold $u = 600$ at which the Bayesian estimate of the tail shape parameter $\hat{\xi} = 0.57$ (*cf.* Table 3) indicates that only a single moment of the underlying P&L distribution is finite. The estimated annual expected excess risk capital based on 250 trading days is also shown as a percentage of the corresponding figure estimated from the full data. Clearly for both threshold levels the more data used in the period of turmoil, the closer our model captures the estimated full-data annual excess capital requirement. Examination of Figure 3 shows visually that while daily losses have settled to early 1997 patterns by the end 1998 about 92% of in-sample annual loss capital provision for 1998 could have been predicted using the statistically determined lower threshold value by 11th September, less than half-way through the turmoil and before the Long Term Capital Management collapse added to volatility.

It is the *severity* of loss that varies between the five chosen 'event' dates, with loss *frequency* playing only a minor role. While the estimated expected excess loss using (24) increases from 232 to 587, the estimated time between exceedances decreases only moderately from about 12 to 9 days. The average number of losses per year[2] exceeding the threshold level $u = 150$ remains approximately at 25; that is, ten trading days on average between excessive losses, which seems to be a reasonable time interval in which to liquidate some risky positions.

However, the estimated excess provision of 16,647 based on the full sample fails to cover actual excess losses over this threshold incurred in the last 250 trading days in the sample of 23,422 – a *deficit* of about 30%.

On the other hand, while we see from Table 4B that only 79% of the full

[2]The Poisson intensity $\hat{\lambda}_U u$ is calculated from equation (24) from the current posterior values for μ, σ and ξ on the MCMC simulation path. This yields an empirical distribution for λ_u from which we select the median estimate $\hat{\lambda}_u$.

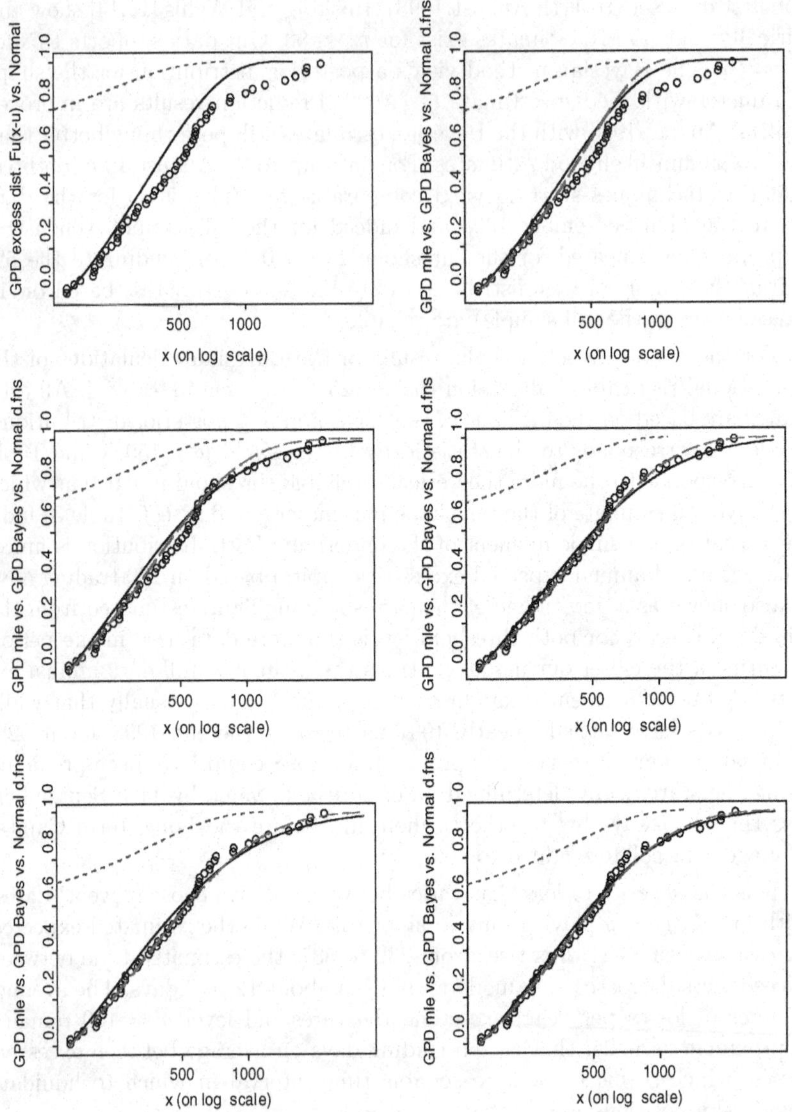

Figure 6: Aggregated P&L with threshold $u = 150$: Fitted GPD excess distribution functions $G_{\xi,\beta}$ based on ML (dashed lines) and Bayesian (solid lines) posterior median estimates of ξ and β vs. normal distribution functions (dashed lines) using data up to the 17th (top-left), 21st (top-right), 25th (middle-left), 28th August (middle-right), 11th September 1998 (bottom-left) and the full sample (bottom-right). Dots represent the empirical distribution function F_u for aggregated losses exceeding u.

Table 4a: Expected excess annual risk capital for the five subsamples and the full-sample based on estimates with $u = 150$.

Data split date	Daily expected Excess beyond u ($u = 150$)	Exponential time gap (in days) $\hat{\lambda}_u^{-1}$ between successive loss excesses	Annualised Poisson intensity $\hat{\lambda}_u$ (expected number of excesses)	Expected excess annual risk capital *(% of the full data estimate)*
17th Aug '98	231.6	11.7	21.4	4,956 *(29.7%)*
21st Aug '98	271.0	11.1	22.5	6,098 *(36.7%)*
25th Aug '98	440.3	10.6	23.6	10,391 *(62.5%)*
28th Aug '98	513.9	10.0	24.9	12,796 *(77%)*
11th Sep '98	586.7	9.6	26.0	15,254 *(91.7%)*
Full sample	517.0	7.7	32.2	16,647 *(100%)*

Table 4b: Expected excess annual risk capital for the five subsamples and the full-sample based on estimates with $u = 600$.

Data split date	Daily expected Excess beyond u ($u = 600$)	Exponential time gap (in days) $\hat{\lambda}_u^{-1}$ between successive loss excesses	Annualised Poisson intensity $\hat{\lambda}_u$ (expected number of excesses)	Expected excess annual risk capital *(% of the full data estimate)*
17th Aug '98	319.9	86.6	2.9	928 *(7.2%)*
21st Aug '98	432.0	69.9	3.6	1,555 *(12%)*
25th Aug '98	933.1	50.1	5	4,666 *(36.4%)*
28th Aug '98	1245.2	38.7	6.4	7,969 *(62.1%)*
11th Sep '98	1459.9	36.2	6.9	10,073 *(78.5%)*
Full sample	1395.4	27.2	9.2	12,838 *(100%)*

sample excess capital provision of 12,838 is covered by 11 September using the more theoretically justified higher threshold, the suggested annual provision at this date compares very favourably with sample excess losses of 8,737 over the last 250 trading days – a *surplus* of about 15% – which might be expected from extreme value theory appropriately applied in predictive mode.

Economic capital for operational risk at business unit level

Having estimated the frequency and severity of the aggregated daily P&L our aim next is to use the hierarchical structure of the Bayesian model for operational risk capital allocation at the level of the four *individual* trading desks. The Bayesian hierarchical MCMC model was applied to the four desks with a fixed loss threshold $u = 130$ for their parameter estimation to ensure

Table 5: Statistical analysis of the aggregated P&L and the four individual P&L data sets: Bayesian estimates of the GPD and Poisson parameters and their resulting risk measures, all based on the medians of the corresponding posterior distributions.

Firm-wide level $u = 150$	Bayes posterior median estimates $\hat{\xi}$	$\hat{\beta}$	Daily severity q-GPD-based 95%	99%	Daily expected excess beyond u	Expected number of excesses beyond u (per annum)	Expected excess annual risk capital
	0.25	340	691.0	1,639.5	517.0	32.2	**16,646**
Business-unit level $u = 130$							
Desk One	0.34	205.2	601.6	1,360.3	365.9	49.3	18,046
Desk Two	0.25	108.1	116.3	324.5	190.4	7.5	1,435
Desk Three	0.24	118.6	179.2	442.0	206.5	13.0	2,688
Desk Four	0.26	106.1	71.2	250.0	192.8	4.8	925
						Total:	**23,094**

a sufficient number of exceedances. The numbers of exceedances (beyond $u =$ 130) are respectively $n_u = 83$, 13, 22 and 8 for desks one, two, three and four, which clearly makes maximum likelihood estimation ill-suited to the task, particularly for desks two and four. The four individual desks estimates for ξ and β as well as for the aggregated P&L data and the annual risk capital (based on 250 trading days) are summarised in Table 5. The GPD-based severity quantile is specified by (18) and expected excess is calculated by (24) and (25).

Expected excess annual risk capital provision at the firm-wide level is less than the sum of the corresponding capital provisions across the four individual desks. Thus the *subadditivity* – or portfolio diversification – property holds under the expected excess loss risk measure [1]. However, in spite of the too low threshold bias discussed above, the sum of the individual desk provisions covers actual firm-wide excess losses of 23,422 to within about 1%. In addition the hierarchical structure of Bayesian method of parameter estimation provides a more transparent risk assessment for the four business units. Based on the estimates of the severity parameters ξ, β and frequency parameter λ_u, we see that desk one is the most risky amongst the four desks. The estimated parameters are given by the respective medians of their posterior marginal distributions[3] as shown in Figure 7.

[3]Boxplot interpretation of posterior marginal parameter distributions: White horizontal line within the whisker of the boxplot indicates the median of the posterior distribution while the whiskers' lower and upper sides represent respectively the 25% and 75% of the

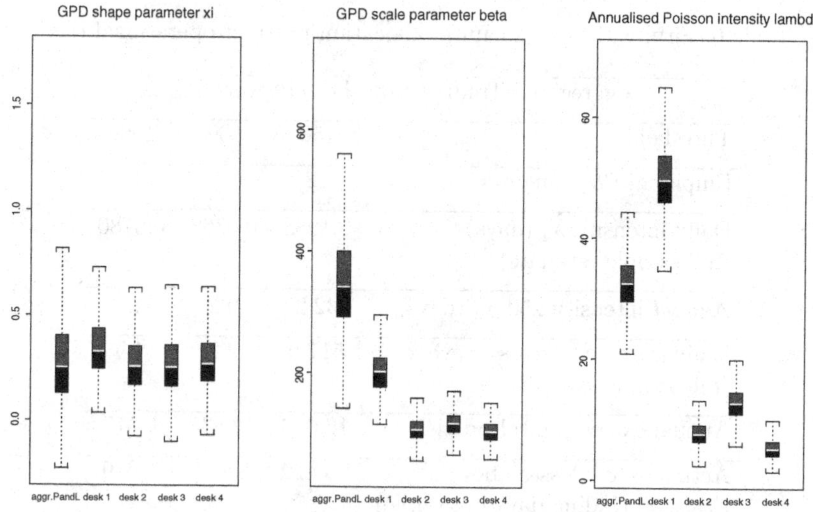

Figure 7: Posterior distributions of the estimated shape ξ and scale β (GPD) parameters, and the annualised Poisson intensity λ_u. The posterior distributions for the aggregated P&L are estimated from losses exceeding threshold $u = 150$ whereas the posterior distributions for the four individual desks are estimated from losses exceeding threshold $u = 130$.

Economic capital for operational risk at firm level

Our example consists of essentially market data with losses due to political events, i.e. operational losses. It is thus important that the unexpected loss threshold is chosen greater than or equal to the combined market and credit VaR threshold. With such a choice the capital allocation will protect against large and rare losses classified as operational. The most problematic aspect of standard VaR methods – underestimation of capital for longer time periods – in this case will be accounted for by exceedances. In our method we have assumed max-stability and therefore only the intensity of the Poisson process is scaled. In Table 6 we summarise the different rules for excess risk capital allocation corresponding to the 18%, 5% and 2.7% quantile thresholds of the empirical P&L distribution and compare them with actual excess losses.

distribution. The lower and upper brackets represent the minimum and maximum values of the distribution.

Table 6: Firm-wide excess capital allocation rules for operational risk.

Aggregated Trading P&L Loss Provision			
Threshold u	150	600	1000
Empirical P&L quantile (%)	18	5	2.7
Daily intensity $\hat{\lambda}_u$ (days) (full sample estimate)	0.1288	0.0368	0.0180
Annual intensity 250 $\hat{\lambda}_u$ (days)	32.2	9.2	4.5
Daily expected excess above u (full sample estimate)	517.0	9.2	4.5
Annual excess capital provision	16,646	12,838	6,877
Actual excess losses above u (last 250 trading days in sample)	23,422	8,737	4,619
Percentage safety margin (%)	–29.0	46.9	48.9

Conclusions and future directions

Losses incurred similar to those of Barings Bank belong to the category of *extreme operational* loss and could have been mitigated through *control and capital allocation*. P&L data, volatility of returns and other factors should be constantly analysed for identification of extremes. Apparent lack of operational loss data suggests an implementation based on Bayesian hierarchical MCMC simulation, which provides us with robust parameter estimates of extreme distributions. When applied at the level of business units Bayesian procedures allow more efficient capital allocation.

In measuring operational risk we propose a framework which allows a consistent integration with market and credit risk capital allocations. Due to fuzzy boundaries between the different risk types, operational risk must be measured as an excess over levels for market and credit risk. Integrated risk management will involve different risk valuations for different business units and by different models. In our model we assume the 'ordering' of thresholds: market $\leq$ credit $\leq$ operational. For integrated risk management further careful adjustments of market and credit thresholds and time re-scaling of intensity should be performed to be comparable with market and credit risk evaluation. These are topics of our current research. Further progress in operational risk modelling depends on cooperation with industry and the wider availability of case study data.

References

1. Artzner P., F. Delbaen, J.M. Eber & D. Heath (1999). Coherent measures of risk. *Mathematical Finance* **9**, 203–228.

2. Basle Committee on Banking Supervision, January (2001). *Operational Risk.* Reprinted as Chapter 6 of this volume.

3. Bernardo J.M. & A.F.M. Smith (1994). *Bayesian Theory.* Wiley.

4. British Bankers' Association (1997). Operational Risk Management Survey.

5. Castillo E. (1988). *Extreme Value Theory in Engineering.* Academic Press.

6. Danielson J. & C. G. de Vries (1997). Tail index and quantile estimation with very high frequency data. *Journal of Empirical Finance* **4**, 241–257.

7. Du Mouchel W.H. (1983). Estimating the stable index α in order to measure tail thickness: a critique. *Annals of Statistics* **11** (4), 1019–1031.

8. Embrechts P., C. Kluppelberg & T. Mikosch (1997). *Modelling Extremal Events.* Springer.

9. Feller W. (1966). *An Introduction to Probability Theory and Its Applications* **2**. Wiley.

10. Galambos J. (1978). *The Asymptotic Theory of Extreme Order Statistics.* Wiley.

11. Gnedenko B.V. & Kolmogorov A. N. (1954). *Limit Distributions for Sums of Independent Random Variables.* Addison-Wesley.

12. Gnedenko B.V. (1941). Limit theorems for the maximal term of a variational series. *Comptes Rendus de l'Academie des Sciences del'URSS* **32**, 7–9.

13. Hosking J.R., J.R. Wallis & E.F. Wood (1985). Estimation of the generalised extreme value distribution by the method of probability-weighted moments. *Technometrics* **27**, 251–261.

14. Hosking J.R. & J.R. Wallis (1987). Parameter and quantile estimation for the generalised Pareto distribution. *Technometrics* **29**, 339–349.

15. Jameson R. (1998). Playing the name game. *Risk*, November, 38–45.

16. Leadbetter M.R., Lindgren G. &. Rootzen H. (1983). *Extremes and Related Properties of Random Sequences and Processes.* Springer.

17. Leadbetter M.R. (1991). On a basis for 'Peaks over Threshold' modeling. *Statistics & Probability Letters* **12**, 357–362.

18. Mandelbrot B.B. (1982). *The Fractal Geometry of Nature,* W.H. Freeman.

19. Medova E.A. (2000). Measuring risk by extreme values. *Risk*, November, 20–26.

20. McNeil A.J. & T. Saladin (1997). The peaks over thresholds method for estimating high quantiles of loss distributions. In *Proceedings of XXVII International ASTIN Colloquium*, 23–43.

21. Ong M.K. (1999). *Internal Credit Risk Models: Capital Allocation and Performance Measurement.* Risk Books, London.

22. Pickands J. (1975). Statistical inference using extreme order statistics, *Annals of Statistics* **3**, 119–131.

23. Samorodnitsky G. & M.S. Taqqu (1994). *Stable Non-Gaussian Random Processes: Stochastic Models with Infinite Variance.* Chapman & Hall.

24. Smith A.F.M. and G.O. Roberts (1993). Bayesian computation via the Gibbs sampler and related Markov chain Monte Carlo methods. *J. Royal Statistical Society* **B 55**, 3–23.

25. Smith R.L. (1985). Threshold methods for sample extremes. In *Statistical Extremes and Applications*, J. Tiago de Oliveira (ed.), NATO ASI Series, 623–638.

26. Smith R.L. (1987). Estimating tails of probability distributions. *Annals of Statistics* **15**, 1174–1207.

27. Smith R.L. (1990). Extreme value theory. In *Handbook of Applicable Mathematics Supplement.* W. Ledermann (Chief ed.), Wiley, 437-472.

28. Smith R.L. (1998). Bayesian and frequentist approaches to parametric predictive insurance. In *Bayesian Statistics* **6**. J.M Bernado, J.O. Berger, A. Dawid, A.F.M. Smith (eds.), Oxford University Press.

29. Smith R.L. (2001). Measuring risk with extreme value theory. This volume, Chapter 8.